Mathematical Problems
in Viscoelasticity

π Pitman Monographs and
Surveys in Pure and Applied Mathematics 35

Mathematical Problems in Viscoelasticity

Michael Renardy, William J Hrusa & John A Nohel
Virginia Polytechnic Institute/Carnegie–Mellon University/University of Wisconsin

Longman
Scientific &
Technical

Copublished in the United States with
John Wiley & Sons, Inc., New York

Longman Scientific & Technical
Longman Group UK Limited
Longman House, Burnt Mill, Harlow
Essex CM20 2JE, England
and Associated Companies throughout the world.

Copublished in the United States with
John Wiley & Sons, Inc., 605 Third Avenue, New York, NY 10158

First published 1987

AMS Subject Classifications: (main) 45K05, 73F15, 76A10
                             (subsidiary) 35L15, 35L67

ISSN 0269–3666

British Library Cataloguing in Publication Data
Renardy, M.
    Mathematical problems in viscoelasticity.
    —(Pitman monographs and surveys in
    pure and applied mathematics, ISSN
    0269-3666; 35)
    1. Viscoelasticity 2. Deformations
    (Mechanics)
    I. Title II. Hrusa, W.J. III. Nohel, J.A.
    620.1'1233   TA418.2
    ISBN 0-582-00320-2

Library of Congress Cataloging-in-Publication Data
Renardy, Michael.
    Mathematical problems in viscoelasticity.
    (Pitman monographs and surveys in pure and applied
mathematics, ISSN 0269-3666 ; 35)
    Bibliography: p.
    Includes index.
    1. Viscoelasticity. 2. Integral equations.
3. Continuum mechanics. I. Hrusa, W. (William)
II. Nohel, John A. III. Title. IV. Series.
QA929.R45   1987   531'.3823   86-21413
ISBN 0-470-20748-5 (USA only)

Printed and Bound in Great Britain at The Bath Press, Avon

Contents

VI. Steady flows of viscoelastic fluids

Preface

The mathematical analysis of initial and boundary value problems in viscoelasticity has progressed significantly during the past ten years. Our purpose in writing this monograph is to present a timely overview rather than a detailed account of the developments, concentrating on qualitative properties of classical solutions.

Most of the material discussed here appears or will appear in research papers. While not new in spirit, certain topics, in particular Sections 4 and 5 of Chapter II, Section 3 of Chapter III, Sections 2 and 5 of Chapter IV and Sections 5 and 6 of Chapter V, do not appear explicitly in the literature.

The individual chapters are essentially self-contained. Chapter I provides a general introduction to the physical background of the subject; the reader may refer to it as needed.

This monograph is addressed to research scientists and advanced graduate students having general familiarity with techniques of modern analysis and partial differential equations. It is not intended to be a textbook.

We are grateful to friends and colleagues including M. Crandall, C. Dafermos, W. Desch, H. Engler, M. Gurtin, K. Hannsgen, D. Joseph, J.U. Kim, R. MacCamy, D. Malkus, V. Mizel, D. Owen, B. Plohr, W. Pritchard, R. Rogers, A. Tzavaras and R. Wheeler for reading various sections of the monograph and offering useful comments and suggestions for improvement. We thank the Mathematical Sciences Divisions of AFOSR, ARO and NSF for supporting our research, and Mrs. Sally Ross for typing large portions of the manuscript.

Michael Renardy, William J. Hrusa and John A. Nohel

Blacksburg, VA, Pittsburgh, PA and Madison, WI

February 1987

Acknowledgements

Thanks are due to Academic Press for permission to reproduce material from the authors' papers in the *Journal of Differential Equations*, volumes **59** and **64** (© Academic Press 1985, 1986).

Notation for function spaces

The following function spaces are used throughout this monograph. We note that derivatives should always be interpreted in the sense of distributions.

A. Sobolev spaces

Let Ω be a domain in $\mathbb{R}^n$ with sufficiently regular boundary $\partial\Omega$. Let $1 \le p \le \infty$, $q = p/(p-1)$, let k be a nonnegative integer, s a real number ≥ 0, let I be an interval in $\mathbb{R}$ and let X be a Banach space. In 1-9 below, the functions may be scalar-, vector- or tensor-valued, depending on context.

1. $L^p(\Omega)$ The space of all (equivalence classes of) measurable functions f on Ω such that $\int_\Omega |f(x)|^p \, dx < \infty$ if p is finite and $\mathrm{ess}-\sup_{x\in\Omega} |f(x)| < \infty$ if $p = \infty$.

2. $W^{k,p}(\Omega)$ The space of all f in $L^p(\Omega)$ whose derivatives through order k are in $L^p(\Omega)$.

3. $W_0^{k,p}(\Omega)$ The completion of $C_0^\infty(\Omega)$ (the set of C^∞-functions with compact support) in $W^{k,p}(\Omega)$ for $1 \le p < \infty$.

4. $W^{-k,q}(\Omega)$ The dual space of $W_0^{k,p}(\Omega)$ via the inner product on $L^2(\Omega)$ for $1 < p < \infty$.

5. $W^{k-1/p,p}(\partial\Omega)$ The trace class of $W^{k,p}(\Omega)$.

6. $H^k(\Omega)$ $= W^{k,2}(\Omega)$.

7. $H_0^k(\Omega)$ $= W_0^{k,2}(\Omega)$.

8. $H^{-k}(\Omega)$ $= W^{-k,2}(\Omega)$.

9. $H^s(\Omega)$ $= [H^m(\Omega), L^2(\Omega)]_{1-s/m}$, where m is any integer $\ge s$ and $[\cdot,\cdot]$ denotes interpolation spaces, see [L6].

10. $L^p(I; X)$ The space of all (equivalence classes of) strongly measurable functions $f : I \to X$ such that $\int_I \|f(t)\|^p \, dt < \infty$ if p is finite and $\mathrm{ess}-\sup_{t\in I} \|f(t)\| < \infty$ if $p = \infty$.

11. $W^{k,p}(I; X)$ The space of all $f \in L^p(I; X)$ whose derivatives through order k (in the sense of vector-valued distributions) are in $L^p(I; X)$.

12. $H^k(I; X)$ $= W^{k,2}(I; X)$.

B. Spaces of continuous functions

Let k be a nonnegative integer and $0 < \theta < 1$.

1. C^k The space of k times continuously differentiable functions; we write C in place of C^0.

2. C^k_b The space of functions which, together with their derivatives through order k, are bounded and continuous.

3. C^k_{bu} The space of functions which, together with their derivatives through order k, are bounded and uniformly continuous.

4. $C^{k,lim}$ The space of k times continuously differentiable functions, defined on an unbounded interval, such that the function and its first k derivatives have limits at infinity.

5. C^θ The space of Hölder continuous functions with exponent θ.

0 Introduction

Continuum mechanics attempts to describe the motions and equilibrium states of deformable bodies. The equations providing this description are of two kinds: **balance laws** and **constitutive assumptions**. Balance laws represent basic physical principles such as conservation of mass and balance of linear momentum and are valid for all bodies, regardless of composition; a constitutive assumption serves to specify the particular material comprising the body under consideration. Throughout this monograph, we deal only with a purely mechanical theory, and thermal effects are not accounted for. This is not because we believe thermal effects are unimportant, but because the simpler, purely mechanical situation is better understood at the present time.

Two types of materials are usually considered in basic texts on continuum mechanics: **elastic materials** and **viscous fluids**. At each material point of an elastic material the stress at the present time depends only on the present value of the strain. Under physically natural assumptions concerning this dependence of stress on strain, the resulting equations of motion are hyperbolic. Such equations propagate singularities in the initial data, and if the stress-strain relation is non-linear, then even initially smooth solutions generally develop singularities in finite time. On the other hand, for an incompressible viscous fluid the stress at a given point is a function of the present value of the velocity gradient at that point (plus an undetermined pressure). The resulting equations of motion are of parabolic type; if the relation between stress and velocity gradient is assumed to be linear, then the classical Navier-Stokes equations are obtained. Here singularities in the initial data are smoothed instantaneously and C^∞-solutions exist locally in time. Further, if the spatial dimension is 1 or 2, then solutions of the Navier-Stokes equations (with Dirichlet boundary conditions) remain of class C^∞ for all positive time. Whether or not any type of singularities can develop in solutions of the three-dimensional Navier-Stokes equations is still an open problem.

Viscoelastic materials have properties between those of elastic materials and viscous fluids. A class of such materials is obtained by letting the stress depend on the present values of both the strain and the velocity gradient. Such a theory can describe, for example, the behavior of compressible gases. Experience shows, however, that other viscoelastic materials, e.g. polymers, suspensions and emulsions, under usual operating conditions, cannot be described in this way. Such materials have **memory**: the stress depends not only on the present values of the strain and/or velocity gradient, but on the entire temporal history of the motion. Typically the memory fades with time: disturbances which occurred in the distant past have less influence on the present stress than those which occurred

1

in the more recent past. The subject of this book is the mathematical analysis of equations which model the motions of materials with memory.

In order to introduce some of the central issues and ideas, while avoiding technical complications, we shall discuss the motion of one-dimensional viscoelastic bodies. Many of the ideas encountered in one dimension have analogues in several dimensions. The description of multi-dimensional motions of viscoelastic materials will be discussed in Chapter I.

Consider the longitudinal motion of a homogeneous one-dimensional body, e.g. a bar of uniform cross-section.[1] We identify the body with an interval $B \subset \mathbb{R}$; for simplicity, we assume that the body occupies B in an unstressed state. The interval B is called a **reference configuration**; a typical point (or particle) in B will be denoted by x. In order to describe the motion, we follow the evolution of points in B. Therefore, the quantities of interest will be regarded as functions of the material point x and the time t.

We assume that the body has unit reference density, and we denote by $u(x,t)$ the displacement at time t of the particle with reference position x (i.e. $x+u(x,t)$ is the position at time t of this particle). The **strain** ϵ is given by

$$\epsilon(x,t) := u_x(x,t), \tag{1}$$

and the equation of balance of linear momentum has the form

$$u_{tt}(x,t) = \sigma_x(x,t) + f(x,t), \ x \in B, \ t \geq 0. \tag{2}$$

Here σ is the **stress** and f is a **body force**; we regard f as prescribed. Roughly speaking, σ is a contact force per unit area, which acts between the two sides of an imagined cross-section of the material, and f is an externally applied force per unit volume (or unit mass), such as gravity. We supplement (2) with a **constitutive assumption** relating the stress to the motion.

If the body is **elastic** then the stress depends on the strain through a constitutive equation of the form

$$\sigma(x,t) = \phi(\epsilon(x,t)), \tag{3}$$

where ϕ is a given smooth function with $\phi(0) = 0$ and $\phi'(0) > 0$. The condition $\phi(0) = 0$ reflects the assumption that the reference configuration be stress-free, while the assumption $\phi'(0) > 0$ means that stress increases with strain,[2] at least near $\epsilon = 0$. The resulting motion is then described by the quasilinear wave equation

$$u_{tt} = \phi(u_x)_x + f, \ x \in B, \ t \geq 0. \tag{4}$$

---

[1] This problem is somewhat artificial, since under realistic conditions a longitudinally deforming bar may experience a variation in its cross-section. More realistic problems in one space dimension will be discussed in Section 4 of Chapter I.

[2] This need not be the case globally; in fact non-monotone stress-strain relations are used to model phase transitions.

Our task is to determine a function u which satisfies (4) together with initial conditions

$$u(x,0) = u_0(x), \ u_t(x,0) = u_1(x), \ x \in B, \tag{5}$$

and suitable boundary conditions if B is not all of the real line. Equation (4) is **hyperbolic** as long as ϕ' is positive. The appropriate analogue of this condition in higher dimensions is the **strong ellipticity condition** which will be introduced in Chapter I.

In the linear case, $\sigma = c^2 \epsilon$, equation (4) reduces to the linear wave equation

$$u_{tt} = c^2 u_{xx} + f. \tag{6}$$

A well-known feature of (6) is that singularities in the data persist for all time and propagate with constant speed c. The amplitude of a singularity neither grows nor decays and, roughly speaking, the solution preserves the smoothness of the data. Indeed, in the special case $B = \mathbb{R}$ and $f \equiv 0$, the unique solution of (6), (5) is given by d'Alembert's formula

$$u(x,t) = \frac{1}{2}(u_0(x+ct) + u_0(x-ct)) + \frac{1}{2c}\int_{x-ct}^{x+ct} u_1(\xi) \, d\xi, \tag{7}$$

which shows that if the initial data are smooth then the solution is smooth for all time. In the nonlinear case, the initial value problem (4), (5) is well-posed locally in time (for an appropriate class of data), but singularities can develop in finite time. Even if the data are of class C^∞ and arbitrarily small, second derivatives of u can become infinite in finite time because the wave speed, $\sqrt{\phi'}$, is not constant. This issue will be discussed in more detail in Chapter II.

For certain materials, the stress depends on the strain ϵ and on the velocity gradient ϵ_t through a constitutive relation of the form[3]

$$\sigma(x,t) = \Psi(\epsilon(x,t), \epsilon_t(x,t)). \tag{8}$$

An instructive model for such a relation is

$$\sigma(x,t) = \phi(\epsilon(x,t)) + \eta\epsilon_t(x,t), \tag{9}$$

where ϕ is as in the elastic case and η is a positive constant. The resulting equation of motion,

$$u_{tt} = \phi(u_x)_x + \eta u_{xtx} + f, \tag{10}$$

is essentially parabolic. In contrast with elasticity, equation (10) has globally defined smooth solutions provided $\phi' > 0$ and the initial data and body force are smooth. Singularities in the initial data are not necessarily smoothed out, but

---

[3] For the linear case, such a theory was proposed by O.E. Meyer [M13] in 1874. The use of this theory to describe viscoelastic solids is criticized by Boltzmann [B11].

there are no propagating singularities. In fact, only singularities that do not move with respect to the material are possible; they result only from singularities in the data and do not develop spontaneously. Stationary singularities have no analogue in elasticity, but, as we shall see in Chapter II, they are present in materials with memory.

In a material with memory the stress at a material point x at time t depends on the entire history of the strain at x. The linearized constitutive relation for small deformations was given in 1874 by Boltzmann [B11]:

$$\sigma(x,t) = \beta\epsilon(x,t) + \int_{-\infty}^{t} m(t-\tau)(\epsilon(x,t) - \epsilon(x,\tau))\, d\tau. \tag{11}$$

Here β is a non-negative constant, and m is a positive, monotone-decreasing function. For (11) to be meaningful, it must be assumed that m is integrable at infinity and that $sm(s)$ is integrable at zero. Boltzmann remarks that a Newtonian viscosity (i.e. a term $\eta\epsilon_t$) can be obtained as a limiting case of (11), by choosing kernels which are large near zero and small elsewhere. Restated in modern terms, Boltzmann's argument is based on setting $m_N(s) = b_N(s)/s$ and choosing a delta sequence for $b_N(s)$; the integral then yields $\epsilon_t(x,t)$ in the limit as $N \to \infty$. In essence, this amounts to setting $m = -\delta'$, where δ is the Dirac delta function. If m in (11) is allowed to be a generalized function and the sign conditions on m and m' are suitably generalized, it can be shown [R9] that m must be a function plus a multiple of $-\delta'$.

If m is integrable, we can write (11) in the alternative form

$$\sigma(x,t) = c^2\epsilon(x,t) - \int_{-\infty}^{t} m(t-\tau)\epsilon(x,\tau)\, d\tau, \tag{12}$$

where $c^2 := \beta + \int_0^\infty m(s)\, ds$. The constant c^2 measures the instantaneous response of stress to strain and is called the **instantaneous stress modulus**; on the other hand, β determines the stress due to a constant strain history and is called the **equilibrium stress modulus**. If we set $\epsilon = 0$ for $t < 0$ and $\epsilon = \epsilon_0$ for $t > 0$, then the stress for $t > 0$ is given by

$$\sigma = \left(\beta + \int_t^\infty m(s)\, ds\right)\epsilon_0 =: G(t)\epsilon_0. \tag{13}$$

An experiment in which the strain is suddenly changed from 0 and held at a constant value is called a **stress relaxation test**; by measuring the stress one obtains the function $G(t)$, which is called the **stress relaxation modulus**. Obviously, $G(0) = c^2$ and $G(\infty) = \beta$. The function $m(t) = -G'(t)$ is called the **memory function**. The assumption that m is positive means that G is monotone-decreasing, i.e. that the stress following a sudden deformation relaxes with increasing time. It follows from (11) and the positivity of m that the stress needed to sustain a strain ϵ is diminished by previous strains of the same sign and increased by previous strains of the opposite sign. The assumption that m

4

is monotone-decreasing (i.e., that G is convex) means that a deformation which occurred in the distant past has less influence on the present stress than one which occurred in the more recent past,[4] i.e. the memory **fades** with time.

Substitution of (11) into the balance of linear momentum yields

$$u_{tt} = \beta u_{xx} + \int_{-\infty}^{t} m(t-\tau)(u_{xx}(x,t) - u_{xx}(x,\tau))\, d\tau + f, \qquad (14)_1$$

or

$$u_{tt} = c^2 u_{xx} - \int_{-\infty}^{t} m(t-\tau) u_{xx}(x,\tau)\, d\tau + f, \qquad (14)_2$$

where the second form is valid only if m is integrable. The history of u up to time $t = 0$ is assumed to be known. It is instructive to compare the behavior of solutions to (14) with that of solutions of the linear wave equation (6). For simplicity, we take $B = \mathbb{R}$, $u(\cdot, t) = 0$ for all $t < 0$, and $f \equiv 0$. However, we impose nontrivial initial conditions at zero, i.e. u is to satisfy the initial conditions (5) as $t \to 0+$. In this problem, two types of singularities can appear in the solution: those which propagate along characteristics, and stationary singularities. If m is smooth on $[0, \infty)$, the propagating singularities move with speed c as they would for the wave equation (6). However, in contrast to the wave equation, the amplitude A of such a singularity is exponentially damped; as we shall see in Chapter II,

$$A(t) = A(0) \exp[\frac{-m(0)t}{2c^2}], \qquad (15)$$

which is similar to the situation for the frictionally damped wave equation

$$u_{tt} + u_t = c^2 u_{xx}. \qquad (16)$$

If m is smooth on $[0, \infty)$, the stationary singularities in the solution are weaker (i.e., they occur in higher order derivatives) than those that propagate. As $t \to \infty$, the amplitude of a stationary singularity decays to zero if $\beta > 0$, but does not decay if $\beta = 0$.

The situation is different when $m(s)$ has a singularity[5] as $s \to 0+$. If we formally set $m(0) = \infty$ in (15) then $A(t)$ becomes zero for positive t, which

---

[4] Boltzmann's original statement reads as follows: "...wobei jedoch eine Dehnung einen um so geringeren Einfluß hat, vor je längerer Zeit sie stattfand; und zwar ist die Kraft, welche zur Erzeugung einer bestimmten Dehnung erforderlich ist, geringer, wenn schon früher eine Deformation im gleichen Sinne Statt hatte."

[5] The possibility of a singular kernel was already considered by Boltzmann [B11], who proposed that $m(s) \sim s^{-1}$ (except for very large s) and fitted such a kernel to torsion measurements on a filament of glass. It is not possible to determine $G(0)$ or $m(0)$ from a stress relaxation test. Nevertheless, it is noteworthy that data from certain experiments are compatible with the possibility of a singular kernel. Oscillatory measurements which determine the Fourier transform $\hat{G}(i\omega)$ in molten polyethylene [L2] show a behavior at large ω that is consistent

suggests that propagating singularities are smoothed out. This is indeed the case, with the degree of smoothing depending on the precise nature of the singularity in m. If m is integrable, it can be shown that disturbances still propagate with the finite wave speed c and hence equation (14) exhibits hyperbolic as well as parabolic behavior. We note, however, that singular kernels do not smooth out stationary singularities. A detailed discussion of linear wave propagation, with particular emphasis on singular kernels, will be given in Chapter II.

Nonlinear materials with memory are described by constitutive relations of the form

$$\sigma(x,t) = \mathcal{G}(\epsilon^t(x,\cdot)), \tag{17}$$

where the **strain history** $\epsilon^t(x,\cdot) : [0,\infty) \to \mathbb{R}$ is defined by

$$\epsilon^t(x,s) := \epsilon(x,t-s), \tag{18}$$

and $\mathcal{G}$ is a **functional** defined on a suitable class of histories. We leave the discussion of constitutive theories of viscoelastic materials to Chapter I, and cite only a specific model which has been studied by a number of authors. We choose this model because it is relatively simple mathematically but yet retains many important qualitative features of more general models. The constitutive relation is

$$\sigma(x,t) = \phi(\epsilon(x,t)) - \int_{-\infty}^{t} m(t-\tau)\psi(\epsilon(x,\tau))\, d\tau, \tag{19}$$

where ϕ and ψ are given smooth functions with

$$\phi(0) = \psi(0) = 0, \ \phi'(0) > 0, \ \psi'(0) > 0, \ \phi'(0) - \psi'(0)\int_0^\infty m(s)\, ds > 0, \tag{20}$$

and m is a positive, monotone-decreasing function. The equation of motion resulting from (19) is

$$u_{tt} = \phi(u_x)_x - \int_{-\infty}^{t} m(t-s)\psi(u_x(x,s))_x\, ds + f, \ x \in B, \ t \geq 0. \tag{21}$$

The function ϕ measures the **instantaneous elastic response** of the material, whereas the **equilibrium elastic response** is characterized by the equilibrium stress function χ defined by

$$\chi(\xi) := \phi(\xi) - \psi(\xi)\int_0^\infty m(s)\, ds. \tag{22}$$

---

with a singular kernel. For certain materials, the values of $G(0)$ computed from wave speed measurements [J7] are much larger than those obtained by extrapolating stress relaxation data, which suggests the use of a singular kernel. In addition, various molecular theories of polymers [D11],[R17],[Z1] have predicted power-type singularities in m. Of course, from a practical point of view, whether or not it is reasonable to consider singular kernels in a given problem depends crucially on the time scales involved.

As initial data for (21) the history of u prior to time $t = 0$, as well as the values of $u(\cdot, 0^+)$ and $u_t(\cdot, 0^+)$, are prescribed. (In general, we do not assume that $u(\cdot, 0^+) = u(\cdot, 0^-)$ or $u_t(\cdot, 0^+) = u_t(\cdot, 0^-)$.) We shall often assume that u vanishes identically for $t < 0$, which leads to the initial value problem

$$u_{tt} = \phi(u_x)_x - \int_0^t m(t-s)\psi(u_x(x,s))_x \; ds + f, \quad x \in B, \; t \geq 0, \qquad (23)$$

$$u(x,0) = u_0(x), \; u_t(x,0) = u_1(x), \quad x \in B; \qquad (24)$$

we supplement (23), (24) with suitable boundary conditions if B is not the entire real line. If the history of u is not zero, its contribution to the memory in (21) can be incorporated into the body force f in (23).

If m is smooth on $[0, \infty)$, then equation (23) is essentially hyperbolic. The integral can be regarded as a lower order perturbation term, since convolution with a smooth kernel acts like a differential operator of order -1. The idea of treating the memory term as a perturbation can be used to prove local existence for the initial value problem (23), (24). Of course, this approach does not use the special structure of (23) and is easily extended to other equations with similar structure. Local existence theorems for viscoelastic materials are discussed in Chapters III and V.

Observe that if $m \equiv 0$ then (21) reduces to the quasilinear wave equation (4). If $m \not\equiv 0$ and appropriate sign conditions hold, the memory term has a damping effect which competes with the tendency of the quasilinear wave equation to produce singularities. Significant information about the competition between the memory effects and the nonlinear elastic response in materials with fading memory was provided by the work of Coleman and Gurtin [C8] on the propagation of acceleration waves into a one-dimensional medium at rest. (An acceleration wave is similar to a shock wave, the difference being that second – rather than first – derivatives of u experience jumps accross the wave front.) For the model equation (21), with m smooth on $[0, \infty)$, $B = \mathbb{R}$, and $f \equiv 0$, the results of [C8] imply that the amplitude A of an acceleration wave satisfies the Riccati-Bernoulli equation

$$\dot{A}(t) = \alpha A^2(t) - \beta A(t), \qquad (25)$$

with material constants α and β given by

$$\alpha = -\frac{\phi''(0)}{2[\phi'(0)]^{3/2}}, \; \beta = \frac{m(0)\psi'(0)}{2\phi'(0)}. \qquad (26)$$

Thus if $\phi''(0) < 0$ and $A(0) < \frac{\beta}{\alpha}$, the corresponding solution $A(t)$ of (25) decays to zero as $t \to \infty$; if $A(0) > \frac{\beta}{\alpha}$, the solution $A(t)$ becomes infinite at a finite time T_0. (The case $\phi''(0) > 0$ is similar.)

This behavior of acceleration waves suggests that if m is smooth on $[0, \infty)$, the initial value problem (23), (24) should have globally defined smooth solutions for data which are sufficiently smooth and small (and compatible with the boundary

7

conditions if B is not $\mathbb{R}$). On the other hand, if $\phi''(0) \neq 0$, then smooth solutions may develop singularities in finite time if the data are suitably large. It will be seen in Chapters II and IV that such results hold for a variety of models of materials with fading memory.

The situation concerning singular kernels is less completely understood. For local existence, the memory term can no longer be treated as a perturbation. Local existence of smooth solutions of (21) has been proved, but the result is not as widely generalizable as for a regular kernel; global existence for smooth and small data has been proved for initial-boundary value problems on finite intervals, but not for the Cauchy problem with $B = \mathbb{R}$. The question of global existence for large data is rather intriguing. On the one hand, singular kernels lead to smoothing in the linearized problem; on the other hand, we shall see in Chapter II that they do not preclude the existence of steady shocks. It is not known whether or not singularities can develop from smooth data. However, there are results on global existence of weak solutions to certain model problems with singular kernels.

The constitutive relation (19) is an example of an **integral model**. Another popular class of constitutive models for viscoelastic materials is the class of **differential models** in which a differential equation for the stress is given. A simple example of a differential constitutive law is

$$\sigma_t + \lambda\sigma = \phi'(\epsilon)\epsilon_t + \lambda\phi(\epsilon) - \psi(\epsilon) \tag{27}$$

where λ is a positive constant and ϕ and ψ are given smooth functions. We note that (27) can also be expressed in integral form. Indeed, under mild assumptions on the behavior of σ and ϵ as $t \to -\infty$, (27) can be written in the form (19) with $m(s) := e^{-\lambda s}$. However, it is not always possible to express a differential model in integral form. Constitutive models for viscoelastic materials will be discussed in more detail in Section 3 of Chapter I.

We now give an outline of the material discussed in this monograph.

In Chapter I, we review some basic concepts from continuum mechanics and introduce the notions required to describe the motion of multi-dimensional bodies. We then discuss the constitutive laws that are used to model viscoelastic materials. This discussion motivates the formulation of specific model equations that will be studied in later chapters. We conclude Chapter I with a discussion of special situations in which the equations of motion lead (exactly or approximately) to one-dimensional problems.

Chapter II is concerned with the propagation and development of singularities in viscoelastic media. We begin by discussing the classification of equations according to types (elliptic, hyperbolic, parabolic) and the possibility of a change of type in multidimensional problems. We then focus on the one-dimensional case and discuss linear wave propagation with particular emphasis on the effect of a singularity in the kernel. For nonlinear problems arising from differential constitutive models and from integral models with smooth kernels we discuss the development of singularities from smooth (but large) initial data. Finally, we study jump conditions that must be satisfied across a shock, and the existence of steady shock and acceleration waves.

Chapter III deals with the problem of local (in time) existence of smooth solutions. The results are based on the contraction mapping principle together with a priori estimates of the energy type. We prove a local existence theorem for Dirichlet initial-boundary value problems for a class of integral models in three dimensions. The principal terms of the equations are similar to those in nonlinear elasticity, and the memory term is treated as a perturbation. We also address the question of how the arguments should be modified when incompressible materials are considered; an existence result for the Dirichlet initial-boundary value problem of incompressible elasticity is established. Finally, we discuss the model problem (21) with a singular kernel. Although some generalizations are possible, the method used relies more essentially on the special form of the equation than in the case of a smooth kernel.

Questions of global existence and asymptotic behavior of smooth solutions are addressed in Chapter IV. We begin with a discussion of one-dimensional model problems; global existence of smooth solutions for problems with small data is established. We first treat a parabolic problem (involving a Newtonian viscosity and a memory term) by means of the implicit function theorem. This method is not applicable to hyperbolic problems, and instead we use a continuation argument based on energy estimates . In order to illustrate the "energy method" we prove a global existence theorem for the quasilinear wave equation with frictional damping. We then consider the initial value problem (23), (24); we treat initial-boundary value problems with B a finite interval, as well as the Cauchy problem with $B = \mathbb{R}$. The latter case is more subtle because the Poincaré inequality cannot be used in the estimates. We then use the energy method to study a three-dimensional problem (with small data); more precisely we establish global existence of spatially periodic motions of a class of fluids called K-BKZ fluids. The chapter concludes with a global existence result for a parabolic problem (in one space dimension) with large initial data; this result is based on energy estimates and a comparison argument.

The global existence results discussed in Chapter IV can also be viewed as proofs for the stability of the equilibrium state. In fluid dynamics, it is of great interest to know when certain flows, such as shearing motions between parallel plates or rotating cylinders, are stable. Eigenvalues for certain specific models of viscoelastic fluids have been computed numerically, but there are essentially no analytical results on stability of flows of viscoelastic fluids within a general class of motions. We remark also that the relation between stability and the eigenvalue spectrum is by no means obvious when the equations of motion are hyperbolic. (See Section 4.4 of [P1].)

Abstract semigroup theory has been a powerful tool in dealing with initial value problems for partial differential equations, and it can also be applied to partial integrodifferential equations arising in viscoelasticity. This will be the subject of Chapter V. We begin with a review of the relevant results from semigroup theory. We then discuss various possibilities of dealing with history dependence in an abstract context. It is possible to represent a differential-delay equation as a differential equation on a history space, and in fact there are many ways to do so.

Some of the options, as well as their advantages and disadvantages, are discussed. After presenting some results on linear viscoelasticity we discuss local existence for one-dimensional problems of parabolic and of hyperbolic type. Finally, we treat three-dimensional problems with incompressibility; we establish local existence theorems for a class of fluids which in a sense perturb Newtonian fluids, and for K-BKZ fluids.

The sixth and final chapter is devoted to steady flows of viscoelastic fluids. Despite the importance of the subject, relatively little is known mathematically about the existence of such flows. For a class of fluids with differential constitutive relations, we prove the existence of slow steady flows which perturb a state of rest. We also address the question of inflow boundary conditions. Because of the memory of the fluid, extra boundary conditions, in addition to those familiar from Newtonian fluid mechanics, are needed at inflow boundaries. The precise nature of such boundary conditions depends highly on the constitutive model, and there seems to be no general principle to formulate them. Although this casts some doubt on the physical significance of the question, the problem is nevertheless important because in numerical simulations it is necessary to truncate infinite domains. We shall characterize inflow boundary conditions for differential constitutive models with a single relaxation mode. It would be of great interest to obtain existence results for steady flows which are not perturbations of rest or rigid motion. Numerical simulations of such flows have encountered great difficulties, and nonexistence has been suspected as a cause. However, no analytical results are known and the numerical evidence is anything but conclusive.

I Continuum mechanics for viscoelastic materials

1. Kinematics and balance laws

The description of the behavior of viscoelastic materials is more complicated than that of either elastic solids or Newtonian fluids. In an elastic solid, the stress is determined by the deformation of the material relative to a fixed reference configuration, while the stress in a Newtonian fluid is determined by the velocity field (and the density, if the fluid is compressible). Viscoelastic materials exhibit behavior between that of classical solids and classical fluids; the stresses in such materials are determined neither by their current state of deformation nor by their current state of motion. Rather, the entire history of the deformation has to be considered; therefore such materials are called **materials with memory**. In this chapter, we provide the necessary background required to understand the modeling of such materials. (For a more extensive discussion, we refer to textbooks on viscoelastic materials. See [B8],[C3],[C6],[H18],[L8],[P3],[P7],[S3],[T5].) The first two sections contain a review of some basic concepts from continuum mechanics and an explanation of our notation. Section 3 is concerned with the formulation of constitutive laws for viscoelastic materials. The basic concepts will be presented in a three-dimensional setting; specializations to one-dimensional motions will be discussed in Section 4. We assume that the reader has some familiarity with continuum mechanics and tensor analysis; for a comprehensive discussion, see e.g. [G18].

By $x = (x^1, x^2, x^3)$ we denote **material** or **Lagrangian coordinates**. One can think of x as the position of a material point in some (fixed) configuration which the body might in principle occupy (or does actually occupy at a certain time). This configuration is then called a **reference configuration**, which we usually denote by Ω. A motion of the body is determined by the positions of material points in space as functions of the reference position x and the time t. Let $y = (y^1, y^2, y^3) = \chi(x, t)$ denote the spatial position at time t of the material point with reference postion x. It will always be assumed that χ is continuous and that $\chi(\cdot, t)$ is invertible for each fixed t; we write $\chi^{-1}(\cdot, t)$ for the inverse of χ with respect to its first argument. Every quantity defined on the body can be regarded either as a function of x and t or as a function of y and t. The latter description of motion is called **Eulerian** or **spatial**.

The **deformation gradient** **F** is defined by

$$\mathbf{F}(x,t) := \frac{\partial \chi(x,t)}{\partial x}; \quad \text{in components:} \quad F_\alpha^i(x,t) := \frac{\partial \chi^i(x,t)}{\partial x^\alpha}. \tag{1}$$

Throughout this monograph, we adopt the summation convention on repeated indices; we use Latin indices for spatial components and Greek indices for components in material coordinates. Tensor quantities will be written in boldface, while scalar and vector quantities will be written in light-face print. The determinant of **F** measures the local change of volume; in order to preclude compression to zero volume, we shall assume that det **F** is strictly positive. This assumption guarantees that $\chi(\cdot,t)$ is locally invertible. (Under certain boundary conditions, one can show that det $\mathbf{F} > 0$ implies global invertibility [B2],[M12].) Many viscoelastic materials are rather easily deformed in a volume preserving way, but show little compressibility. Some attention will therefore be focussed on **incompressible** materials. A body is incompressible if it is possible to choose the reference configuration in such a way that det $\mathbf{F}(x,t) \equiv 1$ in every motion.

The equations of motion will involve time and space derivatives of various quantities. We have to distinguish between $\frac{\partial}{\partial t}|_{x=const.}$ and $\frac{\partial}{\partial t}|_{y=const.}$. The first of these is called a **material** or **Lagrangian time derivative**, while the second is called an **Eulerian time derivative**. For purposes of the present chapter, we use a dot to denote the material time derivative and a prime to denote the Eulerian time derivative. We use the symbols ∇_x and ∇_y to denote gradients in the Lagrangian and Eulerian frames, and we always write $v = \dot{y}$ for the velocity. The material and Eulerian time derivatives of a quantity a are related by

$$\dot{a} = a' + (\dot{y} \cdot \nabla_y)a = a' + (v \cdot \nabla_y)a. \tag{2}$$

For future use, we record the following identities concerning the time derivative of the deformation gradient:

$$\dot{\mathbf{F}} = \nabla_x v = (\nabla_y v)\mathbf{F}, \tag{3$_1$}$$

and

$$(\det \mathbf{F})^\cdot = (\det \mathbf{F}) \, \mathrm{div}_y \, v. \tag{3$_2$}$$

Throughout this book, x will always stand for Lagrangian coordinates and y for Eulerian coordinates; when there is no danger of confusion, we shall omit subscripts on ∇. In the subsequent chapters, we shall not adhere to a rigid convention concerning notation for time derivatives; the symbol $\frac{\partial}{\partial t}$, a subscript t or a dot may denote either a material or an Eulerian time derivative, depending on context, and we shall drop the notation of a prime for the Eulerian time derivative.

Although our spatial coordinates are always Cartesian, we use both upper and lower indices as a matter of convenience. For the gradient of a vector w, we use the convention $(\nabla_y w)_j^i := \frac{\partial w^i}{\partial y^j}$, and $(\nabla_x w)_\alpha^i := \frac{\partial w^i}{\partial x^\alpha}$. As far as matrix products are concerned, the upper index is regarded as the row index.

An objective of continuum mechanics is the formulation and solution of equations that determine the function χ. The equations governing the motion of a body are the equations of mass conservation and balance of momentum. If $\rho_0(x)$ denotes the reference density (i.e. the mass per unit reference volume) and $\rho(y,t)$ the density at time t (i.e. the mass per unit volume in the configuration at time t), the equation of mass conservation reads

$$\rho(\chi(x,t),t) \det \mathbf{F}(x,t) = \rho_0(x). \tag{4}$$

We always assume $\rho_0 > 0$. By taking the material time derivative of (4) and using (3), one obtains

$$\dot{\rho} + \rho \nabla_y \cdot v = 0, \tag{5}$$

or equivalently,

$$\rho' + \nabla_y \cdot (\rho v) = 0. \tag{6}$$

Equation (6) is often called the equation of continuity. For incompressible materials, equation (4) simplifies to

$$\rho(\chi(x,t),t) = \rho_0(x), \tag{7}$$

i.e. $\dot{\rho} = 0$, and hence (5) reduces to

$$\nabla_y \cdot v = 0. \tag{8}$$

The equation of balance of linear momentum involves the concept of **stress**, i.e. force per unit area (see [G18] for more information). For different purposes, it is convenient to use different notions of stress. In Eulerian coordinates, it is natural to use the Cauchy stress tensor $\mathbf{T}$, which measures force per unit area in the present configuration of the body. The balance of linear momentum reads

$$\rho(v' + (v \cdot \nabla_y)v) = \operatorname{div}_y \mathbf{T} + \rho b. \tag{9}$$

Here b denotes a body force per unit mass; we regard b as prescribed. If (9) holds (and there are no extraneous couples), then balance of angular momentum is satisfied if and only if $\mathbf{T}$ is symmetric. Symmetry of $\mathbf{T}$ will be imposed as a restriction on the constitutive relation rather than a constraint on the motion. In Lagrangian coordinates, the equation equivalent to (9) is

$$\rho_0 \ddot{y} = \operatorname{div}_x \mathbf{S} + \rho_0 b. \tag{10}$$

The tensor $\mathbf{S}$ is called the **Piola-Kirchhoff stress**; it measures force per unit area in the reference configuration. The Cauchy and Piola-Kirchhoff stress are related by

$$\mathbf{S}(x,t) = (\det \mathbf{F}(x,t)) \mathbf{T}(\chi(x,t),t) \mathbf{F}^{-T}(x,t). \tag{11}$$

13

Here $\mathbf{F}^{-T}$ denotes the inverse of the transpose of $\mathbf{F}$. The symmetry of $\mathbf{T}$ is equivalent to the relation $\mathbf{SF}^T = \mathbf{FS}^T$. Note that $\mathbf{S}$ is generally not symmetric; its divergence is defined by summation over the second index:

$$[\text{div}_x \ \mathbf{S}]^i (x,t) := \sum_{\alpha=1}^{3} \frac{\partial}{\partial x^\alpha} \mathbf{S}^{i\alpha}. \tag{12}$$

It is also convenient to introduce the **body stress**

$$\mathbf{\Pi}(x,t) := \mathbf{F}^{-1}(x,t)\mathbf{T}(\chi(x,t),t)\mathbf{F}^{-T}(x,t). \tag{13}$$

Note that $\mathbf{\Pi}$ is symmetric.

In order to formulate a dynamical problem, the equations of mass conservation and balance of momentum will be supplemented by a relationship between stress and deformation, and by appropriate boundary and initial conditions. The relation between stress and deformation is called a **constitutive law**. A general discussion of constitutive laws for materials with memory will be given in the next section; specific constitutive models will be discussed in Section 3. The formulation of boundary conditions presents no major problem if the boundary of the domain considered coincides with a material boundary (i.e. if no material moves through the boundary). As in classical elasticity or fluid mechanics, it is reasonable to prescribe either the positions of particles on the boundary as a function of time (displacement boundary conditions) or the **tractions** (i.e. forces per unit area) on the boundary. Let n denote the unit normal of the boundary in the present configuration and m the unit normal of the boundary in the reference configuration. It is natural to consider two types of tractions: $\mathbf{T} \cdot n$ measures force per unit area in the present configuration, while $\mathbf{S} \cdot m$ measures force per unit area in the reference configuration. In fluid mechanics, one often wants to consider problems in which material moves through the boundary. While this presents no problem for Newtonian fluids, the memory of viscoelastic fluids renders the behavior of the fluid inside the domain dependent on the history of its deformation prior to entering the domain. This raises serious questions, which have not been adequately answered, about boundary conditions to be imposed at an inflow boundary, i.e. at a portion of the boundary where material enters the domain. The conditions that are needed at inflow boundaries depend on the constitutive model. A special case will be discussed in Chapter VI.

A motion is called **steady** if the velocity, density and Cauchy stress are independent of time in the Eulerian sense, i.e. they depend only on y. For solutions which are time-independent in the Lagrangian sense (i.e. $\dot{y} = 0$, $\dot{\mathbf{S}} = 0$), there is no difference between the equations governing viscoelastic materials and those governing elastic materials.

14

2. General principles for constitutive laws

Constitutive relations generally involve the deformation gradient $\mathbf{F}$; in particular, the **polar decomposition** of $\mathbf{F}$ is important. For any matrix $\mathbf{F}$ with positive determinant there are unique positive definite symmetric matrices $\mathbf{U}$ and $\mathbf{V}$ and a unique proper orthogonal matrix $\mathbf{Q}$ such that

$$\mathbf{F} = \mathbf{QU} = \mathbf{VQ}. \tag{14}$$

Physically, this means that the deformation is decomposed into stretching and rotation. The tensors $\mathbf{U}$ and $\mathbf{V}$ are called the **right and left stretch tensors**. The **right and left Cauchy-Green tensors** $\mathbf{C}$ and $\mathbf{B}$ are defined by

$$\mathbf{C} := \mathbf{U}^2 = \mathbf{F}^T\mathbf{F}, \text{ and } \mathbf{B} := \mathbf{V}^2 = \mathbf{FF}^T. \tag{15}$$

It is also customary to refer to $\mathbf{C}$ as the **Cauchy strain** and to $\mathbf{B}$ as the **Finger strain**.

For a homogeneous elastic material, the constitutive relation has one of the following equivalent forms:

$$\mathbf{T}(\chi(x,t),t) = \tilde{\mathbf{T}}(\mathbf{F}(x,t)), \tag{16$_1$}$$

$$\mathbf{S}(x,t) = \tilde{\mathbf{S}}(\mathbf{F}(x,t)), \tag{16$_2$}$$

$$\mathbf{\Pi}(x,t) = \tilde{\mathbf{\Pi}}(\mathbf{F}(x,t)), \tag{16$_3$}$$

and the reference density ρ_0 is constant. The equations of motion, in material coordinates, read

$$\rho_0 \ddot{y}^i = \frac{\partial}{\partial x^\alpha} \tilde{S}^{i\alpha} (\nabla_x y) + \rho_0 b^i. \tag{17}$$

If we define the **elasticity tensor** $\mathbf{A}$ by

$$A^{\alpha\beta}_{ij}(\mathbf{F}) := \frac{\partial \tilde{S}^{i\alpha}}{\partial F^j_\beta}(\mathbf{F}), \tag{18}$$

we can rewrite (17) in the form

$$\rho_0 \ddot{y}^i = A^{\alpha\beta}_{ij}(\nabla y)\frac{\partial^2 y^j}{\partial x^\alpha \partial x^\beta} + \rho_0 b^i. \tag{19}$$

The requirement that work be non-negative over closed cycles implies that $\mathbf{A}$ have the following symmetry:

$$A^{\alpha\beta}_{ij} = A^{\beta\alpha}_{ji}. \tag{20}$$

15

If the domain of $\tilde{\mathbf{S}}$ is simply connected, then (20) implies the existence of a **stored energy function** $W = W(\mathbf{F})$ such that[1]

$$\tilde{S}^{i\alpha}(\mathbf{F}) = \frac{\partial W}{\partial F^i_\alpha}(\mathbf{F}). \tag{21}$$

Various generalizations of the one-dimensional notion that "stress increases with strain" have been proposed (cf. [M8] for a summary). In this monograph we employ the **strong ellipticity condition**

$$A^{\alpha\beta}_{ij}(\mathbf{F})\xi^i\xi^j\eta_\alpha\eta_\beta > 0, \ \forall \xi, \eta \in \mathbb{R}^3 \text{ with } |\xi| = |\eta| = 1. \tag{22}$$

This condition is physically reasonable for a large class of elastic materials, and it is exactly the right condition to make initial value problems for (19) (with Dirichlet boundary conditions or on all of space) well-posed locally in time (cf. Section III.2).

Constitutive laws should satisfy the principle of **material objectivity** or **frame indifference**. This principle requires that stress results only from deformations and is unaffected (except for orientation) by rigid rotations of the body. For materials described by (16), this restriction takes the form

$$\mathbf{Q}\tilde{\mathbf{T}}(\mathbf{F})\mathbf{Q}^T = \tilde{\mathbf{T}}(\mathbf{QF}), \ \forall \mathbf{Q} \in \mathrm{SO}(3), \tag{23$_1$}$$

where $\mathrm{SO}(3)$ is the proper orthogonal group. In terms of the Piola-Kirchhoff stress, frame indifference takes the form

$$\mathbf{Q}\tilde{\mathbf{S}}(\mathbf{F}) = \tilde{\mathbf{S}}(\mathbf{QF}), \ \forall \mathbf{Q} \in \mathrm{SO}(3), \tag{23$_2$}$$

and in terms of the body stress

$$\tilde{\mathbf{\Pi}}(\mathbf{F}) = \tilde{\mathbf{\Pi}}(\mathbf{QF}), \ \forall \mathbf{Q} \in \mathrm{SO}(3). \tag{23$_3$}$$

This last relation implies that $\mathbf{\Pi}$ depends only on the right stretch tensor $\mathbf{U}$, or equivalently, only on the Cauchy strain tensor $\mathbf{C}$.

The constitutive equation for a compressible Newtonian fluid is

$$\mathbf{T} = -p(\rho)\mathbf{1} + 2\eta\left(\mathbf{D} - \frac{1}{3}\,\mathrm{div}_y\ v\ \mathbf{1}\right) + \varsigma\ \mathrm{div}_y\ v\ \mathbf{1}, \tag{24}$$

where $\mathbf{D}$ is the symmetric part of the velocity gradient $\nabla_y v$, i.e.

$$\mathbf{D} := \frac{1}{2}\left(\nabla_y v + (\nabla_y v)^T\right), \tag{25}$$

---

[1] Some authors use the term elastic for materials described by (16) without assuming the symmetry condition (20) and then refer to elastic materials which have a stored energy function as hyperelastic. In this monograph, elastic materials will always be assumed to satisfy (20).

η and ς are positive constants (or possibly functions of the density ρ), p denotes the hydrostatic pressure, and $\mathbf{1}$ is the identity tensor.

For **incompressible materials**, we have the constraint det $\mathbf{F} = 1$. Due to this constraint, the stress cannot be completely determined by a constitutive law. The stress tensor in an incompressible material is given by

$$\mathbf{T} = -p\mathbf{1} + \mathbf{T}_E, \tag{26}$$

where $\mathbf{T}_E$, called the **extra stress**, is given by a constitutive relation and the **pressure** p can only be determined by solving the equations of motion. When there is no potential for confusion, we shall often omit the subscript E.

For classical incompressible Newtonian fluids, the constitutive law for the extra stress is

$$\mathbf{T}_E = 2\eta\mathbf{D}, \tag{27}$$

and from (5) and (9) one obtains the Navier-Stokes equations

$$\rho\big(v' + (v \cdot \nabla_y)v\big) = \eta\Delta_y v - \nabla_y p, \tag{28}_1$$

$$\text{div}_y \; v = 0. \tag{28}_2$$

For elastic solids, one usually employs the Lagrangian description of motion, while Eulerian coordinates are generally more convenient for describing Newtonian fluids. As noted earlier, viscoelastic materials have properties between those of elastic solids and viscous fluids, and it is useful to employ both descriptions of motion.

The stress in a viscoelastic material is a functional of the history of the deformation. In this monograph, we consider only **simple materials** [N7]. These are characterized by the condition that the stress depends only on values of the deformation gradient at the same material point x, and not on the deformation gradient at other points. We write the constitutive law for a simple material in the form of a functional of the history:

$$\mathbf{T}(\chi(x,t),t) = \mathcal{F}\big(\mathbf{F}^t(x,\cdot),x\big), \tag{29}$$

where here and throughout a superscript t indicates the history up to time t of a function, i.e.

$$\mathbf{F}^t(x,s) := \mathbf{F}(x,t-s) \quad \forall s \in [0,\infty). \tag{30}$$

Observe that $\mathbf{T}$ depends on the values of $\mathbf{F}$ at all prior times $t-s$, and possibly on the material point x itself.[2] In the present section, we shall discuss only

---

[2] Some authors, e.g. Oldroyd [O3], have proposed to generalize (29) to a relation between $\mathbf{T}$ and the deformation history, which need not be uniquely solvable for $\mathbf{T}$ (with little justification, Oldroyd claims that, on the other hand, the deformation should depend uniquely on the stress history). Specific constitutive laws which allow $\mathbf{T}$ to depend on the deformation history in a nonunique fashion have been proposed by de Gennes [G1] and Hinch [H7] on the basis of molecular theories. We refer to Fan and Bird [F1] for a criticism of the molecular arguments.

purely algebraic conditions on the functional $\mathcal{F}$. We therefore do not specify a space of functions on which $\mathcal{F}$ is defined. In particular, $\mathcal{F}$ may depend on temporal derivatives of the deformation gradient $\mathbf{F}$, and the constitutive relation (29) therefore permits a Newtonian viscosity. For **incompressible simple materials**, the Cauchy stress is given by (26), where the extra stress is given by a constitutive relation as above.[3]

We now introduce some specific tensors related to the motion, which will appear in constitutive laws for simple materials. It is often convenient to consider the gradient of the position of particles at time τ with respect to the positions at some other time t. We define the **relative deformation gradient** $\mathbf{F}_r$ by

$$\mathbf{F}_r(\tau, \chi(x,t), t) := \mathbf{F}(x,\tau)\mathbf{F}^{-1}(x,t), \tag{31}$$

and the **relative Cauchy strain** $\mathbf{C}_r$ by

$$\mathbf{C}_r(\tau, y, t) := \mathbf{F}_r^T(\tau, y, t)\mathbf{F}_r(\tau, y, t). \tag{32}$$

We note that

$$\frac{\partial}{\partial \tau}\mathbf{F}_r(\tau, y, t)\big|_{\tau=t} = \dot{\mathbf{F}}(x,t)\mathbf{F}^{-1}(x,t) = \nabla_y v(y,t), \tag{33}$$

where $y = \chi(x,t)$.

The tensors defined by

$$\mathbf{A}_n(y,t) := \frac{\partial^n}{\partial \tau^n}\mathbf{C}_r(\tau, y, t)\big|_{\tau=t} \tag{34}$$

are known as **Rivlin-Ericksen tensors** [R15], although they appear to have been introduced earlier by Oldroyd [O1]. It is immediately verified that

$$\mathbf{A}_1 = (\nabla_y v)^T + \nabla_y v = 2\mathbf{D}, \tag{35}$$

while the higher Rivlin-Ericksen tensors satisfy the recursion formula

$$\mathbf{A}_{n+1} = \dot{\mathbf{A}}_n + (\nabla_y v)^T \mathbf{A}_n + \mathbf{A}_n(\nabla_y v). \tag{36}$$

For simple materials, the principle of frame indifference requires that

$$\mathbf{Q}(t)\mathcal{F}(\mathbf{F}^t(x,\cdot), x)\mathbf{Q}^T(t) = \mathcal{F}(\mathbf{Q}^t(\cdot)\mathbf{F}^t(x,\cdot), x) \tag{37}$$

---

[3] In certain situations, especially at high pressure, $\mathbf{T}_E$ must be allowed to depend on p. For example, the viscosity of an incompressible Newtonian fluid may depend on the pressure, as already remarked by Stokes [S15]. We refer to [G15] for experiments on the dependence of viscosity on pressure. Some mathematical issues arising from a pressure-dependent viscosity are discussed in [R12]. For a general discussion of material constraints in continuum mechanics see [A9].

for every function $\mathbf{Q}$ taking values in $SO(3)$. Frame indifference takes a particularly simple form when the body stress $\mathbf{\Pi}$ is used. If we define

$$\mathcal{G}\big(\mathbf{F}^t(x,\cdot),x\big) := \mathbf{F}^{-1}(x,t)\mathcal{F}\big(\mathbf{F}^t(x,\cdot),x\big)\mathbf{F}^{-T}(x,t), \qquad (38)$$

then (37) reduces to

$$\mathcal{G}\big(\mathbf{Q}^t(\cdot)\mathbf{F}^t(x,\cdot),x\big) = \mathcal{G}\big(\mathbf{F}^t(x,\cdot),x\big). \qquad (39)$$

In other words, frame indifference requires that $\mathcal{G}$ depend only on the history of the right stretch tensor $\mathbf{U}$ defined in (15). Since $\mathbf{U} = \sqrt{\mathbf{C}}$, we may write

$$\mathcal{G}\big(\mathbf{F}^t(x,\cdot),x\big) = \mathcal{G}\big(\mathbf{U}^t(x,\cdot),x\big) = \mathcal{H}\big(\mathbf{C}^t(x,\cdot),x\big). \qquad (40)$$

The mathematical arguments used in this monograph make no explicit use of consequences of frame indifference; however, we make no assumptions which conflict with frame indifference.

A tensor quantity $\mathbf{P}$ defined on motions of the body is called **objective** if it is transformed to $\mathbf{Q}(t)\mathbf{P}(t)\mathbf{Q}^T(t)$ under a time-dependent proper rotation $\mathbf{Q}(t)$ of the body. Frame indifference requires that the Cauchy stress $\mathbf{T}$ be objective, and consequently, objective tensors play a fundamental role in formulating constitutive laws for $\mathbf{T}$. Examples of such tensors are the left Cauchy-Green tensor $\mathbf{B}$, the symmetric part $\mathbf{D}$ of the velocity gradient, the relative Cauchy strain $\mathbf{C}_r$ and all of the Rivlin-Ericksen tensors. It is important to note that the material time derivative of an objective tensor usually is not objective, because

$$(\mathbf{Q}\mathbf{P}\mathbf{Q}^T)^{\cdot} = \dot{\mathbf{Q}}\mathbf{P}\mathbf{Q}^T + \mathbf{Q}\dot{\mathbf{P}}\mathbf{Q}^T + \mathbf{Q}\mathbf{P}\dot{\mathbf{Q}}^T, \qquad (41)$$

which generally does not equal $\mathbf{Q}\dot{\mathbf{P}}\mathbf{Q}^T$. However, it is possible to define objective time derivatives; we shall encounter some in Section 3.

Throughout this monograph, we consider only materials which are **homogeneous**, i.e. the constitutive functional does not depend explicitly on x. We therefore write

$$\mathbf{T}(\chi(x,t),t) = \mathcal{F}\big(\mathbf{F}^t(x,\cdot)\big) \qquad (42)$$

in place of (29). **Material symmetry** imposes further restrictions on the constitutive law. The symmetry group of a material is the set of all proper unimodular matrices $\mathbf{H}$ for which

$$\mathcal{F}\big(\mathbf{F}^t(x,\cdot)\mathbf{H}\big) = \mathcal{F}\big(\mathbf{F}^t(x,\cdot)\big). \qquad (43)$$

A material is called **isotropic**, if the symmetry group contains[4] $SO(3)$. It can be shown that any frame-indifferent constitutive law for an isotropic simple material can be represented in the form

$$\mathbf{T}_{(E)}(y,t) = \mathcal{J}\big(\mathbf{B}(x,t),\mathbf{C}_r^t(\cdot,y,t)\big), \qquad (44)$$

---

[4] Of course, relations (42) and (43) are not invariant under changes of the reference configuration; we assume that this configuration has been chosen in a way that is adapted to the natural symmetries of the material.

where $y = \chi(x,t)$ and $\mathbf{C}_r^t(s,y,t) := \mathbf{C}_r(t - s, y, t)$, $s \geq 0$. The functional J is isotropic, that is,

$$J\big(\mathbf{QB}(x,t)\mathbf{Q}^T, \mathbf{QC}_r^t(\cdot,y,t)\mathbf{Q}^T\big)$$
$$= \mathbf{Q}J\big(\mathbf{B}(x,t), \mathbf{C}_r^t(\cdot,y,t)\big)\mathbf{Q}^T \quad \forall \mathbf{Q} \in \mathrm{SO}(3). \tag{45}$$

For viscoelastic materials, there is not always a clear distinction between **fluids** and **solids**, and hence there are various definitions in the literature. Noll [N7] calls a material a fluid, if the symmetry group is the proper unimodular group itself. With this definition, every fluid is isotropic, and one can show that a fluid is described by a constitutive law of the form (44) where J depends on $\mathbf{B}$ only through $\det \mathbf{B}$. (Note that for an incompressible fluid $\det \mathbf{B} = 1$.) Since $\mathbf{C}_r = \mathbf{1}$ if the material is always kept at rest, the isotropy of J implies that the stress must in this case also be a multiple of the identity. In other words, a deformation that does not depend on time can produce only an isotropic pressure. This latter property can be used as an alternative definition of a fluid; it allows for a class of fluids wider than Noll's. In three-dimensional problems, the materials we call fluids are fluids in the sense of either definition. In one-dimensional problems, material symmetry is not meaningful. If the one-dimensional motion under consideration involves compression, then there is no distinction between a fluid and a solid. If, however, the one-dimensional problem describes an isochoric (i.e. volume-preserving) motion (see the examples discussed in Section 4), then a time-independent deformation does not affect the stress if the material is a fluid. We shall refer to one-dimensional equations as "models for fluids", if they have this latter property. We use the term solid to mean a material which is not a fluid.

3. Constitutive models for viscoelastic materials

To develop a mathematical theory of existence, uniqueness, stability, etc., we must make mathematical assumptions concerning the functional which relates the stress to the history of the deformation gradient. Constitutive theories for viscoelastic materials have been developed by prescribing a space of functions, called a history space, on which the functional $\mathcal{F}$ in (29) is to be defined. Of course, there are many possible choices for such a space of functions, and a variety of such spaces have been considered. Axiomatic treatments of history spaces, motivated by physical considerations, are given in Coleman and Noll [C5], Coleman and Mizel [C10],[C11], Wang [W2],[W3] and Saut and Joseph [S1].

There has also been a great deal of effort to develop models which specify a more explicit form for the functional $\mathcal{F}$. Many models are motivated by introducing natural modifications into established theories such as finite elasticity, classical fluid dynamics, or linear viscoelasticity. From a physical point of view, the complexity of viscoelastic behavior results from the dynamics of the long chain molecules (in the case of polymers) or the macroscopic particles (in the case of suspensions) of which these materials are composed. Numerous constitutive models are based on attempts to make this idea precise. (See [B9],[B10] for reviews. A discussion of molecular theories goes beyond the scope of this book.) For more information on constitutive models we refer to textbooks on viscoelastic materials, e.g. [B8],[P3],[S3].

In this monograph, we deal primarily with models which specify an explicit form for the functional $\mathcal{F}$, but involve one or more unspecified constitutive functions. In many cases, the mathematical techniques we use would extend to more general constitutive relations. As one would expect, the statement of assumptions becomes more and more complicated as the models increase in generality. Our primary interest in this book is to highlight the methods rather than state the most general results, and we therefore confine ourselves to situations where we feel that both the assumptions and the analysis are reasonably transparent.

There are several restricted classes of motions for which constitutive laws can be characterized rather completely. An example is the class of infinitesimal perturbations of the rest state, where linear viscoelasticity is applicable. Linear viscoelasticity postulates a linear relationship between the stress and the history of the deformation gradient. Such a relationship cannot hold for all motions, because it conflicts with frame indifference. For an isotropic, incompressible linearly viscoelastic medium the constitutive equation has the form

$$\mathbf{T}_E = 2\beta\mathbf{E} + 2\eta\mathbf{D} + 2\int_0^\infty m(s)\big[\mathbf{E}(t) - \mathbf{E}(t-s)\big]\,ds. \tag{46}$$

Here

$$\mathbf{E} = \frac{1}{2}(\mathbf{F} + \mathbf{F}^T) - \mathbf{1} \tag{47}$$

is the linearized strain. As already explained in Chapter O, we assume the sign conditions

$$\beta \geq 0, \eta \geq 0, m \geq 0, m' \leq 0. \tag{48}$$

With u denoting the displacement, the resulting equation of motion is

$$\rho\frac{\partial^2 u}{\partial t^2} = -\nabla p + \beta\Delta u + \eta\Delta\frac{\partial u}{\partial t} + \int_0^\infty m(s)\big[\Delta u(\cdot,t) - \Delta u(\cdot,t-s)\big]\,ds,$$

$$\operatorname{div} u = 0. \tag{49}$$

(For infinitesimal perturbations of the rest state, it is not necessary to distinguish between the Lagrangian and Eulerian description.) In this monograph, we concentrate primarily on nonlinear problems and only certain aspects of linear viscoelasticity will be discussed. For more information on linear viscoelasticity, we refer to the review article of Leitman and Fisher [L4].

Popular models of nonlinear viscoelastic behavior are basically of three kinds:
(i) materials of Rivlin-Ericksen type,
(ii) integral models, and
(iii) differential models.

Rivlin-Ericksen materials [R15] are such that the stress depends only on the present value of $\mathbf{F}$ and on the present values of a finite number of temporal derivatives of $\mathbf{F}$. If frame indifference is taken into account and material isotropy is assumed, this implies that the functional J in (44) depends on the history of $\mathbf{C}_r$ only through the values of a finite number of Rivlin-Ericksen tensors, i.e., J becomes a function of $\mathbf{B}$, $\mathbf{A}_1,...,\mathbf{A}_n$. Such models must be regarded as asymptotic expansions useful in approximating certain special deformations [C5]. For $n > 1$, these models have undesirable properties if one regards them as valid in general deformations [C9],[J4],[R9], and they are unsuited for a discussion of the mathematical problems addressed in this monograph.

Integral models express the stress in terms of the deformation gradient by an expression involving one or several integrals. We remark that expansions in terms of multiple integrals have been proposed as an approximation scheme for a general class of constitutive laws, see Green and Rivlin [G6],[G8] and Green, Rivlin and Spencer [G7].

In this monograph, we shall extensively discuss single-integral models of the form

$$\mathbf{T}(\chi(x,t),t) = \int_0^\infty \mathbf{M}\Big(s, \mathbf{F}(x,t), \mathbf{F}(x,t-s)\Big)\, ds, \qquad (50)$$

or

$$\mathbf{S}(x,t) = \int_0^\infty \mathbf{N}\Big(s, \mathbf{F}(x,t), \mathbf{F}(x,t-s)\Big)\, ds. \qquad (51)$$

where $\mathbf{M}$ and $\mathbf{N}$ are matrix-valued functions. The idea behind such models is an extension of the superposition principle underlying the linear theory of Boltzmann. While the functions $\mathbf{M}$ and $\mathbf{N}$ are allowed to be nonlinear, it is assumed that contributions to the stress associated with different values of the delay time s are superposed in an additive fashion. More specific models of this nature which take account of frame indifference and material symmetry were introduced independently by Kaye [K10] and Bernstein, Kearsley and Zapas [B6]. If $\mathbf{M}$ (or $\mathbf{N}$) is sufficiently regular and well-behaved as $s \to 0$, then the expression giving the stress is meaningful for deformation gradients which have a jump discontinuity at the present time. We refer to materials with this property as materials with **instantaneous elasticity**. Indeed, if we regard the values of $\mathbf{F}$ at all past times $t - s$ as given, then we can regard $\mathbf{T}$ or $\mathbf{S}$ as a function of $\mathbf{F}(x,t)$ alone. The constitutive relation has then the same form as that of an elastic material, and we can define an instantaneous elasticity tensor. It is natural to require this tensor to satisfy the symmetry condition (20) and the strong ellipticity condition (22). Note that the instantaneous elasticity tensor depends on the entire history of $\mathbf{F}$. In Chapter III, we discuss well-posedness (locally in time) of dynamic problems for materials described by (51) which have instantaneous elasticity. Roughly speaking, the idea is to treat such materials as perturbations of elastic materials.

If the material described by (51) is such that the instantaneous elasticity tensor satisfies (20) for all choices of the deformation history, then $\mathbf{N}$ can be expressed as the gradient of a potential. If, in addition, the material is an incompressible fluid, then this potential can be expressed in terms of the invariants of $\mathbf{C}_r$, due to considerations of isotropy. In this case, (50) reduces to

$$\mathbf{T}_E\left(y,t\right) = 2 \int_0^\infty W_{,2}\left(s, I_1, I_2\right) \mathbf{C}_r^{-1}(t - s, y, t) - W_{,3}\left(s, I_1, I_2\right) \mathbf{C}_r(t - s, y, t) \; ds,$$

$$(52)_1$$

where

$$I_1 := \operatorname{tr} \mathbf{C}_r^{-1}(t - s, y, t), \quad I_2 := \operatorname{tr} \mathbf{C}_r(t - s, y, t), \qquad (52)_2$$

and $W_{,2}$, $W_{,3}$ denote the derivatives of the scalar potential W with respect to the second and third arguments. We refer to a material described by (52) as a K-BKZ fluid. It is often assumed that W can be factored into a part depending only on s and a part depending only on I_1 and I_2, i.e. $W(s, I_1, I_2) = m(s)\hat{W}(I_1, I_2)$; in this case we shall speak of a K-BKZ fluid with a **separated kernel**.

Integral models such as (50) may be modified by adding a Newtonian viscous contribution to the stress. This can be further generalized by making the viscosity tensor-valued and dependent on the history of $\mathbf{F}$ in the form of an integral. We remark that, for example, the model of Curtiss and Bird [C13] has such a form.

Differential constitutive laws are such that $\mathbf{T}$ (or $\mathbf{T}_E$) is related to the velocity gradient (and the strain tensor $\mathbf{B}$ in the case of solids) by a system of differential equations. In order to satisfy the principle of frame indifference, the formulation of such differential equations will involve objective time derivatives. Recall that the material time derivative of an objective tensor is usually not objective (see (41)). Some objective time derivatives were introduced by Oldroyd [O1], who also formulated a number of specific differential models [O2]. The recursion formula (36) for the Rivlin-Ericksen tensors expresses $\mathbf{A}_{n+1}$ as a kind of time derivative, called the **lower convected derivative**, of $\mathbf{A}_n$. For any second order tensor $\mathbf{P}$, we define its lower convected derivative $\hat{\mathbf{P}}$ by

$$\hat{\mathbf{P}} := \dot{\mathbf{P}} + (\nabla_y v)^T \mathbf{P} + \mathbf{P} \nabla_y v. \qquad (53)$$

The **upper convected derivative** $\check{\mathbf{P}}$ of $\mathbf{P}$ is defined by

$$\check{\mathbf{P}} := \dot{\mathbf{P}} - (\nabla_y v)\mathbf{P} - \mathbf{P}(\nabla_y v)^T. \qquad (54)$$

One can define a sequence of tensors analogous to the Rivlin-Ericksen tensors with the lower convected derivative replaced by the upper convected derivative in the recursion relation. These tensors are related to $\mathbf{C}_r^{-1}$ in the same way that the Rivlin-Ericksen tensors are related to $\mathbf{C}_r$. Later, we shall also use an "Oldroyd-type" derivative which interpolates between the upper and lower convected derivatives; it is defined by

$$\dot{\mathbf{P}} - \Omega\mathbf{P} + \mathbf{P}\Omega - a(\mathbf{D}\mathbf{P} + \mathbf{P}\mathbf{D}), \qquad (55)$$

23

where a is a constant between -1 and 1, while $\mathbf{D}$ and $\mathbf{\Omega}$ denote the symmetric and antisymmetric parts of $\nabla_y v$. The special case $a = 0$ is known as the **corotational** or **Jaumann derivative**. To see that, for example, the lower convected derivative is objective, we recall (see (41)) that, under a rotation $\mathbf{Q}(t)$, the material derivative of $\mathbf{P}$ transforms to

$$(\mathbf{QPQ}^T)^{\cdot} = \dot{\mathbf{Q}}\mathbf{PQ}^T + \mathbf{Q}\dot{\mathbf{P}}\mathbf{Q}^T + \mathbf{QP}\dot{\mathbf{Q}}^T, \tag{56}$$

while the velocity gradient $\mathbf{L} := \nabla_y v = \dot{\mathbf{F}}\mathbf{F}^{-1}$ transforms to

$$\dot{\mathbf{Q}}\mathbf{Q}^T + \mathbf{QLQ}^T. \tag{57}$$

We therefore find that $\hat{\mathbf{P}}$ transforms to

$$\mathbf{Q}\dot{\mathbf{P}}\mathbf{Q}^T + \mathbf{QP}\dot{\mathbf{Q}}^T + \dot{\mathbf{Q}}\mathbf{PQ}^T + (\mathbf{QL}^T\mathbf{Q}^T + \mathbf{Q}\dot{\mathbf{Q}}^T)\mathbf{QPQ}^T + \mathbf{QPQ}^T(\dot{\mathbf{Q}}\mathbf{Q}^T + \mathbf{QLQ}^T)$$

$$= \mathbf{Q}\hat{\mathbf{P}}\mathbf{Q}^T. \tag{58}$$

In Chapters II and VI, we shall discuss viscoelastic fluids described by the constitutive relation

$$\dot{\mathbf{T}} - \mathbf{\Omega}\mathbf{T} + \mathbf{T}\mathbf{\Omega} - a(\mathbf{DT} + \mathbf{TD}) + \lambda\mathbf{T} = 2\mu\mathbf{D}, \tag{59}$$

which can be regarded either as a special Oldroyd model [O2] or as a special case of the Johnson-Segalman model [J2]. Models such as (59) are nonlinear extensions of the linear theory developed by Maxwell [M10]. The special cases $a = 1$, $a = -1$ and $a = 0$ are known as the upper convected, lower convected and corotational Maxwell models. For other popular differential models, see e.g. [G3],[L5],[O2],[P5],[W4].

Many models can be written in both differential and integral form. For example, the upper convected Maxwell model (equation (59) with $a = 1$) is equivalent to (52) in the special case when $W = \frac{\mu\lambda}{2}e^{-\lambda s}I_1$. We remark, however, that neither the differential nor the integral class of models contains the other.

4. One-dimensional problems

In this monograph significant attention will be devoted to one-dimensional problems. There are two types of three-dimensional motions leading to such problems: shearing motions and elongational motions. Throughout this section, we assume that the material in question is isotropic and incompressible.

The kinematics of a shearing motion is described by

$$y^1 = x^1 + u(x^2, t), \ y^2 = x^2, \ y^3 = x^3. \tag{60}$$

This motion automatically satisfies the incompressibility condition. Since the deformation does not depend on x^1 or x^3, the components of the extra stress **T** do not depend on these coordinates either. Moreover, frame indifference implies that T^{13} and T^{23} vanish. To see this, let

$$\mathbf{Q} = \begin{pmatrix} -1 & 0 & 0 \\ 0 & -1 & 0 \\ 0 & 0 & 1 \end{pmatrix}, \tag{61}$$

and note that

$$\mathbf{QTQ}^T = \begin{pmatrix} T^{11} & T^{12} & -T^{13} \\ T^{12} & T^{22} & -T^{23} \\ -T^{13} & -T^{23} & T^{33} \end{pmatrix}. \tag{62}$$

Since we have

$$\mathbf{F} = \begin{pmatrix} 1 & \frac{\partial}{\partial x^2} u(x^2, t) & 0 \\ 0 & 1 & 0 \\ 0 & 0 & 1 \end{pmatrix}, \tag{63}$$

it is easily verified that $\mathbf{FQ} = \mathbf{QF}$, and by combining (37) and (43), we obtain $\mathbf{QTQ}^T = \mathbf{T}$, which gives the desired result. As a consequence, the third component of $\mathrm{div}_y\, \mathbf{T}$ vanishes, and the second depends only on x^2 and can be balanced by a pressure gradient in the x^2-direction. Thus the equation of motion (9) is trivially satisfied in two components, and the only nontrivial part is in the x^1-direction, where we have

$$\rho \frac{\partial^2 u(x, t)}{\partial t^2} = \frac{\partial T^{12}(x, t)}{\partial x} + \rho b(x, t). \tag{64}$$

Here we have written x for x^2, and b stands for an assigned body force per unit mass pointing in the x^1-direction. Since $x^2 = y^2$, we need not distinguish between the Eulerian and Lagrangian description. The stress component T^{12} is a functional of the history of $\mathbf{F}$, which is a function of u_x. In the case of a fluid, T^{12} is a functional of the history of the relative deformation gradient $\mathbf{F}_r(\tau, x, t)$. The only nonconstant component of $\mathbf{F}_r$ is $F_r^{12} = u_x(x, \tau) - u_x(x, t)$. Single-integral models for K-BKZ fluids lead to equations of the form

$$\rho u_{tt}(x, t) = \int_{-\infty}^t g\left(t - \tau, u_x(x, t) - u_x(x, \tau)\right)_x d\tau + \rho b(x, t), \tag{65}$$

where g is an odd function of the second argument. In Section IV.6, we shall discuss shearing flows of K-BKZ fluids with a Newtonian viscosity. For such materials, the term ηu_{xxt} is added to the right-hand side of (65). For shearing

25

motions of solids, one has to allow a general dependence on $u_x(x,t)$ and $u_x(x,\tau)$ (rather than just on their difference) in the integrand of (65).

Elongational motions are motions dominated by stretching in one particular direction. In contrast to the shearing motions described above, they are usually not exactly described by a one-dimensional equation, but only approximately. An example of exactly one dimensional kinematics would be

$$y^1 = x^1 + u(x^1), \ y^2 = x^2, \ y^3 = x^3; \tag{66}$$

however, such motions are rarely dealt with in applications, and they are impossible in incompressible materials. The stretching of sufficiently thin rods or filaments can be described approximately by a one-dimensional equation when error terms of the order of the thickness are neglected. We refer to the review article of Antman [A6] for the use of projection methods to approximate three-dimensional problems by one-dimensional problems; an asymptotic method which highlights the role of the thinness parameter is described in [N2]. For applications to viscoelastic fluids see [P3],[R4]. We omit the derivation and only discuss the results. We consider a thin rod or filament pointing in the $x = x^1$-direction and consisting of an incompressible material. The thickness of the rod is assumed to be small, say of order δ. We consider motions that, to within order δ, are given by

$$y^1 = x + u(x,t), \ y^2 = \left(1 + u_x(x,t)\right)^{-1/2} x^2, \ y^3 = \left(1 + u_x(x,t)\right)^{-1/2} x^3. \tag{67}$$

The deformation gradient is approximated by

$$\mathbf{F} = \begin{pmatrix} 1 + u_x & 0 & 0 \\ 0 & (1 + u_x)^{-1/2} & 0 \\ 0 & 0 & (1 + u_x)^{-1/2} \end{pmatrix}. \tag{68}$$

We assume that the rod is free of traction on the lateral surface. To leading order, this gives $T^{22} = T^{33} = p$, and the equation of motion (10) becomes

$$\rho u_{tt} = \{(1 + u_x)^{-1}(T^{11} - p)\}_x + \rho b = \{(1 + u_x)^{-1}(T^{11} - T^{22})\}_x + \rho b. \tag{69}$$

The stress components T^{11} and T^{22} are functionals of the history of u_x; in the case of a fluid they are functionals of the history of the relative deformation gradient, which involves the ratio $(1 + u_x(x,\tau))/(1 + u_x(x,t))$. Thus single-integral models for K-BKZ fluids lead to equations of the form

$$\rho u_{tt}(x,t) = \int_{-\infty}^{t} \left\{(1 + u_x(x,t))^{-1} g\left(t - \tau, \frac{1 + u_x(x,\tau)}{1 + u_x(x,t)}\right)\right\}_x d\tau + \rho b(x,t), \tag{70}$$

where $g(\cdot, 1) = 0$.

The equation

$$u_{tt}(x,t) = \int_{-\infty}^{t} g\left(t - \tau, u_{xt}(x,t), u_x(x,t), u_x(x,\tau)\right)_x d\tau + f(x,t) \tag{71}$$

subsumes all of the special cases described above, with or without inclusion of a Newtonian viscosity in the constitutive relation. Note that u_{xt} appears only through its present value. If the material described by (71) is a fluid, we have $g(s, 0, p, p) \equiv 0$.

Many other one-dimensional models have been studied, either as special cases of three-dimensional models or directly as a model for the restricted one-dimensional situation. Some of them will be encountered in the subsequent chapters.

II Hyperbolicity and wave propagation

1. Introduction

Many important features of partial differential equations are intimately related to the "type" (hyperbolic, parabolic or elliptic) of the equations. Two such features are the nature of singularities which can occur in solutions and the nature of the data needed to obtain a well-posed problem. Roughly speaking, the classification of partial differential equations according to types is a characterization of how the differential operator acts on rapidly oscillating functions; for such functions the terms of highest differential order dominate, and consequently only these terms are taken into account in the classification. The classification of the partial integrodifferential equations governing the motions of viscoelastic materials is an important ingredient in their mathematical analysis. We begin with some heuristic remarks which provide ideas that will later lead to rigorous results on existence and uniqueness of solutions for initial value problems and on the propagation and development of singularities.

While the motion of elastic solids is described by wave equations, which are of hyperbolic type, the motion of Newtonian fluids is described by the Navier-Stokes equations, which are parabolic. Viscoelastic materials have intermediate properties, and this is reflected in the mathematical nature of their equations. Some model equations have leading-order terms like those for a viscous fluid and would hence be classified as "parabolic"; others have leading-order terms like those for an elastic solid and would be classified as "hyperbolic", and there are also intermediate possibilities. To illustrate these notions, we consider the equation of one-dimensional linear viscoelasticity (see (49) of Chapter I)

$$u_{tt} = \eta u_{xxt} + \beta u_{xx} + \int_{-\infty}^{t} m(t-\tau)\big(u_{xx}(x,t) - u_{xx}(x,\tau)\big)\,d\tau, \tag{1}$$

where $\eta \geq 0$, $\beta \geq 0$ are constants and the kernel m satisfies $m \geq 0$, $m' \leq 0$. If $\eta > 0$, the highest-order derivative on the right-hand side is u_{xxt}, and the equation is essentially of parabolic type. If $\eta = 0$, then the effect of the memory term depends crucially on the behavior of m near 0. If m is smooth on $[0, \infty)$, then the convolution with m can be regarded as a differential operator of order -1, and the "leading-order" term on the right-hand side of (1) is $(\beta + \int_0^\infty m(s)\,ds)u_{xx}(x,t)$. We therefore say that the equation is hyperbolic. Another way to see this is that, if we differentiate the equation with respect to t, then the convolution term still involves only second derivatives of u whereas the remaining terms in the equation involve third derivatives of u. Singularities in the kernel m lead to behavior that is between hyperbolic and parabolic. In fact, we can think of the first term on the right of (1) as arising from setting m equal to minus the derivative of the delta function. It is natural to consider also the possibility of weaker singularities in m, e.g. $m(t) \sim t^{-\alpha}$ as $t \to 0$.

In Section 2 of this chapter, we discuss a class of differential models for viscoelastic fluids in more than one space dimension. Following Joseph, Renardy and Saut [J5], we find the equations determining characteristics, and we derive criteria for a change of type. There are two ways in which a change of type can occur. First, the equations of motion can lose their evolutionary character; this would, for example, be the case if, in the problem (1) above, $\eta = 0$ and the coefficient $\beta + \int_0^\infty m(s)\, ds$ of u_{xx} becomes negative. The equation then becomes elliptic rather than hyperbolic and the initial value problem becomes ill-posed. A different kind of change of type, analogous to a sonic transition in gas dynamics, arises in the consideration of steady flows. In this case the time-dependent equations do not change type at all, but the velocity of the fluid exceeds the speed of wave propagation, leading to a change of type for the steady flow equations. A satisfactory mathematical theory of existence of steady flows has not yet been developed; in fact, little is known when there is no change of type (see Chapter VI), and nothing when there is.

In Section 3, we discuss linear wave propagation. More specifically, we are interested in the evolution of discontinuities in the initial data. As is well known, hyperbolic equations propagate discontinuities, while parabolic equations lead to smoothing. We shall see that the equations of linear viscoelasticity behave as their type would suggest. Interesting possibilities arise from the "intermediate-type" equations with singular kernels. In particular, we shall see that these models allow wave propagation at finite speed to coexist with smoothing of singularities.

In the nonlinear case, hyperbolic equations not only propagate discontinuities in the initial data, but can also lead to the development of singularities (shocks) in finite time from smooth initial data. Section 4 discusses theorems demonstrating such blow-up in the case of one-dimensional models for viscoelastic materials. Differential models as well as single-integral models with smooth kernels will be considered. It would be interesting to know whether or not such singularities can develop from smooth data for integral models with singular kernels.

Although relatively little is known about the existence of solutions with shocks, Section 5 discusses jump conditions that have to be satisfied across a shock. Such jump conditions can easily be stated for materials with instantaneous elasticity. However, a serious problem arises for materials whose instantaneous behavior is "hypoelastic".

Finally, in Section 6, we discuss the existence of steady shock and acceleration waves for a single-integral model in one dimension. We follow the development of Greenberg [G9] who considered a more general class of materials with fading memory. Although singular kernels lead to smoothing of singularities in the linear case, they do not preclude the existence of steady shocks in the nonlinear case. This makes the question of formation of singularities for equations with singular kernels especially interesting.

2. Characteristics and change of type

The equations governing the motion of viscoelastic materials with differential constitutive relations can be written as systems of quasilinear partial differential equations, for which characteristics (curves or surfaces) are defined in the usual way. For integral models, it is possible to extract "leading-order terms" which yield such a quasilinear system. The governing equations are then classified according to whether or not their characteristics are real. Systems for which all characteristics are real are generally called hyperbolic, while those with no real characteristics are called elliptic. Typically, problems in viscoelasticity are neither hyperbolic nor elliptic; we refer to real characteristics as hyperbolic characteristics and nonreal characteristics as elliptic characteristics. It is also important to note that the notion of a characteristic is not independent of the way the equations are written. For example, trivial characteristics are introduced by differentiating the equations.

In this section, we study a class of differential models for incompressible viscoelastic fluids. We derive equations to determine characteristics and criteria for a change of type. As we shall see, there are always "trivial" hyperbolic characteristics associated with material trajectories (these characteristics are, however, important for the problem of characterizing inflow boundary conditions, see Chapter VI). In addition, there are elliptic characteristics associated with the incompressibility constraint. The remaining characteristics may be either hyperbolic or elliptic. Our treatment essentially follows that of Joseph, Renardy and Saut [J5]; for earlier work, we refer to Rutkevich [R18],[R19], Ultman and Denn [U1] and Luskin [L11].

To review the notions of characteristic and type, we consider the system

$$\mathbf{A}_0(q,y)\frac{\partial q}{\partial t} + \sum_{l=1}^{n} \mathbf{A}_l(q,y)\frac{\partial q}{\partial y^l} = g(q,y), \tag{2}$$

where $q = (q^1, q^2, ..., q^k)$ is a k vector, $\mathbf{A}_0$ and $\mathbf{A}_l$ are $k \times k$ matrices, and $y = (y^1, ..., y^n)$. We assume that $\mathbf{A}_0$, $\mathbf{A}_l$ and g are smooth functions. Characteristic surfaces are those surfaces for which (2) cannot be used to express the normal derivative of q in terms of the tangential derivatives. The Cauchy problem consisting of equation (2) and initial data prescribed on a characteristic surface cannot thus be solved – even formally. Alternatively, characteristic surfaces are those for which the leading-order terms in the equation do not preclude a rapid variation of solutions in the direction normal to the surface; in fact, discontinuities across characteristic surfaces are possible. More precisely, let us consider small perturbations of a given smooth solution Q. If we set $q = Q + \tilde{q}$, and linearize (2) about Q, we obtain,

$$\mathbf{A}_0(Q, y)\frac{\partial \tilde{q}}{\partial t} + \sum_{l=1}^{n} \mathbf{A}_l(Q, y)\frac{\partial \tilde{q}}{\partial y^l}$$

$$= \nabla_q g(Q, y) \cdot \tilde{q} - (\nabla_q \mathbf{A}_0(Q, y) \cdot \tilde{q})\frac{\partial Q}{\partial t} - \sum_{l=1}^{n}(\nabla_q \mathbf{A}_l(Q, y) \cdot \tilde{q})\frac{\partial Q}{\partial y^l}. \qquad (3)$$

Observe that the right-hand side of (3) does not contain any derivatives of $\tilde{q}$. Moreover, in a neighborhood of each point (y, t), we can regard the coefficient matrices $\mathbf{A}_0(Q, y)$ and $\mathbf{A}_l(Q, y)$ as approximately constant. If $\tilde{q}$ oscillates rapidly, then at least one of its derivatives is large compared to $\tilde{q}$, and the left-hand side of (3) will be large compared to the right-hand side. Hence (3) cannot be satisfied unless the derivatives of $\tilde{q}$ are such that they make the left side of (3) zero to leading order with respect to the frequency of oscillation. This is the case if and only if $\tilde{q} = \tilde{q}(\phi(y, t))$, where the scalar-valued function ϕ satisfies

$$\det\left\{\mathbf{A}_0\frac{\partial \phi}{\partial t} + \sum_{l=1}^{n} \mathbf{A}_l\frac{\partial \phi}{\partial y^l}\right\} = 0, \qquad (4)$$

and the direction of $\tilde{q}$ is that of a corresponding nullvector. A surface $\phi(y, t) = 0$, where ϕ satisfies (4), is a **characteristic surface**.

Systems are classified into types according to the nature of their characteristic surfaces. This classification is local in (q, y)-space, i.e. we regard the coefficient matrices $\mathbf{A}_0$ and $\mathbf{A}_l$ to be frozen at the values which they take at a particular point. We call the system (2) **hyperbolic** at (q, y) if $\mathbf{A}_0$ is not singular, if all the roots ν of

$$\det\left(\nu\mathbf{A}_0 + \sum_{l=1}^{n} \alpha_l \mathbf{A}_l\right) = 0 \qquad (5)$$

are real for every choice of the constant vector $\alpha \in \mathbb{R}^n$, and if the eigenvectors defined by

$$\nu\mathbf{A}_0 v + \sum_{l=0}^{n} \alpha_l \mathbf{A}_l v = 0 \qquad (6)$$

form a basis for $\mathbb{R}^k$. If the system is hyperbolic, then there is a maximal number of real characteristic surfaces. If the coefficients in (2) are constant, this means that there is a maximal number of linearly independent wave solutions of the form $q = q(\alpha \cdot y + \nu t)$. At the other extreme, the system (2) is called **elliptic** if the only real solution of equation (5) is $\nu = 0$, $\alpha = 0$, i.e. there are no real characteristic

31

surfaces. The systems that we will study are generally neither hyperbolic nor elliptic; however, many important features of such systems are intimately related to the number of real roots of (5).

It is well known that the initial value problem for an elliptic system is not well posed. We now discuss the notion of evolutionarity which provides a necessary condition for the well-posedness of initial value problems. The system (2) is called **evolutionary** if, for every choice of real parameters α_i, the roots ν of (5) are real. Note that the definitions of hyperbolicity and ellipticity require the matrix $\mathbf{A}_0$ to be nonsingular, while the definition of evolutionarity does not. If a system with constant coefficients fails to be evolutionary, there are spatially periodic initial data which lead to exponentially growing solutions of the form $q = q_0 \exp(i(\alpha \cdot y + \nu t))$; the faster the data oscillate, the faster the solution grows. If the coefficients are variable and evolutionarity fails in the neighborhood of a certain point, then catastrophic growth of solutions due to rapidly varying data is expected to occur locally. This phenomenon is referred to as **Hadamard instability**. In other words, if evolutionarity fails, one expects that the initial value problem is not well posed.

An important example of a system that is evolutionary but not hyperbolic is provided by the Euler equations for incompressible fluids,

$$\rho\left(\frac{\partial v}{\partial t} + (v \cdot \nabla_y)v\right) = -\nabla_y p, \tag{7}_1$$

$$\mathrm{div}_y\, v = 0. \tag{7}_2$$

Observe that the matrix corresponding to $\mathbf{A}_0$ in (2) is singular because there is no time derivative of p.

For steady flows of viscoelastic fluids, the velocity, stress and pressure are independent of time, and we shall be concerned with systems of the form (2) where the unknown q is independent of time (i.e. the term $\frac{\partial q}{\partial t}$ is missing). For such systems, the notion of evolutionarity is not relevant. Nevertheless, it is of interest to characterize the set of $\alpha \in \mathbb{R}^n$ for which (5) is satisfied with $\nu = 0$.

We now discuss the system governing the flows of incompressible fluids described by the constitutive relation (59) of Chapter I. The equations of motion (in Eulerian coordinates and with zero body force) have the form

$$\rho\left(\frac{\partial v}{\partial t} + (v \cdot \nabla)v\right) = \mathrm{div}\ \mathbf{T} - \nabla p, \tag{8}_1$$

$$\mathrm{div}\ v = 0, \tag{8}_2$$

$$\frac{\partial \mathbf{T}}{\partial t} + (v \cdot \nabla)\mathbf{T} - \frac{1+a}{2}(\nabla v \mathbf{T} + \mathbf{T}(\nabla v)^T) + \frac{1-a}{2}(\mathbf{T}\nabla v + (\nabla v)^T \mathbf{T}) + \lambda \mathbf{T}$$

$$= \mu(\nabla v + (\nabla v)^T), \tag{8}_3$$

where μ and λ are positive constants and $-1 \leq a \leq 1$. This system has the general structure of (2) with $n = 3$ and $k = 10$. The ten unknowns are: one for p, three for v and six for the symmetric matrix $\mathbf{T}$.

We begin by considering the special case of one-dimensional simple shearing motions. The flow is in the y^1-direction, with the velocity and stresses depending only on $y^2 = x^2 =: x$. For this special class of motions one can obtain a reduced system which involves only four unknowns: one component of velocity and three components of the extra stress. Let $\sigma := T^{12}$ denote the shear stress and let $\gamma := T^{11}$, $\tau := T^{22}$. Then the equation of motion and the constitutive law $(8)_3$ reduce to

$$\rho v_t = \sigma_x,$$

$$\sigma_t = \frac{1}{2}(\tau - \gamma)v_x + \frac{a}{2}(\tau + \gamma)v_x + \mu v_x - \lambda\sigma, \tag{9}$$

$$\gamma_t = (1 + a)\sigma v_x - \lambda\gamma,$$

$$\tau_t = (a - 1)\sigma v_x - \lambda\tau.$$

The reduced system (9) is hyperbolic at a given state $(v, \sigma, \tau, \gamma)$ if and only if $\frac{1}{2}(\tau - \gamma) + \frac{a}{2}(\tau + \gamma) + \mu > 0$. If this quantity changes sign, a change of type occurs and evolutionarity is lost.

We turn now to the general system (8). We shall analyze the roots ν of equation (5) for given $\alpha \in \mathbb{R}^3$. We consider first the case $\alpha_2 = \alpha_3 = 0$, to which the general case can be reduced by a rotation of the coordinate system. We set $\beta = \nu + v^1\alpha_1$. Equation (5) reads

$$\det \begin{pmatrix} A & B \\ C & D \end{pmatrix} = 0, \tag{$10)_1$}$$

where

$$A = \begin{pmatrix} 0 & \alpha_1 & 0 & 0 \\ \alpha_1 & \rho\beta & 0 & 0 \\ 0 & 0 & \rho\beta & 0 \\ 0 & 0 & 0 & \rho\beta \end{pmatrix}, \tag{$10)_2$}$$

$$B = \begin{pmatrix} 0 & 0 & 0 & 0 & 0 & 0 \\ -\alpha_1 & 0 & 0 & 0 & 0 & 0 \\ 0 & -\alpha_1 & 0 & 0 & 0 & 0 \\ 0 & 0 & -\alpha_1 & 0 & 0 & 0 \end{pmatrix}, \tag{$10)_3$}$$

$$C = \alpha_1 \begin{pmatrix} 0 & -2\mu - 2aT^{11} & (1-a)T^{12} & (1-a)T^{13} \\ 0 & -aT^{12} & -\mu + \frac{1-a}{2}T^{22} - \frac{1+a}{2}T^{11} & \frac{1-a}{2}T^{23} \\ 0 & -aT^{13} & \frac{1-a}{2}T^{23} & -\mu + \frac{1-a}{2}T^{33} - \frac{1+a}{2}T^{11} \\ 0 & 0 & -(1+a)T^{12} & 0 \\ 0 & 0 & -\frac{1+a}{2}T^{13} & -\frac{1+a}{2}T^{12} \\ 0 & 0 & 0 & -(1+a)T^{13} \end{pmatrix}, \tag{$10)_4$}$$

$$D = \begin{pmatrix} \beta & 0 & 0 & 0 & 0 & 0 \\ 0 & \beta & 0 & 0 & 0 & 0 \\ 0 & 0 & \beta & 0 & 0 & 0 \\ 0 & 0 & 0 & \beta & 0 & 0 \\ 0 & 0 & 0 & 0 & \beta & 0 \\ 0 & 0 & 0 & 0 & 0 & \beta \end{pmatrix}. \tag{$10)_5$}$$

After some algebra, (10) reduces to

$$\alpha_1^2 \beta^4 \{\rho\beta^2 - \alpha_1^2 (\mu + \frac{1+a}{2} T^{11} - \frac{1-a}{2} \Lambda_2)\} \{\rho\beta^2 - \alpha_1^2 (\mu + \frac{1+a}{2} T^{11} - \frac{1-a}{2} \Lambda_3)\} = 0.$$

$$(11)_1$$

Here Λ_2 and Λ_3 are the eigenvalues of the matrix

$$\begin{pmatrix} T^{22} & T^{23} \\ T^{23} & T^{33} \end{pmatrix}.$$

$$(11)_2$$

The special cases $a = \pm 1$ and $a = 0$ were derived by Rutkevich [R18], and the two-dimensional version of problem (8) (for general a) was discussed by Joseph, Renardy and Saut [J5].

When α_2 and α_3 are not zero, we must make the following changes in (11): α_1^2 should be replaced by $|\alpha|^2 = \alpha_1^2 + \alpha_2^2 + \alpha_3^2$, β should be replaced by $\nu + v \cdot \alpha$, T^{11} should be replaced by $n \cdot \mathbf{T}n$, where n is the unit vector in the direction of α, and for Λ_2 and Λ_3 we must substitute the two nontrivial eigenvalues of $\mathbf{PTP}$, where $\mathbf{P}$ is the orthogonal projection along n.

As for the Euler equations discussed above, the matrix in (8) which corresponds to $\mathbf{A}_0$ in (2) is singular; hence (8) is neither hyperbolic nor elliptic. To study the evolutionarity of (8), we discuss the nature of the roots of (11). We observe that the factor α_1^2 does not involve ν at all, hence it has no effect on evolutionarity; the corresponding real characteristic surfaces are $t =$const. Indeed, it is evident that the incompressibility condition imposes a constraint on the permissible choices of initial data. The velocity must be divergence free, and initial data for the pressure cannot be prescribed. The factor β^4 always leads to real characteristics; these characteristic surfaces are composed of particle trajectories and they result from the memory of the fluid, which propagates information following the motion of particles. No change of type is associated with these first two factors. However, a change of type and loss of evolutionarity can arise from the two factors in curly brackets in (11). It can be shown that the characteristics associated with these factors are related to the evolution of the vorticity, i.e. the curl of the velocity field. If we apply the divergence operator to (8)$_3$, apply the operator $\partial/\partial t + (v \cdot \nabla)$ to (8)$_1$, insert the expression found for div $[\frac{\partial \mathbf{T}}{\partial t} + (v \cdot \nabla)\mathbf{T}]$ and then take the curl, a straightforward but lengthy calculation yields

$$\rho(\frac{\partial}{\partial t} + (v \cdot \nabla))^2 \, \text{curl } v = \mu\Delta \, \text{curl } v + \frac{1+a}{2} \sum_{i,j} T^{ij} \frac{\partial^2}{\partial y^i \partial y^j} \, \text{curl } v$$

$$+ \frac{1-a}{2} \, \text{curl } \{\mathbf{T} \, \text{curl curl } v\} + \dots$$

$$(12)$$

Here the dots stand for lower-order terms involving only derivatives of v of order less than three. It can be shown that the last two factors of (11) are associated with characteristics of (12).

A loss of evolutionarity occurs if and only if

$$\frac{1-a}{2} \Lambda_{max} - \frac{1+a}{2} \Lambda_{min} > \mu,$$

$$(13)$$

where Λ_{max} and Λ_{min} are the largest and smallest eigenvalues of $\mathbf{T}$. The eigenvalues of $\mathbf{T}$ are not unrestricted. Although mathematically we could assign arbitrary initial values for $\mathbf{T}$, from a physical point of view we think of the material as having obeyed the constitutive law $(8)_3$ for its entire deformation history, and we want only those solutions of $(8)_3$ which behave reasonably as $t \to -\infty$. For example, if the material is not moving at all, then $(8)_3$ leads to $\mathbf{T} = \mathbf{T}_0 e^{-\lambda t}$, but only the solution $\mathbf{T} = 0$ makes physical sense; otherwise, the stress becomes infinite as $t \to -\infty$. It is shown in Johnson and Segalman [J2] that the "reasonable" solution of $(8)_3$ can be expressed as an integral in the following way:

$$\mathbf{T}(y,t) = -\frac{\mu}{a} + \int_{-\infty}^{t} \frac{\mu\lambda}{a} e^{-\lambda(t-\tau)} \mathbf{E}(\tau,y,t)\mathbf{E}(\tau,y,t)^T \ d\tau, \qquad (14)$$

where $\mathbf{E}$ satisfies the differential equation

$$(\frac{\partial}{\partial t} + (v \cdot \nabla))\mathbf{E}(\tau,y,t) - \frac{1+a}{2}\nabla v \mathbf{E} + \frac{1-a}{2}(\nabla v)^T \mathbf{E} = 0 \qquad (15)$$

with the initial condition

$$\mathbf{E}(\tau,y,\tau) = \mathbf{1}. \qquad (16)$$

Noting that $\mathbf{E}\mathbf{E}^T$ is positive definite, we see that all eigenvalues of $\mathbf{T}$ must be greater than $-\frac{\mu}{a}$ if a is positive and less than $-\frac{\mu}{a}$ if a is negative. This rules out loss of evolutionarity in the cases $a = \pm 1$.

It is also of interest to consider steady flows for fluids governed by the constitutive relation $(8)_3$. For such flows, v, p and $\mathbf{T}$ are independent of time, and we simply set $\frac{\partial v}{\partial t}$ and $\frac{\partial \mathbf{T}}{\partial t}$ equal to 0 in (8). In (11), we must then set $\nu = 0$ and hence $\beta = v^1\alpha_1$ (and $\beta = v \cdot \alpha$ when α_2 and α_3 are different from 0). The factor α_1^2 (or, resp., $|\alpha|^2$) now leads to complex characteristics and hence an elliptic part of the system. The factor β^4 leads to characteristic surfaces composed of streamlines. The presence of these characteristics is relevant in the discussion of boundary conditions. The multiplicity four suggests that there should be four associated boundary conditions which need to be imposed at an inflow boundary (in the two-dimensional case, the multiplicity is reduced to two, see [J5]). We shall see in Chapter VI that this is indeed the case. It depends on the velocity as well as the stress whether the characteristics defined by the last two factors in (11) are hyperbolic or elliptic. If the extra stress vanishes, then both factors read $\rho(v^1\alpha_1)^2 - \alpha_1^2\mu$ (or, resp., $\rho(v \cdot \alpha)^2 - |\alpha|^2\mu$). If $|v|^2$ is less than $\frac{\mu}{\rho}$, then the characteristics are elliptic. If $|v|^2$ exceeds $\frac{\mu}{\rho}$, there is a change of type, and hyperbolic characteristics occur. The quantity $\sqrt{\mu/\rho}$ is the speed of propagation of shear waves (see next section): a change of type occurs precisely where the speed of the fluid is faster than the wave speed. If $\mathbf{T} \neq 0$, the wave speed becomes stress-dependent and anisotropic. We emphasize that a change of type in the equations for steady flow does not imply a loss of evolutionarity in the time-dependent problem. This is analogous to the situation in gas dynamics.

In general, viscoelastic materials cannot be described by a set of differential equations. Instead the stress depends on the deformation history through some

35

kind of functional, e.g. by an expression involving integrals. Nevertheless, we can, under appropriate assumptions, extract leading-order terms, in which the highest derivatives appear only through their present values and not through their histories. Characteristics are then defined as above. To demonstrate what we have in mind, let us consider a one-dimensional model problem. Suppose that the stress is given by

$$\sigma(x,t) = \int_{-\infty}^{t} m(t-\tau)h(u_x(x,t), u_x(x,\tau))\, d\tau. \tag{17}$$

If we differentiate (17) with respect to t, we get

$$\sigma_t(x,t) = \int_{-\infty}^{t} m'(t-\tau)h(u_x(x,t), u_x(x,\tau))\, d\tau + m(0)h(u_x(x,t), u_x(x,t))$$

$$+ u_{xt}(x,t)\int_{-\infty}^{t} m(t-\tau)h_{,1}(u_x(x,t), u_x(x,\tau))\, d\tau. \tag{18}$$

In (18), the second derivative u_{xt} occurs only through its present value, and all other terms involve only first derivatives. Thus, if we combine (18) with the equation of motion

$$\rho u_{tt} = \sigma_x, \tag{19}$$

we can identify the curves

$$\frac{dx}{dt} = \pm\sqrt{\frac{1}{\rho}\int_{-\infty}^{t} m(t-\tau)h_{,1}(u_x(x,t), u_x(x,\tau))\, d\tau} \tag{20}$$

as characteristic (to correlate this example with the discussion above, the reader should note that the velocity is now given by u_t). We note that the lines $x = const.$ are also characteristic, resulting from differentiation of equation (19). Obviously, if m and $h_{,1}$ are positive, the equation is hyperbolic. The characteristics depend not only on the instantaneous values of the variables, but also on their histories. The reader should also note that we need sufficient smoothness of the kernel m to justify the calculation leading to (18); it is necessary that $m(t)$ and $m'(t)$ are integrable near zero for the terms on the right-hand side to be finite. If these conditions are not satisfied, we cannot define characteristics in the fashion described, and we shall see in the next section that there are problems with singular kernels for which the wave speed is infinite.

The same idea can be applied to a very general class of viscoelastic materials with instantaneous elasticity, although the equations for the characteristics become very complicated. For an isotropic viscoelastic material, the extra stress is given by a functional of the form (cf. equation (44) in Chapter I)

$$\mathbf{T}(y,t) = J\left\{\mathbf{B}(\chi^{-1}(y,t),t), \mathbf{C}_r^t(\cdot,y,t)\right\}. \tag{21}$$

We assume that this can be differentiated with respect to t using the chain rule, and we denote by $J_{,1}$ and $J_{,2}$ the derivatives of J with respect to its first and second arguments. In this way we obtain the following expression for the material time derivative, denoted by a dot,

$$\dot{\mathbf{T}} = (\frac{\partial}{\partial t} + (v \cdot \nabla_y))\mathbf{T} = J_{,1}\left\{\mathbf{B}, \mathbf{C}_r^t; \dot{\mathbf{B}}\right\} + J_{,2}\left\{\mathbf{B}, \mathbf{C}_r^t; \dot{\mathbf{C}}_r^t\right\}. \tag{22}$$

Using equations $(3)_1$ and (15) of Chapter I, we find that

$$\dot{\mathbf{B}} = \mathbf{L}\mathbf{B} + \mathbf{B}\mathbf{L}^T. \tag{23}$$

Hence the first term on the right of (22) depends linearly on the velocity gradient $\mathbf{L} = \nabla_y v$, with coefficients depending on $\mathbf{B}$ and on the history of $\mathbf{C}_r$. To find $\dot{\mathbf{C}}_r^t(s, y, t)$, let us set $x := \chi^{-1}(y, t)$ and recall that

$$\mathbf{C}_r^t(s, y, t) = \mathbf{F}^{-T}(x, t)\mathbf{F}^T(x, t - s)\mathbf{F}(x, t - s)\mathbf{F}^{-1}(x, t). \tag{24}$$

We use $(3)_1$ of Chapter I to differentiate the outer factors and note that for the inner factors the derivative with respect to t is minus the derivative with respect to s. In this way we obtain

$$\dot{\mathbf{C}}_r^t(s, y, t) = -\mathbf{L}^T(y, t)\mathbf{C}_r^t(s, y, t) - \mathbf{C}_r^t(s, y, t)\mathbf{L}(y, t) - \frac{\partial \mathbf{C}_r^t(s, y, t)}{\partial s}. \tag{25}$$

The first two terms are again linear in the gradient of the velocity field. The last term involves the derivative with respect to s. If the functional is such that it involves smoothing with respect to s, e.g. by convolution with a smooth kernel, then it makes sense to think of this contribution as a term of lower order. In this case we find that, to leading order, $\dot{\mathbf{T}}$ is given by an expression linear in the velocity gradient. In conjunction with $(8)_{1,2}$, we have a first order system for velocities, extra stresses and pressure, for which characteristics can be defined in the usual way.

The implications of a change of type on the behavior of solutions and its relevance to experiments are far from understood. If there is a change of type in the time-dependent equations, the initial value problem becomes ill-posed and there will be a Hadamard-type instability resulting in catastrophic growth of highly oscillatory disturbances. There have been attempts (see e.g. Hunter and Slemrod [H19]) to link such a change of type with the onset of an instability called "melt fracture". This instability occurs when polymers are extruded from a nozzle; it leads to wrinkled surfaces of the extrudate and is sometimes accompanied by (acoustic) noises. For reviews on such instabilities and various efforts to explain them we refer to the articles of Petrie and Denn [P4] and Tordella [T3]. A change of type in the equations for steady flow occurs when the speed of the fluid exceeds the speed for propagation of shear waves. This is analogous to a sonic transition in gas dynamics. The expected consequence would be a qualitative change in the nature of the flow, and the occurrence of surfaces of discontinuity (shocks).

Ultman and Denn [U1] remark that qualitative changes in the behavior of heat transfer and drag coefficients in flows past cylinders occur at fluid speeds which have the same order of magnitude as an estimate for the wave speed, which they obtain from a molecular theory. Yoo, Ahrens and Joseph [Y1] refer to the apparent occurrence of discontinuities in flows into a contraction. Joseph, Matta and Chen [J8] have conducted experiments which suggest that delayed die swell occurs at fluid speeds of the same order of magnitude as the propagation speed of shear waves. To assess the role of change of type in experiments, it is important to have measurements of the speed of wave propagation in viscoelastic materials. See Joseph, Riccius and Arney [J7] for some recent experiments.

Numerical calculations of flows of viscoelastic fluids encounter difficulties when the elasticity of the fluid, generally measured by a dimensionless number called the Weissenberg or Deborah number[1], becomes too high. Change of type seems to be involved in some of these problems [B4],[D13],[S12]; it seems clear, however, that there are also numerous other difficulties not linked to change of type.

3. Linear wave propagation

In this section, we discuss the propagation of singularities in one-dimensional linear viscoelasticity. Primarily, we concentrate on the questions of how a singularity in the initial or boundary data of a problem evolves in time, where the singularities in the solution are located and what their nature is. Some attention will also be given to the long-time asymptotic behavior of solutions. The regularity of the solution is influenced strongly by the behavior of the memory function near 0; singular memory functions lead to interesting features between "hyperbolic" (propagation of singularities) and "parabolic" behavior (smoothing of singularities).

Our discussion of problems with singular memory functions is based on [R5] and [H14]. There is an extensive earlier literature on wave propagation with regular memory functions dealing with problems such as determination of wave speeds and amplitudes of discontinuities, reflection of waves from boundaries and long-time asymptotic behavior. We refer in particular to [A2],[B7],[C4],[F2],[K11], [N1],[T1]. A review of many of these results is given in Christensen's book [C3].

---

[1] For the origin of these terms, we refer to the papers of White [W5] and Reiner [R3].

Rayleigh problem

The so-called Rayleigh problem [R2] concerns the motion of a medium occupying a half-space $x > 0$. The medium is at rest up to time $t = 0$, then the boundary starts to move with a constant (unit) velocity. In linear viscoelasticity, this leads to the following equations (with v denoting the velocity)

$$\rho v_{tt}(x,t) = \eta v_{xxt}(x,t) + \beta v_{xx}(x,t)$$

$$+ \int_{-\infty}^{t} m(t-\tau)(v_{xx}(x,t) - v_{xx}(x,\tau))\, d\tau, \quad x > 0,\ t > 0, \tag{26$_1$}$$

$$v(x,t) = 0 \text{ for } x > 0,\ t \le 0, \tag{26$_2$}$$

$$v(0,t) = 1 \text{ for } t > 0. \tag{26$_3$}$$

The following assumptions are made throughout: $\rho > 0$, $\eta \ge 0$, $\beta \ge 0$, and the memory function m is nonnegative, nonincreasing, integrable at infinity (but not necessarily at 0), and $tm(t)$ is integrable at 0.

Obviously, v is discontinuous when $x = t = 0$, and our main problem is how this discontinuity of the data propagates into the solution. Before we discuss the viscoelastic case, let us recall the very simple solutions for an elastic solid and a Newtonian fluid. In the former case, $(26)_1$ becomes $\rho v_{tt} = \beta v_{xx}$, and the solution is $v = H(t\sqrt{\beta/\rho} - x)$ where H denotes the Heaviside step function. The discontinuity propagates with constant speed $\sqrt{\beta/\rho}$ and constant amplitude, as one would expect for the wave equation. The case of the Newtonian fluid, in which $(26)_1$ reduces to the heat equation $\rho v_t = \eta v_{xx}$, was first solved by Stokes [S16]. The solution is a similarity solution. We set $z = x/\sqrt{t}$ and seek a solution of the form $v(x,t) = \tilde{v}(z)$. The heat equation becomes $-\rho(z/2)\tilde{v}_z = \eta \tilde{v}_{zz}$, and the initial and boundary conditions are satisfied if $\tilde{v}(0) = 1$, $\tilde{v}(\infty) = 0$. A simple calculation yields

$$\tilde{v}(z) = \sqrt{\frac{\rho}{\eta\pi}} \int_{z}^{\infty} \exp\left(-\frac{w^2\rho}{4\eta}\right) dw. \tag{27}$$

Clearly this solution is non-zero and analytic everywhere in the quarter plane $x \ge 0$, $t > 0$. The speed of propagation is infinite and the initial discontinuity has been smoothed out. This behavior is typical of parabolic equations.

To avoid trivial inconsistencies, we assume from now on that m does not vanish identically. The results concerning the propagation of the singularity in (26) can be summarized roughly as follows:

(i) If $\eta > 0$, the Newtonian viscosity dominates. The singularity is smoothed and the speed of propagation is infinite. In fact, if m is completely monotone (see definition below), the solution is analytic in the quarter plane $x \ge 0$, $t > 0$.

(ii) If $\eta = 0$ and m is smooth on $[0,\infty)$, the singularity propagates with finite speed

$$c := \sqrt{\left(\beta + \int_{0}^{\infty} m(\tau)\, d\tau\right)/\rho}, \tag{28}$$

39

and its amplitude decays exponentially. We have $v \equiv 0$ for $x > ct$, and v is smooth for $x < ct$.

(iii) If m has a non-integrable singularity at 0, the behavior is similar to the case $\eta > 0$; in particular, if m is completely monotone, then v is analytic in the quarter plane $x \geq 0$, $t > 0$.

(iv) If $\eta = 0$, m is integrable at 0 and $m(0) = +\infty$, interesting types of intermediate behavior occur. The speed of propagation is finite. However, the singularity at $x = t = 0$ is smoothed in a degree depending on the strength of the singularity in m.

Solution by Laplace transform

To solve (26), we employ the Laplace transform with respect to the time variable. We denote the Laplace transform by a hat:

$$\hat{m}(\lambda) := \int_0^\infty e^{-\lambda t} m(t) \ dt, \tag{29}_1$$

$$\hat{v}(x, \lambda) := \int_0^\infty e^{-\lambda t} v(x, t) \ dt. \tag{29}_2$$

Transforming (26), we obtain

$$\rho \lambda^2 \hat{v}(x, \lambda) = (\eta \lambda + \beta + \hat{m}(0) - \hat{m}(\lambda)) \hat{v}_{xx}(x, \lambda), \quad \hat{v}(0, \lambda) = \frac{1}{\lambda}. \tag{30}$$

If m is not integrable, then $\hat{m}(0)$ and $\hat{m}(\lambda)$ are not well-defined individually, and we must use the interpretation

$$\hat{m}(0) - \hat{m}(\lambda) := \int_0^\infty m(t)(1 - e^{-\lambda t}) \ dt, \tag{31}$$

which is meaningful for Re $\lambda \geq 0$. We also note that it follows from the monotonicity assumptions on m that, for Re $\lambda \geq 0$, $\lambda \neq 0$, we have

$$\text{Re } (\hat{m}(0) - \hat{m}(\lambda)) > 0, \ \text{sgn } (\text{Im } (\hat{m}(0) - \hat{m}(\lambda))) = \text{sgn Im } \lambda. \tag{32}$$

The solution of (30) is given by

$$\hat{v}(x, \lambda) = \frac{1}{\lambda} e^{-\lambda x \sqrt{\rho / (\eta \lambda + \beta + \hat{m}(0) - \hat{m}(\lambda))}}, \tag{33}$$

and v itself is given by the inverse Laplace transform

$$v(x, t) = \frac{1}{2\pi i} \int_{\gamma - i\infty}^{\gamma + i\infty} \frac{1}{\lambda} e^{\lambda t} e^{-\lambda x \sqrt{\rho / (\eta \lambda + \beta + \hat{m}(0) - \hat{m}(\lambda))}} \ d\lambda, \tag{34}$$

where γ is any positive real number. The properties (32) guarantee that $\hat{v}(x, \cdot)$ is square integrable along the line $(\gamma - i\infty, \gamma + i\infty)$, and hence the inverse transform is meaningful.

40

The regularity of the solution (34) depends on the behavior of $\hat{v}(x, \lambda)$ as $\lambda \to \infty$. The faster the transform decays to zero, the more often is the solution differentiable with respect to t and x.

If $\eta = 0$ and $m \in L^1(0, \infty)$, then the solution of (26) propagates with finite speed; more precisely, $v(x, t) = 0$ for $x > ct$ where c is given by (28). Indeed, in this case the Riemann-Lebesgue lemma guarantees that $\hat{m}(\lambda) \to 0$ as $\lambda \to \infty$, $\text{Re } \lambda \geq 0$. Consequently if $\eta = 0$ and $x > ct$ the contour of integration in (34) can be closed by a circle on the right. Since the integrand has no singularities with $\text{Re } \lambda > 0$, it follows from Cauchy's theorem that $v(x, t) = 0$. We note that one cannot expect the speed of propagation to be finite if $\eta > 0$ or if m has a nonintegrable singularity at zero.

A class of kernels popular in rheology is the class of completely monotone kernels. The kernel m is said to be **completely monotone** if it is of class C^∞ on $(0, \infty)$ and $(-1)^k m^{(k)}(t) \geq 0$ for all $k \in \mathbb{N}$, $t > 0$. Every completely monotone function is analytic; in fact, a classical theorem of Bernstein asserts that m is completely monotone if and only if there is a positive, locally finite measure μ such that

$$m(t) = \int_0^\infty e^{-\alpha t} \, d\mu(\alpha) \quad \forall t > 0. \tag{35}$$

The Laplace transform of m is then given by

$$\hat{m}(\lambda) = \int_0^\infty \frac{1}{\alpha + \lambda} \, d\mu(\alpha), \tag{36$_1$}$$

again with $\hat{m}(0) - \hat{m}(\lambda)$ interpreted in the obvious way:

$$\hat{m}(0) - \hat{m}(\lambda) = \int_0^\infty \frac{\lambda}{\alpha(\alpha + \lambda)} \, d\mu(\alpha). \tag{36$_2$}$$

Our integrability assumptions on m translate into the integrability of $(1/\alpha)d\mu(\alpha)$ at $\alpha = 0$ and integrability of $(1/\alpha^2)d\mu(\alpha)$ at $\alpha = \infty$.

The case $\eta > 0$

If $\eta \neq 0$, then for large λ the term $\eta\lambda$ dominates the expression under the square root in (34). This can be seen as follows: Let

$$M(t) := \int_t^\infty m(\tau) \, d\tau. \tag{37$_1$}$$

Then M is integrable near 0, and we can define its Laplace transform

$$\hat{M}(\lambda) := \int_0^\infty M(t)e^{-\lambda t} \, dt \tag{37$_2$}$$

for $\text{Re } \lambda > 0$. According to the Riemann-Lebesgue lemma, $\hat{M}(\lambda)$ tends to zero as $\text{Im } \lambda \to \infty$ for fixed $\text{Re } \lambda > 0$. On the other hand, a simple integration by parts in (37)$_2$ shows that

$$\hat{M}(\lambda) = \frac{1}{\lambda}(\hat{m}(0) - \hat{m}(\lambda)). \tag{38}$$

Hence the term $\eta\lambda$ dominates over $\hat{m}(0) - \hat{m}(\lambda)$. The integrand in (34) decays like $\exp\left(-x\sqrt{\frac{\lambda\rho}{\eta}}\right)$, i.e. faster than any negative power of λ. Hence we may differentiate any number of times with respect to t and x under the integral, and the integral remains absolutely convergent (uniformly for t and $x > 0$ in compact sets). The solution is of class C^∞ in the quarter plane $x > 0$, $t \geq 0$. If more is known about the kernel m, we can make further conclusions. If m is completely monotone, then $\hat{m}(0) - \hat{m}(\lambda)$ is defined for every nonzero λ off the negative real axis and it is $o(\lambda)$ as $\lambda \to \infty$ in any sector excluding the negative real axis. Moreover, its imaginary part has the same sign as Im λ. As a consequence, we find that the contour in (34) can, for $t > 0$, be deformed into a path as indicated in Fig. 1, which extends into the left half plane. The integral along this modified path remains convergent even if t and x are extended into a region of the complex plane; it is an analytic function of t and x for $t > 0$, $x \geq 0$. We thus find that as far as the propagation of singularities is concerned, the equation for $\eta \neq 0$ has properties similar to the heat equation, as would of course be expected. The same contour deformation also works if $\eta = 0$ and $\hat{m}(0) - \hat{m}(\lambda)$ tends to ∞ as $\lambda \to \infty$. This is the case if $1/\alpha \, d\mu(\alpha)$ is not integrable at infinity, i.e. if m is not integrable at 0.

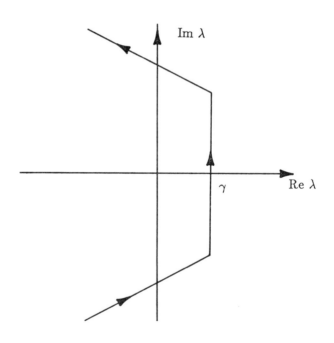

Fig. 1: Deformed contour of integration

Regular kernels

Let us now assume, on the other hand, that $\eta = 0$ and that $m \in L^1(0, \infty)$. Recall that in this case $v(x, t) = 0$ for $x > ct$, and hence we have the "hyperbolic" property of a finite wave speed. It is of interest to study the behavior of the solution near the wave front $x = ct$. (We note that if m is completely monotone then the contour of Fig. 1 can be used to show that v is analytic behind the wave front, i.e. for $0 \leq x < ct$.)

We now consider the case where m is a smooth kernel. More precisely, we assume that $m \in C^\infty[0, \infty)$ and that all derivatives of m are integrable. (The last assumption serves only to simplify the discussion; since the solution at time t depends only on the behavior of m up to time t and on the integral of m, we can always find a kernel with all derivatives integrable that produces the same solution up to time t as the given m). Under these assumptions, the transform $\hat{m}(\lambda)$ has an asymptotic expansion in powers of $\frac{1}{\lambda}$ as $\lambda \to \infty$:

$$\hat{m}(\lambda) = \frac{m(0)}{\lambda} + \frac{m'(0)}{\lambda^2} + \dots \tag{39}$$

When we insert this expression into the exponential in (34), we find

$$e^{-\lambda x \sqrt{\rho/(\beta + \hat{m}(0) - \hat{m}(\lambda))}} = e^{-\lambda x/c} e^{-m(0)x/(2c^3\rho)} + O\left(\frac{1}{\lambda}\right). \tag{40}$$

When the integration in (34) is carried out, the first term of (40) leads to $e^{-m(0)x/(2c^3\rho)} \times H(t - x/c)$ and the second leads to a continuous function of x and t, since the integral converges absolutely and uniformly. Hence the solution is discontinuous across the wave front, and the amplitude of the discontinuity decays exponentially. Using the fact that $x = ct$ at the wave front, we find that the amplitude of the discontinuity is

$$A(t) := \exp\left(\frac{-m(0)t}{2c^2\rho}\right). \tag{41}$$

Elsewhere the solution is continuous. If we proceed further with the asymptotic expansion in (39), we can make the remainder term in the analogue of (40) smaller than any negative power of λ. In this fashion, we obtain correction terms to (40) leading to functions which have jumps in derivatives across the wave front, and a remainder term for which these derivatives are continuous. This argument shows that the solution is in fact of class C^∞ everywhere except at the wave front. If m is completely monotone, then we can use the deformed contour of Fig. 1 provided $t > x/c$, and hence the solution is analytic behind the wave. Of course it is also analytic (namely zero) ahead of the wave.

We next discuss how this picture is changed when m has a singularity at 0. Some suggestions can already be inferred from the results above. If m is not integrable at 0, then c is infinite, and we do not expect a finite propagation speed. If, however, m is integrable, but $m(0) = \infty$, then c remains finite, but the amplitude of the discontinuity across the wave becomes zero. This suggests the interesting possibility that a finite wave speed may coexist with some kind of smoothing in this case. We shall see below that this is exactly what happens.

Examples with singular kernels

We now discuss the regularity of the solution v of (26) for various kinds of singular kernels in the case when $\eta = 0$. Clearly, the crucial point is the behavior of $\hat{m}(\lambda)$ as $\lambda \to \infty$. Although there is no precise correspondence, this behavior is roughly related to the behavior of $m(t)$ as $t \to 0$. (There are some rigorous results [D10] which say that the behavior of $m(t)$ as $t \to 0+$ determines the behavior of $\hat{m}(\lambda)$ as $\lambda \to \infty$ in sectors of the form $-\frac{\pi}{2} + \varepsilon < \arg \lambda < \frac{\pi}{2} - \varepsilon$ with $\varepsilon > 0$. However, such results are not adequate for our purposes here.) The examples of singular kernels studied below are infinite sums of exponentials; kernels of this structure appear in some rheological models that are motivated by molecular considerations. In order to study the behavior of $m(t)$ as $t \to 0+$ and of $\hat{m}(\lambda)$ as $\lambda \to \infty$ we shall make use of the simple calculus inequality

$$\left| \int_0^\infty f(\nu) \, d\nu - \sum_{n=1}^\infty f(n) \right| \leq \int_0^\infty |f'(\nu)| \, d\nu. \tag{42}$$

We note that (42) remains valid if the lower limit in both integrals is replaced by 1.

A. Power-type singularities

The first class of singular kernels that we shall look at is given by

$$m(t) = \sum_{n=1}^\infty e^{-n^\alpha t}. \tag{43}$$

In order to insure that $t m(t)$ is integrable at zero, we must have $\alpha > \frac{1}{2}$. The behavior of $m(t)$ as $t \to 0$ can be determined from the inequalities

$$\int_1^\infty e^{-\nu^\alpha t} \, d\nu < m(t) < \int_0^\infty e^{-\nu^\alpha t} \, d\nu. \tag{44}$$

The difference between the two bounds is $\int_0^1 \exp(-\nu^\alpha t) \, d\nu$, which tends to 1 as $t \to 0+$. The substitution $\phi = t^{1/\alpha} \nu$ yields

$$\int_0^\infty e^{-\nu^\alpha t} \, d\nu = t^{-1/\alpha} \int_0^\infty e^{-\phi^\alpha} \, d\phi \tag{45}$$

and consequently m behaves like $t^{-1/\alpha}$ as $t \to 0+$. Accordingly, m is integrable if and only if $\alpha > 1$.

If $\frac{1}{2} < \alpha \le 1$, then $\hat{m}(0)$ and $\hat{m}(\lambda)$ are not well-defined individually; however, using the interpretation (31) we have

$$\hat{m}(0) - \hat{m}(\lambda) = \sum_{n=1}^{\infty} \frac{\lambda}{n^\alpha (n^\alpha + \lambda)}. \tag{46}$$

To analyze the behavior as $\lambda \to \infty$ in (46), we treat separately the cases $\alpha = 1$ and $\frac{1}{2} < \alpha < 1$. For $\alpha = 1$ we have

$$\hat{m}(0) - \hat{m}(\lambda) = \sum_{n=1}^{\infty} \frac{1}{n} - \frac{1}{n + \lambda}$$

$$= \lim_{N \to \infty} \sum_{n=1}^{N} \frac{1}{n} - \sum_{n=1}^{N} \frac{1}{n + \lambda}. \tag{47}$$

When N is large, $\sum_{n=1}^{N} \frac{1}{n}$ behaves like $\ln N + C$, where C is Euler's constant. Moreover, using (42) with $f(\nu) = (\nu + \lambda)^{-1}$, we find that $\sum_{n=1}^{N} \frac{1}{n+\lambda} = \ln(N + \lambda) - \ln \lambda + O(|\lambda|^{-1})$. It therefore follows from (47) that

$$\hat{m}(0) - \hat{m}(\lambda) = \ln \lambda + C + O(\frac{1}{|\lambda|}). \tag{48}$$

Suppose now that $\frac{1}{2} < \alpha < 1$. In this case, we use (42) with $f(\nu) = \frac{\lambda}{\nu^\alpha (\nu^\alpha + \lambda)}$ and with the lower bound in the integrals replaced by 1. We find that, to leading order

$$\hat{m}(0) - \hat{m}(\lambda) = \int_0^{\infty} \frac{\lambda}{\nu^\alpha (\nu^\alpha + \lambda)} \, d\nu = \lambda^{1/\alpha - 1} \frac{\pi}{\alpha} \operatorname{cosec} \frac{(1 - \alpha)\pi}{\alpha}. \tag{49}$$

Using (48) and (49), we find that when $\frac{1}{2} < \alpha \le 1$, the integrand in (34) decays faster than any negative power of λ along the path of integration; moreover, the contour deformation of Fig. 1 can be carried out for every $x \ge 0$, $t > 0$. The solution is analytic in the quarter plane $x \ge 0$, $t > 0$.

For $\alpha > 1$, we have

$$\hat{m}(\lambda) = \sum_{n=1}^{\infty} \frac{1}{n^\alpha + \lambda}, \quad \hat{m}(0) = \varsigma(\alpha), \tag{50}$$

where ς denotes the Riemann zeta function. Employing (42) with $f(\nu) = \frac{1}{\nu^\alpha + \lambda}$ and setting $\lambda = \gamma + i\psi$, we find

$$\int_0^{\infty} |f'(\nu)| \, d\nu = \frac{1}{\psi} (\frac{\pi}{2} - \arctan \frac{\gamma}{\psi}). \tag{51}$$

Since [G5]

$$\int_0^{\infty} \frac{1}{\lambda + \nu^\alpha} \, d\nu = \frac{\pi}{\alpha} \lambda^{1/\alpha - 1} \operatorname{cosec} \frac{\pi}{\alpha}, \tag{52}$$

we conclude that, as $\lambda \to \infty$ in any sector excluding the negative half-axis,

$$\hat{m}(\lambda) = \frac{\pi}{\alpha} \operatorname{cosec} \frac{\pi}{\alpha} \lambda^{1/\alpha - 1} + O\left(\frac{1}{\lambda}\right). \tag{53}$$

Inserting (53) into (34), we find that the integrand decays to zero like $\exp(-\kappa x |\lambda|^{1/\alpha})$ with $\kappa > 0$, i.e. the integrand decays faster than any power of $|\lambda|^{-1}$. Hence the solution is of class C^∞ for $x > 0$, $t \geq 0$. Since m is integrable when $\alpha > 1$, the solution propagates with finite speed. Moreover, since m is completely monotone, the contour of Fig. 1 can be used to show that the solution is analytic behind the wave, i.e. for $x < ct$. At the wave front, the solution is obviously not analytic, since it is zero on one side and non-zero on the other, but failure to be analytic is the only way in which the solution is singular.

B. Logarithmic singularity

The reader may suspect that a weaker singularity in the kernel produces a lesser degree of smoothing in the solution, and we shall see that this is indeed so. As an example of a kernel with a logarithmic singularity, we use

$$m(t) = \sum_{n=1}^{\infty} e^{-e^n t}. \tag{54}$$

To verify that this kernel behaves like a logarithm near zero, we first note that (42) implies that the asymptotic behavior as $t \to 0+$ will not change if the sum is replaced by an integral (the difference is order 1). Moreover, we have

$$\int_0^\infty e^{-e^\nu t}\, d\nu = \int_0^\infty e^{-e^{\nu + \ln t}}\, d\nu = \int_{\ln t}^\infty e^{-e^\nu}\, d\nu$$

$$= \int_0^\infty e^{-e^\nu}\, d\nu + \int_{\ln t}^0 e^{-e^\nu}\, d\nu. \tag{55}$$

Since $\exp(-e^\nu)$ tends to one as $\nu \to -\infty$, the last term in (55) behaves like $|\ln t|$ as $t \to 0+$. The Laplace transform of m is given by

$$\hat{m}(\lambda) = \sum_{n=1}^{\infty} \frac{1}{\lambda + e^n}. \tag{56}$$

Using (42) we find that, as $\lambda \to \infty$ in any sector excluding the negative half-axis,

$$\hat{m}(\lambda) = \int_0^\infty \frac{d\nu}{\lambda + e^\nu} + O(|\lambda|^{-1}) = \frac{1}{\lambda} \ln(1 + \lambda) + O(|\lambda|^{-1}), \tag{57}$$

from which we obtain

$$\lambda \sqrt{\frac{\rho}{\beta + \hat{m}(0) - \hat{m}(\lambda)}} = \frac{\lambda}{c} + \frac{\ln \lambda}{2c^3 \rho} + O(1), \tag{58}$$

46

and hence

$$\exp\left(-\lambda x\sqrt{\frac{\rho}{\beta + \hat{m}(0) - \hat{m}(\lambda)}}\right) = e^{-\lambda x/c}\lambda^{-x/(2c^3\rho)}\exp[O(1)]. \tag{59}$$

As $\lambda \to \infty$, this decays like a negative power of λ with exponent proportional to x. Hence the smoothness of the solution across the wave front $x = ct$ increases in a manner proportional to x (or, equivalently, proportional to t). At each positive time, the solution is continuous across the wave.

C. Log-log singularity

As an example of an even weaker singularity, let us consider the kernel

$$m(t) = \sum_{n=1}^{\infty}\exp(-te^{e^n}). \tag{60}$$

Again the behavior as $t \to 0+$ is like that of the corresponding integral. We write this integral as follows

$$\int_0^{\infty}\exp(-te^{e^{\nu}})\,d\nu = \int_0^{\ln|\ln t|}\exp\left[-e^{e^{\nu}+\ln t}\right]d\nu + \int_{\ln|\ln t|}^{\infty}\exp\left[-e^{e^{\nu}+\ln t}\right]d\nu. \tag{61}$$

For $t < 1$, the integrand in the first term on the right ranges between e^{-et} and e^{-1}. Since both these bounds are positive, the term has a singularity of order $\ln|\ln t|$ as $t \to 0+$. The second integral on the right of (61) becomes, after setting $\mu = \nu - \ln|\ln t|$,

$$\int_0^{\infty}\exp\left[-e^{(e^{\mu}-1)|\ln t|}\right]d\mu, \tag{62}$$

which tends to zero as $t \to 0+$. Hence m has a log-log singularity at zero. The transform of m is given by

$$\hat{m}(\lambda) = \sum_{n=1}^{\infty}\frac{1}{\lambda + \exp(e^n)}, \tag{63}$$

and, within an error of order $O(|\lambda|^{-1})$, this is approximated by the integral

$$I(\lambda) = \int_0^{\infty}\frac{d\nu}{\lambda + \exp(e^{\nu})}. \tag{64}$$

We substitute $w = \exp(e^{\nu})$, and obtain, for real λ,

$$I(\lambda) = \int_e^{\infty}\frac{dw}{(\lambda + w)w\ln w} = \int_e^{\lambda}\frac{dw}{(\lambda + w)w\ln w} + \int_{\lambda}^{\infty}\frac{dw}{(\lambda + w)w\ln w}. \tag{65}$$

47

In each of the integrals on the right of (65) we expand $(\lambda + w)^{-1}$ in a geometric series. This yields

$$I(\lambda) = \frac{1}{\lambda} \sum_{n=0}^{\infty} (-1)^n \lambda^{-n} \int_e^{\lambda} \frac{w^{n-1}}{\ln w} \, dw + \sum_{n=0}^{\infty} (-1)^n \lambda^n \int_{\lambda}^{\infty} \frac{dw}{w^{n+2} \ln w}$$

$$= \frac{1}{\lambda} \ln \ln \lambda + \frac{1}{\lambda} \sum_{n=1}^{\infty} (-1)^n \lambda^{-n} \int_1^{\ln \lambda} \frac{e^{nv}}{v} \, dv + \sum_{n=0}^{\infty} (-1)^n \lambda^n \int_{\ln \lambda}^{\infty} \frac{dv}{v e^{(n+1)v}}$$

$$= \frac{1}{\lambda} \ln \ln \lambda + \sum_{n=1}^{\infty} \frac{(-1)^n}{\lambda^{n+1}} [\text{Ei}(n \ln \lambda) - \text{Ei}(n)] + \sum_{n=0}^{\infty} (-1)^n \lambda^n \, \text{E}_1 ((n+1) \ln \lambda). \quad (66)$$

We use the asymptotic formula for the exponential integral for large argument [A1], and obtain

$$\text{E}_1 ((n+1) \ln \lambda) = \frac{\lambda^{-(n+1)}}{(n+1) \ln \lambda} (1 + O(\frac{1}{n|\ln \lambda|})),$$

$$\text{Ei}(n) = \frac{e^n}{n} (1 + O(\frac{1}{n})),$$

$$\text{Ei}(n \ln \lambda) = \frac{\lambda^n}{n \ln \lambda} (1 + O(\frac{1}{n|\ln \lambda|})). \quad (67)$$

From this we can conclude that, at least for large $|\lambda|$, the series in (66) converge even if λ is not real, and by the identity theorem for analytic functions (66) is valid. We see moreover that

$$I(\lambda) = \frac{1}{\lambda} \ln \ln \lambda + O(\frac{1}{|\lambda (\ln \lambda)^2|}). \quad (68)$$

Hence the exponential in (34) decays like a negative power of $|\ln \lambda|$, with exponent proportional to x. This leads to an "infinitesimal" gain of regularity, but we can conclude that the solution is continuous across the wave for large enough x. The leading part in $I(\lambda)$ yields (we set $a = \frac{x}{2c^3 \rho}$) the following term in the solution:

$$\int_{\gamma - i\infty}^{\gamma + i\infty} e^{\lambda (t - x/c)} \frac{1}{\lambda (\ln \lambda)^a} \, d\lambda, \quad (69)$$

and for $t - x/c > 0$ this equals [B3]

$$\frac{1}{\Gamma(a)} \int_0^{\infty} \frac{(t - x/c)^s s^{a-1}}{\Gamma(s+1)} \, ds. \quad (70)$$

It is evident that, for $a > 0$, this expression tends to zero as $t - x/c \to 0$, and is hence continuous. The error estimate

$$\hat{m}(\lambda) - I(\lambda) = O(|\lambda|^{-1}) \quad (71)$$

is not good enough to infer the continuity of the solution across the wave front unless x is sufficiently large, and a more refined argument is required. Such an argument was given by Prüß [P9] for a general class of kernels. We describe his result below.

General singular kernels

We recall that, if m is completely monotone and has a non-integrable singularity at 0, then the solution is analytic in the quarter plane $x \geq 0$, $t > 0$. More recently, Prüß has considered the Rayleigh problem with a general class of kernels having integrable singularities. He assumes that $\log M$ (with M defined by $(37)_1$) is convex and proves (joint) continuity of the solution across the wave front. (We note that in order to convert his equation to ours, one should make the identifications $m = -a''$, $\eta = a(0)$, $\beta = a'(\infty)$.) The essential part of Prüß's argument is to show that, for each fixed x, the function $v(x, \cdot)$ is of bounded variation (and actually monotone increasing; for completely monotone m this is also shown by Pipkin [P7], p. 50). Therefore, it must have a limit as $t \to \frac{x}{c}+$, which can easily be shown to be zero. This implies continuity in t for each x. The joint continuity is obtained by Helly's selection theorem. The monotonicity of $v(x, \cdot)$ is shown by proving that

$$\lambda \to \exp\left\{-\lambda x \sqrt{\frac{\rho}{\beta + \hat{m}(0) - \hat{m}(\lambda)}}\right\} \tag{72}$$

is completely monotone and hence the Laplace transform of a positive measure. See [P9] for details.

As the examples above show, the degree of smoothing due to a singular kernel depends crucially on the strength of the singularity in m. It is easy to show that, if m is convex and $m(0) = \infty$, then $\lambda \operatorname{Im} \hat{m}(\lambda) \to \infty$ as λ approaches infinity on a parallel to the imaginary axis. Hence the integrand in (34) decays faster than $1/\lambda$, and in some sense the regularity of the solution is better than that of the data (the transform of a step jump behaves like $1/\lambda$). Hannsgen and Wheeler [H3] have proved a theorem on compactness of evolution operators for initial value problems on bounded domains, assuming that m is convex and $m(0) = \infty$. Also under the assumption that m is convex, Desch and Grimmer [D9] have shown that C^∞-smoothing occurs in the Rayleigh problem if and only if the singularity in the kernel is stronger than logarithmic. The proof of this is based on an estimate of $\lambda \operatorname{Im} \hat{m}(\lambda)$ as $\lambda \to \infty$ in terms of the behavior of $m(t)$ as $t \to 0+$. This estimate exploits the convexity of the kernel.

An open problem concerns the regularity of the solution behind the wave. If m is singular at 0, but C^∞ elsewhere, one expects that the solution should also be of class C^∞ behind the wave (in fact we have demonstrated this above if m is not singular). The case of completely monotone kernels (including all the examples above) is trivial, and in [H14] we give a proof under the assumption that, for $t \to 0$, the growth of every derivative of m is limited by some negative power of t.

49

Shock layers

It is also of interest to look at singularly perturbed problems. We discuss only the simple case of an elastic solid perturbed by a small Newtonian viscosity. For simplicity we set $\rho = \beta = 1$. Equation $(26)_1$ becomes

$$v_{tt} = v_{xx} + \varepsilon v_{xxt},\tag{73}$$

and the solution (34) reduces to

$$v(x,t) = \frac{1}{2\pi i}\int_{\gamma-i\infty}^{\gamma+i\infty}\frac{1}{\lambda}e^{\lambda t}e^{-\lambda x/\sqrt{1+\varepsilon\lambda}}\,d\lambda.\tag{74}$$

Certainly one expects that the solution must in some sense converge to the Heaviside function $H(t-x)$ as $\varepsilon \to 0$; since, however, the solution is continuous for every positive ε, the convergence cannot be uniform, and there must be a transition layer near $x = t$. It is the width and structure of this transition layer which we want to determine. The form of (74) suggests the substitution $\varepsilon\lambda = \sigma$, leading to

$$v(x,t) = \frac{1}{2\pi i}\int_{\gamma-i\infty}^{\gamma+i\infty}\frac{1}{\sigma}e^{\sigma t/\varepsilon}e^{-\sigma x/(\varepsilon\sqrt{1+\sigma})}\,d\sigma.\tag{75}$$

This is the same as the solution for $\varepsilon = 1$ with arguments $\frac{x}{\varepsilon}$ and $\frac{t}{\varepsilon}$. To proceed further, we set $t = x + \xi\sqrt{\varepsilon x}$. We also shift the integration contour to the imaginary axis, with a "dent" to the right at the origin. We now obtain

$$v(x, x + \xi\sqrt{\varepsilon x}) = \frac{1}{2\pi i}\int_{-i\infty}^{i\infty}\frac{1}{\sigma}e^{\sigma\xi\sqrt{x/\varepsilon}}e^{\sigma(x/\varepsilon)(1-1/\sqrt{1+\sigma})}\,d\sigma.\tag{76}$$

If x is positive and ε approaches zero, the integrand in (76) becomes very small when $|\sigma| \gg \sqrt{\frac{\varepsilon}{x}}\sqrt{\ln\frac{x}{\varepsilon}}$, and the infinite integral can be approximated by an integral with finite bounds of this order of magnitude. Within these bounds, we can use the approximation $\frac{1}{\sqrt{1+\sigma}} \sim 1 - \frac{\sigma}{2}$, and the second exponential in the integrand becomes $\exp\frac{\sigma^2 x}{2\varepsilon}$. With this new integrand we can extend the bounds to infinity and make only a small error. Thus we find the following uniform approximation for v:

$$\frac{1}{2\pi i}\int_{-i\infty}^{i\infty}\frac{1}{\sigma}e^{\sigma^2 x/(2\varepsilon)}e^{\sigma\xi\sqrt{x/\varepsilon}}\,d\sigma$$

$$= \frac{1}{2\pi i}\int_{-i\infty}^{i\infty}\frac{1}{\phi}e^{\phi^2/2}e^{\phi\xi}\,d\phi$$

$$= \frac{1}{2} + \frac{1}{2\pi}\int_{-\infty}^{\infty}e^{-\theta^2/2}\frac{\sin\xi\theta}{\theta}\,d\theta = \frac{1}{\sqrt{2\pi}}\int_{-\infty}^{\xi}e^{-\psi^2/2}\,d\psi.\tag{77}$$

As $\xi \to \pm\infty$, this approaches the limits 1 and 0, respectively, i.e. for large $|\xi|$ the solution is close to that of the unperturbed problem. The width of the transition layer is of order $\sqrt{\varepsilon x}$ and its shape is like that of the Newtonian solution (27). It would be of interest to have a similar analysis for the smoothing mechanism introduced by singular kernels rather than Newtonian viscosity.

Asymptotic behavior as $t \to \infty$

We now study the asymptotic behavior as $t \to \infty$ of the solution (34) of the Rayleigh problem (26) by following the development of Chu [C4]. We will exclude the trivial case $\eta = 0$, $m = 0$ in (26). In addition to the basic hypotheses that m is nonnegative, nonincreasing, and integrable at infinity, we assume for simplicity that

(i) $m \in C^\infty [0, \infty)$ with $m^{(k)} \in L^1(0, \infty)$ for every $k \in \mathbb{N}$;
(ii) $\hat{m}(\lambda)$ is C^2-smooth on Re $\lambda \geq 0$ (this will be the case if $t^2 m(t) \in L^1(0, \infty)$).

Since the integrand in (34) has no singularities in the half-plane Re $\lambda > 0$ and only a simple pole at $\lambda = 0$ on the imaginary axis, we shift the contour to the path Γ consisting of the imaginary axis indented by a semicircle in the right half-plane centered at the origin and having radius $\delta > 0$. We decompose Γ into three parts $\Gamma_1, \Gamma_2, \Gamma_3$. Let $M >> \delta$ (the choices δ small and M large will be discussed in the analysis). Γ_1 is the part of Γ from $-i\infty$ to $-iM$ and from iM to $i\infty$; Γ_2 is the part of Γ from $-iM$ to $-i\delta$ and from $i\delta$ to iM; Γ_3 is the semicircle from $-i\delta$ to $i\delta$. We define

$$I_k := \frac{1}{2\pi i} \int_{\Gamma_k} \frac{1}{\lambda} e^{\lambda t} e^{-\lambda x \sqrt{\rho/(\eta \lambda + \beta + \hat{m}(0) - \hat{m}(\lambda))}} \, d\lambda, \quad k = 1, 2, 3, \qquad (78)$$

where $x > 0$, $t > 0$. In the analysis of each integral I_k we will distinguish two cases: (a) $x > 0$ fixed and $t \to \infty$, (b) both x and t large.

We begin by analyzing I_1. If $\eta > 0$, it is easily seen using the expansion (39) that on Γ_1

$$e^{-\lambda x \sqrt{\rho/(\eta \lambda + \beta + \hat{m}(0) - \hat{m}(\lambda))}} = e^{-\sqrt{\lambda \rho/\eta} \, x + O(x/|\lambda|^{1/2})}. \qquad (79)_1$$

For $\frac{x}{M^{1/2}}$ sufficiently small, we use the approximation $\exp(O(\frac{x}{|\lambda|^{1/2}})) = 1 + O(\frac{x}{|\lambda|^{1/2}})$, $|\lambda| \geq M$, in $(79)_1$, and we find that I_1 is approximated by the integral

$$\frac{1}{2\pi i} \int_{\Gamma_1} \frac{1}{\lambda} e^{\lambda t - \sqrt{\frac{\lambda \rho}{\eta}} x} \, d\lambda, \qquad (79)_2$$

to within terms of order $O(\frac{x}{M^{1/2}})$. By deforming the contour Γ_1 into the left half-plane, it follows easily that the integral $(79)_2$ tends to zero as $t \to \infty$, uniformly for x in compact sets. The error term is then made arbitrarily small, uniformly for x in compact sets, by choosing M sufficiently large. If x is large, then we can use $(79)_1$ directly (without approximating the term $\exp(O(\frac{x}{|\lambda|^{1/2}}))$) to see that I_1 is small (uniformly in t) for large M.

If $\eta = 0$, we use formula $(40)_1$ to obtain

$$e^{-\lambda x \sqrt{\rho/(\beta + \hat{m}(0) - \hat{m}(\lambda))}} = e^{-\lambda x/c} e^{-m(0)x/2c^3 \rho} e^{O(x/|\lambda|)}, \qquad (80)$$

51

where c is the constant defined in (28). We approximate $\exp(O(x/|\lambda|))$ by $1 + O(x/|\lambda|)$ if λ is large compared to x. Again it is easy to show from (80) that I_1 is small for both cases (a) and (b) if M is chosen large enough.

We next discuss the integral I_2 in (78). Since the imaginary part of $\eta\lambda + \beta + \hat{m}(0) - \hat{m}(\lambda)$ has the same sign as the imaginary part of λ, one shows that the integrand in I_2 tends to zero as $x \to \infty$. It then follows that I_2 tends to zero as $x \to \infty$, uniformly for $t \in [0, \infty)$. In fact, we can make δ small (of order $x^{-(1/2-\epsilon)}$ for $\beta \neq 0$ and of order $x^{-(2-\epsilon)}$ for $\beta = 0$) if x is large. On the other hand, if x is fixed one integrates by parts and finds that, for any fixed δ and M, I_2 tends to zero like $\frac{1}{t}$ as $t \to \infty$. Therefore the integral I_2 becomes small in both cases (a) and (b).

Finally, we analyze the integral I_3 in (78). If $\eta > 0$, $\beta > 0$, the integrand can be written in the form

$$\frac{1}{\lambda} e^{\lambda t} e^{-\lambda x \sqrt{\rho/\beta}} e^{\lambda^2 x \sqrt{\rho/\beta^3}(\eta - \hat{m}'(0))/2} e^{O(|\lambda|^3 x)}. \tag{81}_1$$

We have $\lambda = \delta e^{i\theta}$, $-\frac{\pi}{2} \leq \theta \leq \frac{\pi}{2}$, and we can choose δ such that $\delta^3 x$ is small. Therefore, we can use the approximation $\exp(O(\delta^3 x)) = 1 + O(\delta^3 x)$ and drop the $O(\delta^3 x)$-term. Then I_3 is approximated by

$$\frac{1}{2\pi i} \int_{\Gamma_3} \frac{1}{\lambda} e^{\lambda t} e^{-\lambda x \sqrt{\rho/\beta}} e^{\lambda^2 x \sqrt{\rho/\beta^3}(\eta - \hat{m}'(0))/2} \, d\lambda. \tag{81}_2$$

Observe that by hypothesis $(-\hat{m}'(0)) = \int_0^\infty t m(t) \, dt > 0$, and therefore $(81)_2$ also approximates I_3 when $\eta = 0$. If $\beta = 0$, one easily shows that the integrand in I_3 is approximated by

$$\frac{1}{\lambda} e^{\lambda t} e^{-x \sqrt{\lambda \rho/(\eta - \hat{m}'(0))}} e^{O(|\lambda|^{3/2} x)}, \tag{82}_1$$

and since we can choose δ such that $\delta^{3/2} x$ is small, we can replace the last factor by 1. Therefore, I_3 is approximated by

$$\frac{1}{2\pi i} \int_{\Gamma_3} \frac{1}{\lambda} e^{\lambda t} e^{-x \sqrt{\lambda \rho/(\eta - \hat{m}'(0))}} \, d\lambda. \tag{82}_2$$

The above analysis shows that the only significant contribution to the solution v comes from I_3 in the form of the integrals $(81)_2$ if $\beta > 0$, or $(82)_2$ if $\beta = 0$. One can check that if $\eta \geq 0$, $\beta > 0$, the contribution $(81)_2$ is exactly the same as that of a material with elastic modulus β and a Newtonian viscosity $\eta - \hat{m}'(0)$. We can therefore summarize the asymptotic behavior of the solution of the Rayleigh problem as follows:

The behavior of the solution (34) as $t \to \infty$ is the same as that of a material with elastic modulus β and a Newtonian viscosity $\eta - \hat{m}'(0)$.

Using the result of the discussion of transition layers in the previous subsection, the asymptotic behavior as $t \to \infty$ is characterized by a wave moving at speed $\sqrt{\beta/\rho}$ and a transition layer having a width of order $\sqrt{t}$.

We remark that the only reason for requiring assumption (i) regarding m is to obtain an estimate for $\hat{m}(\lambda)$ as $\lambda \to \infty$ in a right half-plane (essential use of this hypothesis is only made in analyzing I_1). For specific singular kernels, such as those discussed earlier in this section, there is no difficulty in carrying out a similar analysis provided one has explicit information on $\hat{m}(0) - \hat{m}(\lambda)$ as $\lambda \to \infty$ in a right half-plane. However, the problem of establishing the asymptotic result for a general class of singular kernels is open.

Stationary singularities

So far we have looked at the evolution of a singularity introduced at the boundary of the domain. For the discussion of general initial value problems it is important to look also at the propagation of singularities that originate in the interior of the domain. Let us consider a one-dimensional linearly viscoelastic medium occupying the whole real line. With u denoting the displacement, the motion is described by

$$\rho u_{tt}(x,t) = \eta u_{xxt}(x,t) + \beta u_{xx}(x,t)$$

$$+ \int_{-\infty}^{t} m(t-\tau)(u_{xx}(x,t) - u_{xx}(x,\tau))\, d\tau, \ x \in \mathbb{R}, \ t > 0. \tag{83}$$

We assume that the history up to time $t = 0$ is again given by 0, but that non-zero initial data for u and u_t with a discontinuity are prescribed at $t = 0$:

$$u(x,t) = 0 \text{ for } x \in \mathbb{R}, \ t < 0; \ u(x,0+) = a \text{ sgn } x, \ u_t(x,0+) = b \text{ sgn } x. \tag{84}$$

The Laplace transform leads to

$$\rho\lambda^2 \hat{u}(x,\lambda) - a\rho\lambda \text{ sgn } x - b\rho \text{ sgn}x = -2\eta a\delta'(x) + (\eta\lambda + \beta + \hat{m}(0) - \hat{m}(\lambda))\hat{u}_{xx}(x,\lambda). \tag{85}$$

We note that we do not get the term $-2\eta a\delta'(x)$ on the right hand side if we approximate the Newtonian viscosity by an exponential memory and then pass to the limit of zero relaxation time. There is a problem of exchange of limits involved here. The initial conditions (84) mean that u and u_t change discontinuously at $t = 0$, and we should really think of this discontinuous transition as a limit of smooth transitions taking place very rapidly. The limits of zero transition time and zero relaxation time of the fluid cannot be interchanged, and the difference manifests itself in the term $-2\eta a\delta'(x)$. For the following discussion, we assume that $\eta = 0$, in which case the solution of (85) is given by

$$\hat{u}(x,\lambda) = (a\lambda + b)\left(\frac{1}{\lambda^2}\text{sgn } x - \frac{1}{\lambda^2}H(x)\exp(-\lambda x/\sqrt{(\beta + \hat{m}(0) - \hat{m}(\lambda))/\rho})\right.$$

$$\left. + \frac{1}{\lambda^2}H(-x)\exp(\lambda x/\sqrt{(\beta + \hat{m}(0) - \hat{m}(\lambda))/\rho})\right). \tag{86}$$

53

The last two contributions correspond to a superposition of the solution of the Rayleigh problem and its time integral. Thus the nature of the solution for $x > 0$ or $x < 0$ is as discussed before. The new feature arising in the present case is the behavior at $x = 0$. At this point all three terms in (86) have discontinuities, but these discontinuities cancel. The second derivative, however, is discontinuous. Hence u_{xx} is discontinuous at $x = 0$. We denote the jump in u_{xx} by $J(t) = u_{xx}(0+,t) - u_{xx}(0-,t)$. We conclude from (86) that the transform of J is given by

$$\hat{J}(\lambda) = \frac{-2(a\lambda + b)\rho}{\beta + \hat{m}(0) - \hat{m}(\lambda)}. \tag{87}$$

Here we have regarded u as a function of x taking values in the space of tempered distributions (with respect to t). We recall, however, that it follows from the discussion of the Rayleigh problem that for $t > 0$ the jump $J(t)$ is defined in the classical sense for a wide class of kernels (such as smooth kernels and completely monotone kernels).

There are special cases in which $J(t)$ vanishes for every $t > 0$. These cases occur if $\hat{J}$ is the transform of a distribution that is supported only at 0. According to a well-known theorem, this is the case if and only if $\hat{J}$ is a polynomial. Equation (87) yields

$$\hat{m}(0) - \hat{m}(\lambda) = -\beta - \frac{2(a\lambda + b)\rho}{\hat{J}(\lambda)}. \tag{88}$$

If $\hat{J}$ were a polynomial of degree higher than one, then $\hat{m}(0) - \hat{m}(\lambda)$ would tend to $-\beta$ as $\lambda \to \infty$. This is incompatible with the positivity of m. Hence, if $\hat{J}$ is a polynomial, it is of degree one, and it is easy to conclude that in this case m must be an exponential. If this is the case, then there are certain combinations of a and b for which J vanishes for every positive time.

As $t \to \infty$, we expect that the limit of J is equal to the limit of $\lambda\hat{J}(\lambda)$ as $\lambda \to 0$, which is zero if $\beta \neq 0$ and $2b\rho/\hat{m}'(0)$ if $\beta = 0$, i.e., the jump would relax to zero for a solid, but would persist in a fluid, as would physically be expected. It is clear that complex variable methods can be used to prove these conjectures, if sufficiently strong assumptions on m are introduced. We chose instead to use a result of Jordan, Staffans and Wheeler [J3], which requires much weaker assumptions. The result is a theorem of Wiener-Levy type which says that certain analytic expressions formed with Laplace transforms of L^1-functions are again transforms of L^1-functions. We can use this to prove that a function tends to zero by showing that both the function and its derivative are in L^1.

The theorem of Jordan, Staffans and Wheeler says that a function defined on the closed right half plane is the transform of an L^1-function if it vanishes at infinity and is locally analytic. Here a function ϕ is called locally analytic if, in the neighborhood of every point in the closed right half plane, it can be represented in the form

$$\phi(\lambda) = \psi(\lambda, \hat{\mu}_1(\lambda), ..., \hat{\mu}_k(\lambda)), \tag{89}$$

where ψ is analytic and the $\hat{\mu}_i(\lambda)$ are Laplace transforms of measures with finite total variation, and if in a neighborhood of infinity it can be represented in the form

$$\phi(\lambda) = \psi(\frac{1}{\lambda}, \hat{a}_1(\lambda), ..., \hat{a}_n(\lambda), \frac{\hat{\mu}_1(\lambda)}{\lambda}, ..., \frac{\hat{\mu}_k(\lambda)}{\lambda}), \tag{90}$$

where again ψ is analytic, the $\hat{a}_i$ are transforms of L^1-functions and the $\hat{\mu}_i$ are transforms of measures with finite total variation.

We want to apply this to the jump J in the case of a smooth kernel m with integrable derivatives. In this case, $\hat{J}$ differs only by a polynomial (which does not contribute to $J(t)$ for $t > 0$) from

$$\frac{2(a\lambda + b)\rho(-\hat{m}(\lambda))(\beta + \hat{m}(0)) + 2a\rho m(0)(\beta + \hat{m}(0) - \hat{m}(\lambda))}{(\beta + \hat{m}(0))^2 (\beta + \hat{m}(0) - \hat{m}(\lambda))}. \tag{91}$$

It is clear that this expression is locally analytic except possibly at zero (at infinity, use the fact that $\lambda\hat{m}(\lambda) = \widehat{m'}(\lambda) + m(0)$). At zero, there is no problem if $\beta \neq 0$, but if $\beta = 0$, then the denominator vanishes. In this case, the numerator also vanishes if $b = 0$, and we use (38) to obtain

$$\frac{-2a\rho\hat{m}(\lambda)\hat{m}(0) + 2a\rho m(0)\hat{M}(\lambda)}{\hat{m}(0)^2 \hat{M}(\lambda)}. \tag{92}$$

Since M is nonnegative, nonincreasing and convex, it can be shown that $1/\hat{M}(\lambda)$ is locally analytic at zero even if M is not integrable at infinity (see [J3], p. 775). Hence $\hat{J}$ is locally analytic at zero and $J \in L^1(0, \infty)$. Clearly $\widehat{J'}(\lambda) = \lambda\hat{J}(\lambda) - J(0)$ is locally analytic at zero, and hence $J' \in L^1(0, \infty)$. If $\beta = 0$ and b is non-zero, we can still show that J' is in L^1. Compared to the previous case, the Laplace transform of J' contains an additional term

$$-\frac{2b\rho\hat{m}(\lambda)}{\hat{m}(0)\hat{M}(\lambda)}, \tag{93}$$

which is locally analytic at zero. Since J' is integrable, J has a limit at infinity. It is easy to show that this limit must equal $\lim_{\lambda \to 0+} \lambda\hat{J}(\lambda) = -2b\rho/\hat{M}(0)$, where the last term denotes zero if M is not integrable. Finally, we remark that the argument given here fails for singular kernels because we cannot expect that J and J' are integrable at 0. If one wants to extend these results to singular kernels, one must therefore somehow separate the behavior at zero from that at infinity.

In summary, we find that the discontinuity in the initial data leads to two propagating waves and a stationary singularity in the solution. Singularities in the kernel lead to smoothing of the propagating waves, but they do not smooth the stationary discontinuity. This stationary discontinuity is in the second derivative u_{xx}, because we have introduced a discontinuity only in the instantaneous values of u and u_t; we would, however, get a discontinuity in u itself if we also allowed the history to have spatial discontinuities. For general initial data this means that we cannot expect any spatial smoothing to occur, if the histories are discontinuous; if the discontinuity is only in the initial data, we gain at most two derivatives. We do, however, gain some temporal smoothing (see e.g. [G16],[P8] for related results on abstract Volterra equations). In the nonlinear case, the propagating waves can lead to development of shocks (see next section). Since these propagating waves are smoothed out by singular kernels, the linear theory might lead us to expect that these kernels lead to solutions that remain smooth. We shall see in Section 6, however, that steady shocks can exist in the nonlinear case even if the kernel is singular.

4. Development of singularities

It is well known that nonlinear hyperbolic equations not only propagate singularities, but can also lead to their development from smooth initial data. As a simple example, consider Burgers' equation

$$u_t + uu_x = 0, \ x \in \mathbb{R}, \ t > 0, \ u(x,0) = u_0(x). \tag{94}$$

Suppose that u is a smooth solution of (94) on $\mathbb{R} \times [0,T)$ for some $T > 0$. For each $\xi \in \mathbb{R}$, the characteristic through $(\xi,0)$ is the curve $x = \tilde{x}(\xi,t)$, $t \geq 0$, where $\tilde{x}(\xi,t)$ is the unique solution of

$$\frac{d}{dt}\tilde{x}(\xi,t) = u\big(\tilde{x}(\xi,t),t\big),$$

$$\tilde{x}(\xi,0) = \xi. \tag{95}$$

We note that

$$\frac{d}{dt}u\big(\tilde{x}(\xi,t),t\big) = u_t\big(\tilde{x}(\xi,t),t\big) + u\big(\tilde{x}(\xi,t),t\big)u_x\big(\tilde{x}(\xi,t),t\big) = 0, \tag{96}$$

and hence u is constant along characteristics, i.e.

$$u\big(\tilde{x}(\xi,t),t\big) \equiv u_0(\xi). \tag{97}$$

Moreover, the characteristic curves are straight lines, and they can be determined directly from u_0, i.e.

$$\tilde{x}(\xi,t) = \xi + tu_0(\xi), \ t \geq 0. \tag{98}$$

The relations (97) and (98) can be used to provide a complete description of the solution of (94). If u_0 is smooth, then the solution of (94) remains smooth for as long as the characteristics do not cross. In particular, the solution remains smooth globally in time if and only if

$$u_0'(\xi) \geq 0 \quad \forall \xi \in \mathbb{R}. \tag{99}$$

f the initial datum does not satisfy (99), then a jump discontinuity in u develops n finite time.

The development of a singularity can also be seen by monitoring the evolution f u_x along characteristics. (The solution of (94) retains the full smoothness of he initial datum as long as u_x remains bounded). We note that

$$\frac{d}{dt} u_x\big(\tilde{x}(\xi,t),t\big) = -u_x^2\big(\tilde{x}(\xi,t),t\big), \tag{100}$$

s can be verified by differentiating equation (94) with respect to x. Since the solution of $\dot{y} = -y^2$, $y(0) = y_0$ becomes infinite in finite time if $y_0 < 0$, we onclude that the solution of (94) will develop a singularity if $u_0'(x_0) < 0$ for ome x_0. Observe that singularities can develop for arbitrarily small initial data; iowever, the time it takes to form a singularity increases as the datum becomes mall.

If, instead of (94), we consider the problem

$$u_t + uu_x + \alpha u = 0, \quad u(x,0) = u_0(x) \tag{101}$$

with $\alpha > 0$, the characteristics are still given by (95), but in place of (96) we have

$$\frac{d}{dt} u\big(\tilde{x}(\xi,t),t\big) = -\alpha u\big(\tilde{x}(\xi,t),t\big). \tag{102}$$

Of course, the characteristic curves for (101) generally are not straight lines. Using energy methods, it can be shown for a class of equations including (101) that the olution retains the full smoothness of u_0 for as long as u_x remains bounded (see e.g. [M5]). The evolution of u_x along characteristics is governed by

$$\frac{d}{dt} u_x\big(\tilde{x}(\xi,t),t\big) = -u_x^2\big(\tilde{x}(\xi,t),t\big) - \alpha u_x\big(\tilde{x}(\xi,t),t\big). \tag{103}$$

Since the solution $y(t)$ of

$$\dot{y} = -y^2 - \alpha y, \quad y(0) = y_0 \tag{104}$$

exists for all $t \geq 0$ if and only if $y_0 \geq -\alpha$, one can conclude that, for smooth u_0, the solution of (101) remains smooth globally in time if and only if

$$u_0'(x) \geq -\alpha \quad \forall x \in \mathbb{R}. \tag{105}$$

In particular, (105) holds if $\|u_0'\|_{L^\infty} \leq \alpha$. This type of behavior is typical for quasilinear hyperbolic equations with damping.

The quasilinear wave equation

A blow-up result for the quasilinear wave equation

$$u_{tt} = \phi(u_x)_x, \quad x \in \mathbb{R}, \ t > 0, \tag{106}$$

was proved by Lax [L3], using Riemann invariants. (Lax's argument requires that ϕ be either convex or concave; a blow-up result for the case when ϕ has an inflection point was first established by MacCamy and Mizel [M1].) Anticipating our treatment of problems in viscoelasticity, we transform (106) to a system in a nonconventional fashion. We assume that $\phi' > 0$. Setting $v = u_t$, $\sigma = \phi(u_x)$, we find that (106) becomes

$$v_t = \sigma_x, \quad \sigma_t = \Phi(\sigma)v_x, \tag{107}$$

where

$$\Phi(\sigma) := \phi'(\phi^{-1}(\sigma)). \tag{108}$$

We assume that Φ and the initial data for v and σ are of class C^∞ and that the initial data are bounded. This guarantees the existence of a local solution of class C^∞ which can be continued as long as first derivatives of v and σ remain bounded. The two families of characteristics for (107) are given by $\frac{dx}{dt} = \pm\sqrt{\Phi(\sigma)}$, and the **Riemann invariants** are

$$r := v + \Psi(\sigma), \quad s := v - \Psi(\sigma), \quad \text{where } \Psi(\sigma) := \int^\sigma \frac{dw}{\sqrt{\Phi(w)}}. \tag{109}$$

The significance of the Riemann invariants is that they are constant along characteristics. Since (107) is hyperbolic, i.e. Φ is strictly positive, the mapping $(v, \sigma) \to (r, s)$ is invertible. If we rewrite (107) in terms of r and s we obtain

$$r_t = \lambda(r - s)r_x, \quad s_t = -\lambda(r - s)s_x. \tag{110}$$

Here we have set $\lambda(r - s) := \sqrt{\Phi(\Psi^{-1}(\frac{r-s}{2}))}$. It follows that r is constant along the characteristic curves $dx/dt = -\lambda$ and s is constant along $dx/dt = \lambda$. Consequently, if we consider the Cauchy problem associated with (110), then, as long as a smooth solution exists, r and s cannot take values other than those which they had initially. Let us, on the other hand, look at the evolution of r_x and s_x. If we differentiate (110) with respect to x, we obtain

$$r_{xt} - \lambda(r - s)r_{xx} = \lambda'(r - s)(r_x^2 - r_x s_x),$$

$$s_{xt} + \lambda(r - s)s_{xx} = \lambda'(r - s)(s_x^2 - r_x s_x). \tag{111}$$

The product terms $r_x s_x$ are inconvenient; we can transform them away by setting $\rho := \lambda^{1/2}r_x$, $\eta := \lambda^{1/2}s_x$. We then obtain the system

$$\rho_t - \lambda(r - s)\rho_x = \lambda^{-1/2}(r - s)\lambda'(r - s)\rho^2,$$

$$\eta_t + \lambda(r - s)\eta_x = \lambda^{-1/2}(r - s)\lambda'(r - s)\eta^2. \tag{112}$$

Hence the derivatives of ρ and η along characteristics are proportional to ρ^2 and η^2, respectively. From this we can easily conclude that (except for $\lambda' \equiv 0$), there are solutions which are not globally smooth.

58

Theorem II.1:

Assume that $\inf_{x_1,x_2 \in \mathbb{R}} \lambda'(r(x_1,0) - s(x_2,0)) > 0$ *and that* $m :=$ $\max\left[\sup_x \rho(x,0), \sup_x \eta(x,0)\right] > 0$. *Let*

$$A := \sup_{x_1,x_2} \lambda^{-1/2} \lambda'(r(x_1,0) - s(x_2,0)), \quad B := \inf_{x_1,x_2} \lambda^{-1/2} \lambda'(r(x_1,0) - s(x_2,0)).$$
(113)

Then at least one of ρ *and* η *becomes unbounded at a time between* $(mA)^{-1}$ *and* $(mB)^{-1}$.

An analogous theorem holds if λ' is negative. We note that the sign of λ' is related to that of ϕ'', as is easily verified from the definition of λ. Moreover, it is easy to express the conditions which guarantee blow-up directly in terms of the initial data for (107) (or (106)).

The proof of Theorem II.1 follows from the calculations above and the following simple remark.

Remark II.2:

Consider the ODE $\dot{z}(t) = a(t)z^2(t)$ with initial condition $z(0) = m > 0$. If $0 < B \le a(t) \le A$, then z becomes infinite at a time between $(mA)^{-1}$ and $(mB)^{-1}$, as can easily be seen by solving the equation explicitly. Let t_c be the time when the solution of the ODE becomes infinite. Then at times prior to t_c we have the estimate $\frac{1}{A(t_c - t)} < z(t) < \frac{1}{B(t_c - t)}$.

Consider now the damped quasilinear wave equation,

$$u_{tt} + u_t = \phi(u_x)_x, \quad x \in \mathbb{R}, \ t > 0.$$
(114)

As before, we let $v := u_t$, $\sigma := \phi(u_x)$ and we define r, s, and λ in the same way as above. Instead of (110) we now obtain

$$r_t - \lambda(r-s)r_x = -\frac{r+s}{2}, \quad s_t + \lambda(r-s)s_x = -\frac{r+s}{2}.$$
(115)

The quantities $\rho := \lambda^{1/2} r_x$ and $\eta := \lambda^{1/2} s_x$, with r and s given by (109), satisfy

$$\rho_t - \lambda(r-s)\rho_x = \lambda^{-1/2}(r-s)\lambda'(r-s)\rho^2 - \frac{\rho+\eta}{2},$$

$$\eta_t + \lambda(r-s)\eta_x = \lambda^{-1/2}(r-s)\lambda'(r-s)\eta^2 - \frac{\rho+\eta}{2}.$$
(116)

We note that r and s are no longer constant along characteristics, but it is possible to obtain bounds for them and hence for the coefficients in (116). One can still establish blow-up for solutions of (116), but one must now assume that the initial data are sufficiently large, so that the quadratic terms on the right of (116) dominate the linear terms. More precisely, one takes the data such that $\sup_{x \in \mathbb{R}} (|r(x,0)| + |s(x,0)|)$ is sufficiently small and $\rho(x,0)$ or $\eta(x,0)$ is sufficiently

59

large. Results of this nature for damped hyperbolic systems were obtained b
Kosiński [K15]. For smooth and sufficiently small data, Theorem IV.3 establishe
global existence of classical solutions of equation (114).

Certain one-dimensional models in viscoelasticity are similar to the friction
ally damped wave equation. They look like the quasilinear wave equation whe
only terms of the highest differential order are considered, and hence one expect
solutions to develop singularities when the terms of highest differential order ar
dominant. This is expected to be the case when the gradients of the initial dat
are steep, and the hyperbolic system leads to blow-up in a very short time. If o
the other hand, the initial data and their derivatives are small, then the dissipa
tive mechanism may overcome the tendency to form singularities. We shall see i
Chapter IV that this is indeed the case.

Acceleration waves

The propagation of acceleration waves into a medium at rest (see Colemar
Gurtin and Herrera [C7], Coleman and Gurtin [C8]) exemplifies the behavio
discussed in the preceding paragraph. For simplicity we consider an equation c
motion of the form

$$u_{tt} = \int_{-\infty}^{t} m(t-\tau)h(u_x(x,t), u_x(x,\tau))_x \, d\tau. \tag{117}$$

The analysis of [C7] and [C8] was carried out for a more general functional depen
dence of the stress on the strain history. We consider a solution of (117) whic
is of class C^2 on the complement of a smooth curve $t = \gamma(x)$, across which u
u_t and u_x are continuous, but the second derivatives experience a jump. Such
singularity is called an acceleration wave. We consider an acceleration wave prop
agating into an undeformed medium at rest; we assume that $u = 0$ for $t < \gamma(x)$
By $[w]$ we denote the jump of a quantity w across the curve $t = \gamma(x)$. It follow
from the assumed continuity of u_t and u_x that

$$[u_{xx}] = -\gamma'(x)[u_{xt}] = (\gamma'(x))^2 [u_{tt}]. \tag{118}$$

On the other hand, the equation of motion yields

$$u_{tt} = \left\{ h_{,1}(0,0) \int_0^\infty m(s) \, ds \right\} u_{xx} \tag{119}$$

for $t = \gamma(x)+$. Here $h_{,1}$ is the derivative of h with respect to the first argument
The quantity

$$c^2 := h_{,1}(0,0) \int_0^\infty m(s) \, ds \tag{120}$$

is assumed to be positive. By combining (118) and (119), we find that $\gamma'(x) = c^{-1}$
Hence acceleration waves propagating into an undeformed medium at rest do s
with constant speed, even though the equation is nonlinear.

60

Differentiation of (117) with respect to t yields

$$u_{ttt} = m(0)h(u_x(x,t), u_x(x,t))_x + \int_{-\infty}^{t} m'(t-\tau)h(u_x(x,t), u_x(x,\tau))_x \, d\tau$$

$$+ \int_{-\infty}^{t} m(t-\tau)h(u_x(x,t), u_x(x,\tau))_{xt} \, d\tau. \tag{121}$$

For $t = \gamma(x)+$, (121) simplifies to

$$u_{ttt} = c^2 u_{xxt} + \left\{ h_{,11}(0,0) \int_0^\infty m(s) \, ds \right\} u_{xx} u_{xt} + h_{,2}(0,0)m(0)u_{xx}. \tag{122}$$

Let $\frac{d}{dt}$ denote the derivative along the wavefront: $\frac{d}{dt} = \frac{\partial}{\partial t} + c\frac{\partial}{\partial x}$. We find

$$\frac{d}{dt}[u_{tt}] = [u_{ttt}] + c[u_{xtt}], \quad \frac{d}{dt}[u_{xt}] = [u_{xtt}] + c[u_{xxt}], \tag{123}$$

which, in conjunction with the relation $[u_{tt}] = -c[u_{xt}]$, yields

$$[u_{ttt}] - c^2[u_{xxt}] = 2\frac{d}{dt}[u_{tt}]. \tag{124}$$

Hence equation (122) yields

$$\frac{d}{dt}[u_{tt}] = \alpha[u_{tt}]^2 - \beta[u_{tt}], \tag{125}$$

where

$$\alpha := -\frac{1}{2c^3}\left\{ h_{,11}(0,0) \int_0^\infty m(s) \, ds \right\},$$

$$\beta := -\frac{1}{2c^2} h_{,2}(0,0)m(0). \tag{126}$$

Observe that (125) is an ordinary differential equation for the jump in accelera-tion. If β is positive, solutions of (125) with small initial data will decay to zero, while solutions with large initial data having the same sign as α become infinite in finite time. This suggests that the discontinuity in acceleration develops into a stronger type of singularity, presumably a shock (i.e. a discontinuity in u_t and u_x). We refer to the review article of Chen [C2] for a formal study of shock waves in viscoelastic solids.

We now discuss the development of singularities from smooth data for several problems in viscoelasticity. The method we use is an adaptation of Lax's method. Although we are not dealing with a hyperbolic 2×2-system and there are no Riemann invariants which are constant on characteristics, we have a 2×2-system "to leading order" and there is something like approximate Riemann invariants. We shall discuss both differential and integral models. The method described here was applied to a special problem in [N6]. A very similar approach was developed by Dafermos [D7]. For related work we refer to Dafermos [D6], Gripenberg [G17], Hattori [H4], Malek-Madani and Nohel [M6], and Rammaha [R1].

Differential models

We first discuss one-dimensional motions of materials with differential constitutive relations. For simplicity we assume that there is no external body force and that the density is 1 so that the equation of motion is

$$v_t = \sigma_x. \tag{127}$$

Two particular models were studied by Slemrod ([S7], appendix to [J6]). In the first, we have

$$\sigma(x,t) = h\left(\int_{-\infty}^{t} e^{-\lambda(t-s)} v_x(x,s) \, ds\right), \tag{128}$$

and in the second

$$\sigma(x,t) = \int_{-\infty}^{t} e^{-\lambda(t-s)} h(v_x(x,s)) \, ds. \tag{129}$$

Both models can be reduced to the quasilinear wave equation with linear frictional damping. For the first model, we set

$$z(x,t) := \int_{-\infty}^{t} e^{-\lambda(t-s)} v(x,s) \, ds, \tag{130}$$

which leads to the differential equation

$$z_t = -\lambda z + v, \tag{131}$$

and the equation of motion becomes

$$z_{tt} + \lambda z_t = h(z_x)_x. \tag{132}$$

For the second model, we obtain

$$\sigma_t = -\lambda \sigma + h(v_x), \tag{133}$$

and in conjunction with the equation of motion, we find

$$v_{tt} + \lambda v_t = h(v_x)_x. \tag{134}$$

Under appropriate assumptions on h and the data, second derivatives of z (and hence first derivatives of v and σ) become infinite in finite time for the first model; for the second model, second derivatives of v (and of σ) blow up. We shall find a similar result in a more general context below.

62

We consider systems of the form

$$v_t = \sigma_x,$$

$$\sigma_t = a(\sigma, w)v_x + b(\sigma, w), \qquad (135)$$

$$w_t = c(\sigma, w).$$

Here v and σ are scalar quantities (identified with velocity and stress), while w takes values in $\mathbb{R}^k$. As we shall see, a general class of differential constitutive laws leads to equations of motion which can be put in the canonical form (135).

We consider equation (135) for $x \in \mathbb{R}$ with initial data close (in the L^∞-sense) to a constant state $(\bar{v}, \bar{\sigma}, \bar{w})$. We assume that the initial data and the functions a, b and c are of class C^∞. Moreover, it is assumed that (at least in a neighborhood of $(\bar{\sigma}, \bar{w})$) the system (135) is hyperbolic, i.e. $a > 0$, and genuinely nonlinear, i.e. $a_\sigma \neq 0$. It is well known (see [M5]) that a C^∞-solution to the Cauchy problem associated with (135) exists and can be continued as long as (135) remains hyperbolic and v, σ, w and their first derivatives remain bounded. The characteristic speeds of the system (135) are $\pm\sqrt{a(\sigma, w)}$ and 0; the multiplicity of 0 is equal to the dimension k of the vector w. The development of singularities for (135) can be studied using the general method of John [J1] for hyperbolic systems; this has been carried out in a special example (F. John, private communication). However, the system is highly degenerate, which makes a simpler approach possible. The x-derivative of w is not present, which suggests treating (135) as a perturbation of a 2×2-system and using a variant of Lax's method.

We shall show that by choosing initial data with steep gradients we can obtain solutions such that (v, σ, w) remain close to the constants $(\bar{v}, \bar{\sigma}, \bar{w})$, but the first derivatives of v and σ become infinite in finite time. We can make the blow-up time arbitrarily short by making the gradients of the initial data steep enough.

We now introduce "approximate Riemann invariants" for (135). By this we mean the quantities that would be Riemann invariants if w could be treated as a parameter and a depended only on σ. According to (109), these approximate Riemann invariants are given by

$$r = v + \Psi(\sigma, w), \quad s = v - \Psi(\sigma, w), \quad \text{where } \Psi(\sigma, w) = \int^\sigma \frac{1}{\sqrt{a(\tau, w)}} \, d\tau. \qquad (136)$$

Since the derivative of Ψ with respect to σ is positive, we can invert the relationship (136): we have $v = \frac{r+s}{2}$ and σ is a function of $r - s$ and w. Computation of the derivatives of r, s and w along the characteristics yields

$$r^{\cdot} = -\tilde{g}(r, s, w) \cdot w_x + \tilde{h}(r, s, w),$$

$$s' = -\tilde{g}(r, s, w) \cdot w_x - \tilde{h}(r, s, w), \qquad (137)$$

$$w_t = \tilde{c}(r, s, w),$$

where

$$` = \frac{\partial}{\partial t} - \sqrt{a}\frac{\partial}{\partial x}, \quad ' = \frac{\partial}{\partial t} + \sqrt{a}\frac{\partial}{\partial x}, \tag{138}$$

and

$$\tilde{g}(r,s,w) = \nabla_w \Psi(\sigma,w)\sqrt{a(\sigma,w)},$$

$$\tilde{h}(r,s,w) = \frac{b(\sigma,y)}{\sqrt{a(\sigma,w)}} + \nabla_w \Psi(\sigma,w) \cdot c(\sigma,w), \tag{139}$$

$$\tilde{c}(r,s,w) = c(\sigma,w).$$

Observe that $r`$ and s' are not zero as in (110), but the right-hand sides of (137) do not contain any derivatives of r and s, only the x-derivative of w is present. We shall use this feature below to obtain estimates on the right-hand sides which can be used to control the growth of r and s.

In analogy with Lax's method, we set $\rho := a^{1/4}(v_x + a^{-1/2}\sigma_x)$, $\eta := a^{1/4}(v_x - a^{-1/2}\sigma_x)$. Note that if w were a constant parameter, then $v_x \pm a^{-1/2}\sigma_x$ would be the x-derivatives of r and s. We differentiate ρ, η and w_x along characteristics, and use equation (135) to obtain a system of the form

$$\rho` = \frac{a_\sigma(\sigma,w)}{4a^{1/4}(\sigma,w)}\rho^2 + O(|\rho||w_x| + |\eta||w_x| + |\rho| + |\eta| + |w_x|),$$

$$\eta' = \frac{a_\sigma(\sigma,w)}{4a^{1/4}(\sigma,w)}\eta^2 + O(|\rho||w_x| + |\eta||w_x| + |\rho| + |\eta| + |w_x|), \tag{140}$$

$$w_{xt} = O(|\rho| + |\eta| + |w_x|).$$

Here a term like $O(|\rho||w_x|)$ stands for an expression bilinear in ρ and w_x with coefficients depending only on $\sigma = \sigma(r - s, w)$ and w. It is straightforward (although tedious) to obtain explicit expressions for the right-hand sides in (140); however, these expressions are quite complicated and we feel that the present form of (140) is more transparent. The first terms on the right of (140)$_1$ and (140)$_2$ correspond to those in (112), and if these terms dominate, we can expect blow-up of ρ or η to occur.

We now choose the initial data in such a way that the following conditions are satisfied for $t = 0$: σ and w are (in the L^∞-sense) close to the constant values $\bar{\sigma}$ and $\bar{w}$. Recall that we have assumed that $a_\sigma(\bar{\sigma},\bar{w}) \neq 0$, w.l.o.g. let it be positive. On the other hand we require that ρ and η at time $t = 0$ are such that at least one of them has a large positive maximum, while the maxima of $-\rho$, $-\eta$ and $|w_x|$ are not too large. If (σ,w) remains in a sufficiently small neighborhood U of $(\bar{\sigma},\bar{w})$, then a and a_σ remain positive and bounded away from 0, and the coefficients in the O-terms of (140) remain bounded. Moreover, we have a positive lower bound γ_0 for $a_\sigma/(4a^{1/4})$. For $t \geq 0$, let

$$h(t) := \max[\sup_x \rho(x,t), \sup_x \eta(x,t)]. \tag{141}$$

As long as $(\sigma, w) \in U$, $h(t)$ is large and $\sup_x |w_x| << h(t)$, we find from (140) that

$$(\frac{d}{dt})_+ h(t) \geq \gamma_1 (h(t))^2, \quad \text{and} \quad \sup_x |w_{xt}| \leq \kappa h(t) << (h(t))^2 \qquad (142)$$

with appropriate positive constants γ_1 and κ. It therefore follows that $\sup_x |w_x|$ remains small compared to $h(t)$, and that $h(t)$ becomes infinite in finite time, which can be made arbitrarily short by choosing the initial value of h large enough. We have to make sure that (σ, w) remains in U while h becomes infinite. For this, we first note that there is also a constant γ_2 such that $(\frac{d}{dt})_+ h(t) \leq \gamma_2 (h(t))^2$, and hence we have an estimate of the form $\frac{c_1}{t_c - t} \leq h(t) \leq \frac{c_2}{t_c - t}$, where t_c is the blow-up time. (See Remark II.2.) The third equation of (140) implies that $|w_x|$ grows at most logarithmically as $t \to t_c$. Since the logarithm is integrable, it follows from (137) that r, s and w remain bounded and in fact close to their initial data provided t_c is small enough. Since we can make t_c arbitrarily small, we can make sure that (σ, w) will stay in U while h becomes infinite. Hence we have proved that there are initial data leading to blow-up.

We summarize this result in the following theorem.

Theorem II.3:

Let $T > 0$ and the constant state $(\bar{v}, \bar{\sigma}, \bar{w})$ be given. Assume that a is positive and bounded away from 0 on a neighborhood U of $(\bar{\sigma}, \bar{w})$ and that $a_\sigma(\bar{\sigma}, \bar{w}) \neq 0$. Then in every L^∞-neighborhood of $(\bar{v}, \bar{\sigma}, \bar{w})$, there are C^∞-functions which, when chosen as initial data for (135), yield a local classical solution with the property that (σ, w) remains in U and the maximal time t_c of existence does not exceed T. As $t \to t_c$, first derivatives of v and σ become unbounded.

We now apply this theorem to some model equations in viscoelasticity.

Example II.4: Maxwell-type solid

Consider the constitutive relation

$$\sigma(x, t) = \phi(\epsilon(x, t)) - \int_{-\infty}^{t} e^{-\mu(t-\tau)} \psi(\epsilon(x, \tau)) \, d\tau, \qquad (143)$$

which is a special case of equation (19) of Chapter O. Here ϵ is the strain. Differentiation of (143) with respect to t shows that σ satisfies the differential equation

$$\sigma_t + \mu\sigma = \phi'(\epsilon)v_x + \mu\phi(\epsilon) - \psi(\epsilon), \qquad (144)$$

where v is the velocity. The motion is therefore governed by the system

$$v_t = \sigma_x,$$

$$\sigma_t = \phi'(\epsilon)v_x - \mu\sigma + \mu\phi(\epsilon) - \psi(\epsilon), \qquad (145)$$

$$\epsilon_t = v_x.$$

We can apply Theorem II.3 to show that, given any constant state $(\bar{v}, \bar{\sigma}, \bar{\epsilon})$ such that $\phi'(\bar{\epsilon}) > 0$, $\phi''(\bar{\epsilon}) \neq 0$, there are smooth initial data in any L^∞-neighborhood of that state such that the solution of (145) develops a singularity in finite time. The system as it stands is not in the form (135). However, we can transform it to that form as follows. In a neighborhood of $(\bar{v}, \bar{\sigma}, \bar{\epsilon})$, we set

$$w := \sigma - \phi(\epsilon), \tag{146}$$

which can be inverted to give

$$\epsilon = \phi^{-1}(\sigma - w). \tag{147}$$

The system (145) then assumes the form (135) with

$$a(\sigma, w) = \phi'(\phi^{-1}(\sigma - w)),$$

$$b(\sigma, w) = c(\sigma, w) = -\mu w - \psi(\phi^{-1}(\sigma - w)). \tag{148}$$

It is easy to see that

$$a_\sigma(\sigma, w) = \frac{\phi''(\phi^{-1}(\sigma - w))}{\phi'(\phi^{-1}(\sigma - w))}. \tag{149}$$

Similar considerations apply to the multi-mode Maxwell model

$$\sigma(x, t) = \phi(\epsilon(x, t)) - \int_{-\infty}^{t} \sum_{i=1}^{n} e^{-\mu_i(t-\tau)} \psi_i(\epsilon(x, \tau)) \, d\tau. \tag{150}$$

We set

$$w^1(x, t) := \sigma(x, t) - \phi(\epsilon(x, t)), \tag{151}_1$$

and

$$w^{i+1}(x, t) := \int_{-\infty}^{t} e^{-\mu_i(t-\tau)} \psi_i(\epsilon(x, \tau)) \, d\tau, \quad i = 1, ..., n. \tag{151}_2$$

The motion is governed by the system

$$v_t = \sigma_x,$$

$$\sigma_t = \phi'(\phi^{-1}(\sigma - w^1)) v_x - \sum_{i=1}^{n} \psi_i(\phi^{-1}(\sigma - w^1)) + \sum_{i=1}^{n} \mu_i w^{i+1},$$

$$w_t^1 = -\sum_{i=1}^{n} \psi_i(\phi^{-1}(\sigma - w^1)) + \sum_{i=1}^{n} \mu_i w^{i+1}, \tag{152}$$

$$w_t^{i+1} = \psi_i(\phi^{-1}(\sigma - w^1)) - \mu_i w^{i+1}.$$

The conditions needed to obtain blow-up are the same as in the case of a single relaxation mode: $\phi'(\bar{\epsilon}) > 0$ and $\phi''(\bar{\epsilon}) \neq 0$.

66

Example II.5: Shearing flow of a Johnson-Segalman fluid

We consider shearing flows of a fluid described by the differential constitutive relation (59) of Chapter I. It is assumed that the flow is in the y^1-direction, with the velocity and stresses only depending on $y^2 = x^2 =: x$. Let $\sigma := T^{12}$ denote the shear stress and let $\gamma := T^{11}$, $\tau := T^{22}$. Then the equation of motion and the constitutive law (59) of Chapter I reduce to the following system:

$$v_t = \sigma_x,$$

$$\sigma_t = \frac{1}{2}(\tau - \gamma)v_x + \frac{a}{2}(\tau + \gamma)v_x + \mu v_x - \lambda\sigma, \tag{153}$$

$$\gamma_t = (1 + a)\sigma v_x - \lambda\gamma,$$

$$\tau_t = (a - 1)\sigma v_x - \lambda\tau.$$

We can simplify this to a system for three variables by introducing $Z(\tau, \gamma) := \frac{1}{2}(\tau - \gamma) + \frac{a}{2}(\tau + \gamma)$; we then obtain

$$v_t = \sigma_x,$$

$$\sigma_t = (Z + \mu)v_x - \lambda\sigma, \tag{154}$$

$$Z_t = (a^2 - 1)\sigma v_x - \lambda Z.$$

The system (154) is hyperbolic provided $Z > -\mu$. In order to obtain a genuinely nonlinear system, we must assume that $a \neq \pm 1$. Since the case $|a| > 1$ is not physically reasonable for this model, let $-1 < a < 1$. We set

$$w := \exp\left(\frac{1}{2}(Z + \mu)^2\right)\exp\left(\frac{1 - a^2}{2}\sigma^2\right). \tag{155}$$

In the range $Z \geq -\mu$, we can obviously express Z as a function of σ and w: $Z = \tilde{Z}(\sigma, w)$. We now obtain the new system

$$v_t = \sigma_x,$$

$$\sigma_t = \tilde{Z}(\sigma, w)v_x - \lambda\sigma, \tag{156}$$

$$w_t = -\lambda(1 - a^2)\sigma^2 w - \lambda(\tilde{Z}(\sigma, w) + \mu)\tilde{Z}(\sigma, w)w.$$

This system has the structure of (135) and is genuinely nonlinear if $\tilde{Z}_\sigma \neq 0$, which is the case if $\sigma \neq 0$.

We now discuss a general strategy to transform differential models to the canonical form (135). We assume that σ is a given function of a vector quantity $z = (z_1, z_2, ..., z_n)$, say $\sigma = \phi(z)$, and that z is coupled to v by a system of differential equations of the form

$$z_t = \psi(z, v_x). \tag{157}$$

The motion is therefore governed by

$$v_t = \phi(z)_x,$$

$$z_t = \psi(z, v_x). \tag{158}$$

We are interested in solutions where z is close to a constant state $\bar{z}$. If $\nabla\phi(\bar{z}) \neq 0$, then there is a local coordinate transformation which makes σ one of the coordinates. Utilizing this transformation, the system (158) can be written in the form

$$v_t = \sigma_x,$$

$$\sigma_t = \alpha(\sigma, Z, v_x), \tag{159}$$

$$Z_t = \beta(\sigma, Z, v_x),$$

where Z is an $(n-1)$-vector. The characteristic speeds of this system are $\pm\sqrt{\frac{\partial \alpha}{\partial v_x}}$ and 0; the multiplicity of 0 is $n-1$. We assume that $\frac{\partial \alpha}{\partial v_x}$ is positive, so that the system is hyperbolic. We distinguish between the cases when α and β are linear in v_x (as in the examples above) and when they are nonlinear (as in Slemrod's second model (129) and some models of Oldroyd [O2]).

Let us assume that α and β are linear in v_x, so that (159) can be written in the form

$$v_t = \sigma_x,$$

$$\sigma_t = \tilde{\alpha}(\sigma, Z)v_x + \gamma(\sigma, Z), \tag{160}$$

$$Z_t = \tilde{\beta}(\sigma, Z)v_x + \delta(\sigma, Z).$$

Next we introduce another local change of coordinates that will remove the $\tilde{\beta}$-term from the equation. Let $(\bar{\sigma}, \bar{Z})$ be a point in $\mathbb{R}^n$ such that $\tilde{\alpha}(\bar{\sigma}, \bar{Z}) \neq 0$. Then any solution of the system of ordinary differential equations

$$\sigma' = \tilde{\alpha}(\sigma, Z),$$

$$Z' = \tilde{\beta}(\sigma, Z) \tag{161}$$

with initial data $(\hat{\sigma}, \hat{Z})$ in a sufficiently small neighborhood of $(\bar{\sigma}, \bar{Z})$ will cross the hyperplane $\sigma = \bar{\sigma}$; let the point of intersection be $(\bar{\sigma}, w(\hat{\sigma}, \hat{Z}))$. Obviously, we have $w(\bar{\sigma}, Z) = Z$ and the function $w(\sigma, Z)$ is an integral of the system (161), i.e. $w_\sigma \tilde{\alpha} + \nabla_Z w \cdot \tilde{\beta} = 0$. We can invert the relationship between w and Z and express Z as a function of σ and w: $Z = Z^*(\sigma, w)$. The system (160) now assumes the form (135) with

$$a(\sigma, w) = \tilde{\alpha}(\sigma, Z^*(\sigma, w)),$$

$$b(\sigma, w) = \gamma(\sigma, Z^*(\sigma, w)), \tag{162}$$

$$c(\sigma, w) = w_\sigma(\sigma, Z^*(\sigma, w))\gamma(\sigma, Z^*(\sigma, w)) + \nabla_Z w(\sigma, Z^*(\sigma, w)) \cdot \delta(\sigma, Z^*(\sigma, w)).$$

Assume now that α and β are nonlinear in v_x. If the system is hyperbolic, i.e. $\partial\alpha/\partial v_x > 0$, we can express v_x as a function of α, σ and Z; let us write

$v_x = g(\alpha, \sigma, Z)$. We introduce α as a new variable in place of v_x, and we also set $\gamma = v_t = \sigma_x$. We can then rewrite (159) as follows:

$$\gamma_t = \alpha_x,$$

$$\alpha_t = \frac{\partial \alpha}{\partial \sigma}(\sigma, Z, g(\alpha, \sigma, Z))\alpha + \nabla_Z \alpha(\sigma, Z, g(\alpha, \sigma, Z)) \cdot \beta(\sigma, Z, g(\alpha, \sigma, Z))$$

$$+ \frac{\partial \alpha}{\partial v_x}(\sigma, Z, g(\alpha, \sigma, Z))\gamma_x, \tag{163}$$

$$\sigma_t = \alpha,$$

$$Z_t = \beta(\sigma, Z, g(\alpha, \sigma, Z)).$$

This system has exactly the same structure as (135) if we let γ take the role of v, α the role of σ and (σ, Z) the role of w. Hence the analysis of Theorem II.3 applies provided we assume genuine nonlinearity, which now takes the form $\partial^2 \alpha / (\partial v_x)^2 \neq 0$. There is, however, a difference between the results for (160) and (163). For (160), the first derivatives of v and σ blow up, but for (163) the first derivatives of α and γ (which are equal to second derivatives of v and σ) blow up. These features generalize those found by Slemrod in the special case.

Integral models

The same technique can be applied to integral models. Let us consider a problem of the form

$$u_{tt}(x,t) = \int_{-\infty}^{t} m(t - \tau) h(u_x(x,t), u_x(x,\tau))_x \, d\tau. \tag{164}$$

We assume hyperbolicity and genuine nonlinearity, i.e. $m > 0$, $h_{,1} > 0$ and $h_{,11} \neq 0$. We set $v = u_t$, $\epsilon = u_x$. Then (164) can be written in the form

$$\epsilon_t(x,t) = v_x(x,t),$$

$$v_t(x,t) = \int_{-\infty}^{t} m(t - \tau) h_{,1}(\epsilon(x,t), \epsilon(x,\tau)) \, d\tau \, \epsilon_x(x,t) \tag{165}$$

$$+ \int_{-\infty}^{t} m(t - \tau) h_{,2}(\epsilon(x,t), \epsilon(x,\tau)) \epsilon_x(x,\tau) \, d\tau.$$

The following result is a special case of a theorem of Dafermos [D7].

Theorem II.6:

Assume that h is of class C^3, that m and m' are continuous and integrable on $[0, \infty)$; moreover, assume that $m > 0$ on $[0, \infty)$, $h_{,1}(0,0) > 0$, $h_{,11}(0,0) > 0$. Consider (165) together with the initial conditions

$$\epsilon(x,\tau) = 0, \ x \in \mathbb{R}, \ \tau \leq 0, \ v(x,0) = v_0(x), \ x \in \mathbb{R}. \tag{166}$$

Given any numbers $N, T > 0$, there is a positive number δ and a positive number $M(\delta, N, T)$ such that when v_0 is a C^2-function with compact support in $\mathbb{R}$ which satisfies

$$\max_{x \in \mathbb{R}} |v_0(x)| < \delta, \quad \min_{x \in \mathbb{R}} v_0'(x) > -N, \quad \max_{x \in \mathbb{R}} v_0'(x) > M, \qquad (167)$$

then the length of the maximal time interval of existence for a classical solution of (165), (166) does not exceed T.

The proof given by Dafermos uses a method which is similar in spirit (but not identical) to the one used for proving Theorem II.3. We close this section by sketching how the proof of Theorem II.3 could be adapted to treat equation (165). We introduce approximate Riemann invariants by $r = v + \Psi$, $s = v - \Psi$, where Ψ is defined by

$$\Psi(x, t) := \int^{\epsilon(x, t)} \sqrt{\int_{-\infty}^{t} m(t - \tau) h_{,1}(\varsigma, \epsilon(x, \tau)) \, d\tau} \, d\varsigma. \qquad (168)$$

After some algebra, it is found that

$$r_t = A r_x + P_1, \quad s_t = -A s_x + P_2, \qquad (169)$$

where

$$A := \sqrt{\int_{-\infty}^{t} m(t - \tau) h_{,1}(\epsilon(x, t), \epsilon(x, \tau)) \, d\tau}, \qquad (170)$$

and P_1 and P_2 are terms which involve ϵ and v but depend on derivatives only through integrals of of $\epsilon_x(x, \tau)$. These integrals play the role of w_x in (137). If ϵ_x can be shown to blow up like $\frac{1}{t_c - t}$, then its integral behaves like a logarithm, and r and s can be bounded in a similar fashion as before. We need to introduce appropriate analogues of ρ and η. It turns out that the same choice as for (135) also works in the present example. Let σ be the stress

$$\sigma(x, t) = \int_{-\infty}^{t} m(t - \tau) h(\epsilon(x, t), \epsilon(x, \tau)) \, d\tau. \qquad (171)$$

Then we set as before $\rho := A^{1/2}(v_x + \sigma_x / A)$ and $\eta := A^{1/2}(v_x - \sigma_x / A)$. A blow-up result can now be proved by an argument analogous to the proof of Theorem II.3. The equations for ρ and η (corresponding to the first two equations of (140)) now contain terms involving ϵ_x. To deal with these, one notes that differentiation of (171) with respect to x yields a linear integral relationship between ϵ_x and σ_x. For small times one finds that, up to perturbations that can be controlled, $\epsilon_x \sim A^{-2} \sigma_x$.

Although no such result has yet been proved, the close similarity of the equations with hyperbolic conservation laws suggests that the outcome of the type of blow-up discussed above is a shock solution with discontinuities in the derivatives. The numerical results of Markowich and Renardy [M7] do in fact give such solutions. The equation studied in [M7] is

$$u_{tt}(x,t) = \phi(u_x(x,t))_x - \int_0^t m(t-s)\psi(u_x(x,s))_x\ ds, \qquad (172)$$

and the computations were done with the special choices

$$\phi(\epsilon) = \psi(\epsilon) = 2\epsilon + 5\epsilon^2 + 25\epsilon^3, \quad m(s) = 0.4e^{-s} + 0.2e^{-2s}, \qquad (173)$$

and the initial data

$$u_x(x,0) = \delta(1 - 3x - x^2 + x^3)e^{-x^2/2}, \quad u_t(x,0) = \delta(1 - x^2)e^{-x^2/2}. \qquad (174)$$

For small values of δ, the computations lead to smooth solutions in agreement with analytical results to be discussed in Chapter IV. For larger values of δ, shocks develop, and the development of a shock becomes more rapid as δ is increased. The following figures show the computed u_x (called V) and u_t (called W) at various times for $\delta = 1$.

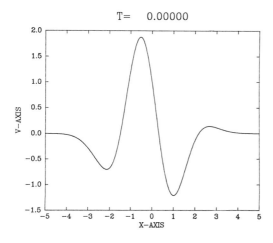

Fig. 2a: Initial data at t=0

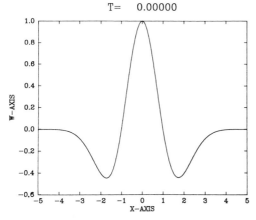

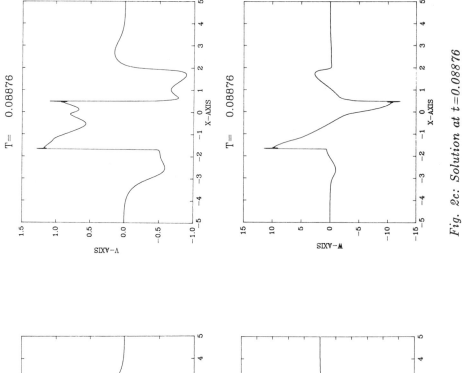

Fig. 2c: Solution at t=0.08876

Fig. 2b: Solution at t=0.05663

5. Shock conditions, instantaneous elasticity, and hypoelasticity

In Section 4, we showed that several one-dimensional models governing the motion of viscoelastic materials lead to development of singularities in finite time from smooth but large initial data. While no rigorous theory exists, numerical evidence and the analogy with hyperbolic conservation laws suggest that the solution can be continued beyond the blow-up time in such a way that the velocity and stress will be discontinuous across one or more smooth curves. Solutions with such discontinuities are called shock solutions, and the associated curves are called shocks.

The theory of existence and uniqueness of discontinuous solutions of nonlinear hyperbolic systems of partial differential equations is far from complete. It is well known that weak solutions of nonlinear hyperbolic equations are in general not unique. In order to single out a physically relevant solution of a given initial value problem, an admissibility criterion must be imposed. A variety of admissibility criteria have been proposed, but the uniqueness issue has not yet been completely settled - even for equations without memory. We shall not discuss admissibility criteria further.

For a single first order equation with memory (in one space dimension), Greenberg and Hsiao [G14] and MacCamy [M4] have discussed the Riemann problem and constructed solutions with shocks. The existence of steady shock and acceleration waves in one dimension for a class of materials with memory will be discussed in Section 6.

Our objective in this section is to discuss jump conditions which must hold at a shock in order that the equations of motion be satisfied in the weak sense. These conditions are analogues of the Rankine-Hugoniot conditions for hyperbolic equations without memory. Let us assume that a solution sustains jump discontinuities in the deformation gradient, velocity and stress across a smooth surface given by $\phi(x,t) = 0$. We assume that the solution is smooth elsewhere, at least in some neighborhood of the surface. We consider the equation of motion in Lagrangian coordinates (with zero body force),

$$\rho_0 v_t = \operatorname{div}_x \mathbf{S}, \tag{175}$$

where $\mathbf{S}$ denotes the Piola-Kirchhoff stress. The weak form of (175) requires that across the surface $\phi(x,t) = 0$ the relation

$$\rho_0 \phi_t [v] = [\mathbf{S}] \nabla_x \phi \tag{176}$$

must be satisfied where $[\cdot]$ denotes the jump of a quantity across the shock. Another jump relation results from the kinematic identity

$$\mathbf{F}_t = \nabla_x v. \tag{177}$$

73

The weak form of this equation yields

$$[\mathbf{F}]\phi_t = [v] * \nabla_x \phi, \tag{178}$$

with $*$ denoting the dyadic product. The relations (176) and (178) do not form a complete set of jump conditions. What is lacking is a relation between $[\mathbf{S}]$ (or the jump in the Piola-Kirchhoff extra stress if the material is incompressible) and $[\mathbf{F}]$. Such a relation can come only from the constitutive law. We note that, for a homogeneous elastic material obeying the constitutive law $(16)_2$ in Chapter I, the appropriate condition is $[\mathbf{S}] = [\tilde{\mathbf{S}}(\mathbf{F})]$.

Before obtaining a jump condition from the constitutive law for a viscoelastic material, one must answer the following question: Does the constitutive law provide a well-defined expression for the stress when the history of the deformation gradient involves a jump? In more precise mathematical terms, suppose the deformation gradient changes rapidly from one value to another. As we make the transition more and more rapid, does the corresponding stress converge to a unique limit depending only on the initial and final states, but not on the values taken by the deformation gradient during the transition? For some constitutive laws, this property holds if we impose a mild restriction, e.g. boundedness of $\mathbf{F}$ and $\mathbf{F}^{-1}$ during the transition. Viscoelastic materials satisfying the above property are those with instantaneous elasticity (see the discussion following (51) in Section 3, Chapter I).

Problems involving Newtonian viscosity or non-integrable kernels will lead to infinite stresses during the transition. This causes no difficulty, since such models are not expected to lead to the development of shocks from smooth data. A genuine problem arises with certain equations for which singularities develop, but for which the limit of the stress in a rapid change of the deformation gradient is not well defined. An example is given by the Johnson-Segalman fluid (see $(8)_3$, Section 2). In one-dimensional motions, we have shown in Example II.5 (Section 4) that singularities develop in finite time for large data; one expects that this is also the case in general three-dimensional motions. To determine the shock conditions for (8), observe that in a rapid transition $\mathbf{T}$ becomes discontinuous and ∇v becomes a δ-function, and their product makes no sense. In a rapid transition, the term $\lambda \mathbf{T}$ in $(8)_3$ is of lower order and hence irrelevant. If this term is dropped, then $(8)_3$ falls within a class of constitutive laws studied by Truesdell [T4], which are called **hypoelastic**. Following Truesdell and Noll [T5], we call a simple material hypoelastic if the constitutive law for $\mathbf{T}$ satisfies the following two restrictions:

(i) For every nondecreasing function $g : [0, \infty) \to \mathbb{R}$ such that $g(0) = 0$ and $\lim_{s \to \infty} g(s) = \infty$, we have

$$\mathbf{T}(\chi(x,t),t) = \mathcal{F}(\mathbf{F}^t(x,\cdot)) = \mathcal{F}(\mathbf{F}^t(x,g(\cdot))).$$

(ii) There exists a tensor-valued function $\mathbf{f}$ such that

$$\dot{\mathbf{T}} = \mathbf{f}(\mathbf{T}, \nabla_y v)$$

for every possible motion of the material. Here $\dot{\mathbf{T}}$ denotes the material time derivative of $\mathbf{T}$.

In physical terms, (i) states that the stress at time t depends only on the order in which the body has occupied its past configurations, but not on the time-rate at which these past configuration were traversed. It can be shown that the function f in item (ii) must be linear with respect to $\nabla_y v$; this is clearly the case in $(8)_3$. For hypoelastic materials, the stress after a transition from one state of deformation to another does, in general, depend on the structure of the transition. However, Bernstein [B5] has shown that if certain integrability conditions are satisfied, then this dependence on the structure of the transition disappears. For the constitutive relations $(8)_3$, these integrability conditions are satisfied if and only if $a = \pm 1$; in this case it is well known that the Johnson-Segalman model reduces to a special case of a K-BKZ material (recall that the latter has instantaneous elasticity). We conclude that if $a \neq \pm 1$ in a Johnson-Segalman fluid, then there is a serious difficulty in defining a shock solution.

An issue closely related to shock conditions and hypoelasticity concerns the validity of the Lodge-Meissner relation [L7]. This relation correlates the shear and normal stresses following a sudden shearing deformation. In the context of elasticity, this correlation has long been known, see Rivlin [R14]. The extension to viscoelasticity can be obtained by an argument which follows the lines of Rivlin [R16]. We consider a material which is at rest $(y = x)$ for $t < 0$ and is then instantaneously deformed in the following way:

$$y^1 = x^1 + s_0 x^2, \ y^2 = x^2, \ y^3 = x^3. \tag{179}$$

For $t > 0$, the material is again held at rest in this new configuration. We assume the material is isotropic, and the constitutive law (44) in Chapter I, Section 2 holds. For $t > 0$, the tensor $\mathbf{B}$ is given by

$$\mathbf{B} = \begin{pmatrix} s_0^2 + 1 & s_0 & 0 \\ s_0 & 1 & 0 \\ 0 & 0 & 1 \end{pmatrix}, \tag{180}$$

and the relative Cauchy tensor $\mathbf{C}_r(\tau, y, t)$ is equal to the identity for $\tau > 0$ and equal to $\mathbf{B}^{-1}$ for $\tau < 0$. If the constitutive relation can be extended to deformation histories with step discontinuities, then the extra Cauchy stress $\mathbf{T}_E$ must be a function of $\mathbf{B}$, and because of invariance requirements this function must be isotropic. Hence $\mathbf{T}_E$ can be represented in the form

$$\mathbf{T}_E = \gamma_1(s_0, t)\mathbf{B} + \gamma_2(s_0, t)\mathbf{B}^{-1} + \gamma_3(s_0, t)\mathbf{1}. \tag{181}$$

Therefore, we find that

$$T^{11} - T^{22} = (\gamma_1 - \gamma_2)s_0^2,$$

$$T^{12} = (\gamma_1 - \gamma_2)s_0, \tag{182}$$

and the ratio $(T^{11} - T^{22})/(T^{12} s_0)$ must be 1. This is known as the Lodge-Meissner relation.

For the Johnson-Segalman fluid with $-1 < a < 1$, the stress following a sudden transition depends on the transition structure. If, however, it is assumed that the motion is a one-dimensional simple shearing motion during the transition period, then unique limits for the stresses exist. Nevertheless, the above argument fails because it is based on assumptions concerning the full three-dimensional constitutive behavior. In fact, it is well known that the Lodge-Meissner relation is not satisfied for the Johnson-Segalman fluid; one obtains instead (see Lodge [L9]),

$$\frac{T^{11} - T^{22}}{s_0 T^{12}} = \frac{2}{\kappa s_0} \tan(\frac{1}{2}\kappa s_0), \tag{183}$$

where $\kappa = \sqrt{1 - a^2}$ (note that we recover the Lodge-Meissner relation for $a = \pm 1$). The Lodge-Meissner relation is important because it provides a distinction between elastic and hypoelastic behavior which can be checked experimentally. For a review of experiments and associated problems we refer to Lodge [L9]. The limited experimental evidence seems to indicate that the Lodge-Meissner relation is at least approximately valid for certain polymers.

6. Steady shock and acceleration waves in one dimension

There is at present no general theory establishing the existence of weak solutions of the equations governing motions of materials with memory. Pipkin [P6] constructed explicit solutions involving travelling shock and acceleration waves for a specific model. Motivated by Pipkin's result, Greenberg [G9] proved the existence of travelling shock waves for a general class of materials with fading memory. In this section, we shall prove such a result for a single-integral model. This specialization considerably simplifies the statement of assumptions as well as the proofs of the theorems, but at the same time it highlights many of the essential features.

The equation we consider is

$$u_{tt}(x,t) = \int_0^\infty m(s)h\big(u_x(x,t), u_x(x,t-s)\big)_x \, ds. \tag{184}$$

We seek a solution in the form of a wave travelling to the left, i.e. $u(x,t) = \tilde{u}(z)$, where $z = x + ct$ for some (constant) positive wave speed c. In this case (184) reduces to

$$c^2 \tilde{u}_{zz}(z) = \int_0^\infty m(s)h\big(\tilde{u}_z(z), \tilde{u}_z(z-cs)\big)_z \, ds. \tag{185}$$

Without loss of generality we assume $h(0,0) = 0$. We are interested in solutions of (185) which vanish identically for $z < 0$. We can integrate (185) with respect to z and obtain

$$c^2 \epsilon(z) = \int_0^\infty m(s) h\big(\epsilon(z), \epsilon(z - cs)\big) \, ds, \tag{186}$$

where we have set $\epsilon := \tilde{u}_z$ to simplify notation. Obviously, $\epsilon = 0$ is a trivial solution. We look for nontrivial solutions which experience a jump in ϵ at $z = 0$. The equation that must be satisfied by $\epsilon(0+)$ is

$$c^2 \epsilon = M h(\epsilon, 0), \quad \text{where } M := \int_0^\infty m(s) \, ds. \tag{187}$$

We seek solutions with $\epsilon(0+)$ negative. If the problem (184) is interpreted as modelling the longitudinal deformation of a one-dimensional rod, the condition $\epsilon(0+) < 0$ means that we are dealing with compressive shocks. Only solutions for which $\epsilon > -1$ make physical sense in this case, and the hypotheses on the function h will be formulated accordingly.

We now make the following assumptions:
(i) m is strictly positive, continuous and integrable on $[0, \infty)$.
(ii) The function h is of class C^1 on $(-1, 0] \times (-1, 0]$; moreover, on $(-1, 0] \times (-1, 0]$, we have $h_{,1} > 0$, $h_{,2} < 0$, and h is strictly concave with respect to the first argument. Moreover, we have $\lim_{\epsilon \to -1} h(\epsilon, 0) = -\infty$, $\lim_{\epsilon \to -1} h(\epsilon, \epsilon) = -\infty$.
(iii) The function $\epsilon \to h(\epsilon, \epsilon)$ is monotone increasing and strictly concave on $(-1, 0]$.

These assumptions guarantee that equation (187) has no solutions in $(-1, 0)$ if

$$c^2 \leq G(0) := M h_{,1}(0, 0). \tag{188}$$

If, on the other hand, $c^2 > G(0)$, then (187) has exactly one solution in $(-1, 0)$, which we denote by $\epsilon_0(c)$. From the assumed concavity of h it follows that

$$c^2 < M h_{,1}\big(\epsilon_0(c), 0\big). \tag{189}$$

If $c^2 > G(0)$, then the assumptions also imply that the equation

$$c^2 \epsilon = M h(\epsilon, \epsilon) \tag{190}$$

has a unique solution in $(-1, 0)$, which we denote by $\epsilon_\infty(c)$. One easily sees that $\epsilon_\infty(c) < \epsilon_0(c)$.

Theorem II.7:

Assume that (i)-(iii) hold and that $c^2 > G(0)$. Then there is a unique solution ϵ to equation (186) which vanishes identically on $(-\infty, 0)$, is continuously differentiable and monotone decreasing on $(0, \infty)$ and satisfies $\lim_{z \to 0+} \epsilon(z) = \epsilon_0(c)$, $\lim_{z \to \infty} \epsilon(z) = \epsilon_\infty(z)$.

Proof:

Using (189) and a straightforward contraction argument one easily obtains the existence of a local solution on some interval $(-\infty, Z]$ which has the property that ϵ vanishes identically on $(-\infty, 0)$ and $\epsilon(0+) = \epsilon_0(c)$. We shall now show that this local solution can be continued globally, is monotone decreasing for $t > 0$ with a strictly negative derivative, and always remains larger than $\epsilon_\infty(c)$.

We first note that for $z > 0$ we can rewrite equation (186) in the form

$$c^2 \epsilon(z) = \int_0^{z/c} m(s)h\big(\epsilon(z), \epsilon(z - cs)\big) \, ds + \int_{z/c}^\infty m(s)h\big(\epsilon(z), 0\big) \, ds. \qquad (191)$$

Differentiation with respect to z yields

$$c^2 \epsilon_z(z) = \frac{1}{c} m(z/c) h\big(\epsilon(z), \epsilon(0+)\big) - \frac{1}{c} m(z/c) h\big(\epsilon(z), 0\big)$$

$$+ \left(\int_0^{z/c} m(s)h_{,1}\big(\epsilon(z), \epsilon(z - cs)\big) \, ds + \int_{z/c}^\infty m(s)h_{,1}\big(\epsilon(z), 0\big) \, ds \right) \epsilon_z(z)$$

$$+ \int_0^{z/c} m(s)h_{,2}\big(\epsilon(z), \epsilon(z - cs)\big) \epsilon_z(z - cs) \, ds. \qquad (192)$$

Upon setting $z = 0+$, this reduces to

$$c^2 \epsilon_z(0+) = \frac{1}{c} m(0) h\big(\epsilon_0(c), \epsilon_0(c)\big) - \frac{1}{c} m(0) h\big(\epsilon_0(c), 0\big) + M h_{,1}\big(\epsilon_0(c), 0\big) \epsilon_z(0+), \qquad (193)$$

and in conjunction with (189) and assumption (ii) it follows that $\epsilon_z(0+)$ is negative. Hence $\epsilon(z)$ is monotone decreasing at least initially.

Suppose there is a finite maximal number Z such that up to $z = Z$ we have a solution which is monotone decreasing and stays larger than $\epsilon_\infty(c)$. Upon using (191) at $z = Z$ and assumption (ii), we find that

$$c^2 \epsilon(Z) = \int_0^{Z/c} m(s)h\big(\epsilon(Z), \epsilon(Z - cs)\big) \, ds + \int_{Z/c}^\infty m(s)h\big(\epsilon(Z), 0\big) \, ds$$

$$> \int_0^{Z/c} m(s) \Big(h\big(\epsilon(Z), \epsilon(Z - cs)\big) - h\big(0, \epsilon(Z - cs)\big) \Big) \, ds + \int_{Z/c}^\infty m(s)h\big(\epsilon(Z), 0\big) \, ds$$

$$> \epsilon(Z) \left\{ \int_0^{Z/c} m(s)h_{,1}\big(\epsilon(Z), \epsilon(Z - cs)\big) \, ds + \int_{Z/c}^\infty m(s)h_{,1}\big(\epsilon(Z), 0\big) \, ds \right\}. \qquad (194)$$

Hence we have

$$c^2 < \int_0^{Z/c} m(s)h_{,1}\big(\epsilon(Z), \epsilon(Z - cs)\big) \, ds + \int_{Z/c}^\infty m(s)h_{,1}\big(\epsilon(Z), 0\big) \, ds. \qquad (195)$$

This implies that the solution can be continued beyond Z, and using (192) we find that $\epsilon_z(Z)$ is negative, i.e. the solution remains monotone decreasing beyond Z. Finally, we note that by assumption (ii)

$$c^2 \epsilon_\infty = \int_0^\infty m(s) h(\epsilon_\infty, \epsilon_\infty)\, ds$$

$$> \int_0^{Z/c} m(s) h(\epsilon_\infty, \epsilon(Z - cs))\, ds + \int_{Z/c}^\infty m(s) h(\epsilon_\infty, 0)\, ds, \qquad (196)$$

and hence $\epsilon(Z)$ cannot be ϵ_∞.

We conclude that the solution can be continued globally, and because it is monotone decreasing and bounded below, it must have a limit at ∞. It is evident that this limit must satisfy equation (190), and hence it must be ϵ_∞. This concludes the proof.

∎

Finally, we look at the interesting limiting case $c^2 = G(0)$. We make the same assumptions as before except we assume that h is of class C^3 rather than just C^1 and that $m \in C^1[0, \infty)$. We now look for acceleration waves, i.e. solutions ϵ that vanish on $(-\infty, 0)$, satisfy $\epsilon(0+) = 0$, but have a nonvanishing derivative as $z \to 0+$. That is, for small positive z, ϵ is proportional to z, and equation (191) assumes the form

$$c^2 \epsilon(z) = \int_0^{z/c} m(s) \Big[h_{,1}(0,0)\epsilon(z) + h_{,2}(0,0)\epsilon(z - cs) \Big]\, ds$$

$$+ \int_{z/c}^\infty m(s)\, ds \times \Big[h_{,1}(0,0)\epsilon(z) + \frac{1}{2} h_{,11}(0,0)\epsilon(z)^2 \Big] + O(z^3). \qquad (197)$$

Noting that $c^2 = G(0)$, it follows that if solutions of the desired type exist, we must have

$$\frac{m(0)h_{,2}(0,0)\epsilon_z(0+)}{2c} + \frac{1}{2} M h_{,11}(0,0)[\epsilon_z(0+)]^2 = 0. \qquad (198)$$

We have already assumed that $h_{,2} < 0$, and if we add the assumption that $h_{,11}(0,0)$ is strictly negative, then equation (198) has a negative solution, let us denote it by ϵ_0'. A contraction argument can again be used to establish the existence of a local solution to equation (198) for which $\epsilon_z(0+) = \epsilon_0'$. Global continuation can then be shown as above. In this way we obtain the following theorem.

Theorem II.8:

Assume that (i)-(iii) hold and that in addition h is of class C^3, $m \in C^1[0, \infty)$, and $h_{,11}(0,0)$ is strictly negative. Moreover, assume that $c^2 = G(0)$. Then there exists a unique solution to equation (186) which vanishes identically for $z \in (-\infty, 0)$, is continuously differentiable and monotone decreasing on $(0, \infty)$, and satisfies $\lim_{z \to 0+} \epsilon(z) = 0$, $\lim_{z \to 0+} \epsilon_z(z) = \epsilon_0'$, and $\lim_{z \to \infty} \epsilon(z) = \epsilon_\infty(c)$.

Remarks II.9:

1. Note that the value of $\epsilon_z(0+)$ found above agrees precisely with the critical amplitude for acceleration waves as discussed in Section 4 (note that we are now dealing with the jump in u_{xx} rather than in u_{tt} and that the wave is moving in the opposite direction). The analysis of Section 4 suggests that the steady acceleration wave of Theorem II.8 is not stable. On the other hand, steady shocks are observed in experiments (see e.g. Nunziato et al. [N8], Walsh [W1]) and appear to be quite stable.

2. If the kernel m has an integrable singularity at 0, the proof of Theorem II.7 remains valid. The only difference is that $\epsilon_z(0+) = -\infty$; the expression for $\epsilon(0+)$ remains unchanged. Recall from Section 3 that for a linear problem with a singular kernel it is not possible to have solutions with shocks. We note that some of the shock profiles shown in [N8],[W1] appear to have vertical tangents behind the shock. This suggests the possibility of modeling by a singular kernel.

III Local existence results based on energy methods

1. Energy methods

In this chapter we use the so-called energy method to establish local (in time) existence of smooth solutions to several model problems in nonlinear viscoelasticity. Our main purpose is to illustrate the method rather than to obtain results of the greatest possible generality. The basic idea is to construct the solution via an iteration scheme which involves solving a sequence of linearized problems. The solvability of the linearized problems as well as convergence of the iterates is proved through the use of a priori bounds that follow from "energy estimates". A lowest-order energy estimate is typically obtained by considering the physical energy of the system. However, bounds obtained from the physical energy do not control norms of sufficiently high order to be useful in an existence proof. It is therefore necessary to differentiate the equation of motion and to consider analogues of the energy for the differentiated equations.

The energy approach seems best suited to problems of hyperbolic type, although it works for many parabolic problems as well. However, it does not generally yield optimal results in parabolic situations. The method is usually cumbersome to employ, but it has the advantage of being rather versatile. In particular, it is applicable to hyperbolic problems involving boundary conditions; many other methods are not well suited to such problems. In Chapter V we shall discuss the application of abstract operator theoretic methods known as semigroup theory. The problems discussed there include parabolic boundary value problems and hyperbolic initial value problems on all of space. Kato [K8],[K9] has shown that the semigroup approach can be adapted to hyperbolic boundary value problems, but we shall not pursue this issue here.

The chapter is organized as follows. In Section 2, we review existence results concerning the Dirichlet initial-boundary value problem for compressible elastic materials. Then, in Section 3, we show how the procedure outlined in Section 2 can be adapted to establish an existence theorem for a class of compressible viscoelastic materials of integral type with smooth memory functions; for such materials the memory term can be treated as a perturbation. The energy method can also be applied to problems involving incompressible materials. We illustrate this in Section 4 by giving an existence proof for incompressible elastic materials;

an approach similar to that of Section 3 can be used to extend these results to viscoelastic materials. For viscoelastic materials with singular memory functions, the situation is more complicated and the iteration scheme of Section 3 cannot be used. Instead our existence proof is based on an iteration scheme which involves the memory term in a more essential way. In contrast with Section 3, our results for singular memory functions are so far limited to a more special class of model equations.

The local existence theorems discussed here concern the full dynamic equations with inertia. We remark that there is also an extensive literature on quasistatic problems, i.e. problems where the inertial terms are neglected. Such problems essentially involve solving elliptic equations. See e.g. Babuška and Hlaváček [B1] for an early reference.

2. Review of the elastic case

For a homogeneous elastic body with reference configuration $\Omega \subset \mathbb{R}^3$ and unit reference density, the equation of motion reads

$$\ddot{u}^i = A^{\alpha\beta}_{ij}(\nabla u)\frac{\partial^2 u^j}{\partial x^\alpha \partial x^\beta} + f^i, \quad x \in \Omega, \ t \geq 0, \tag{1}$$

where $u(x,t)$ is the displacement at time t of the particle with reference position x, $\mathbf{A}$ is the elasticity tensor, and $f = f(x,t)$ is an assigned body force. (This is the same equation as (19) in Chapter I. However, we have formulated it in terms of the displacement $u = y - x$ rather than the deformation y.) We seek a solution u subject to given initial conditions

$$u(x,0) = u_0(x), \ \dot{u}(x,0) = u_1(x), \ x \in \Omega, \tag{2}$$

and Dirichlet boundary conditions

$$u|_{\partial\Omega} = 0 \tag{3}$$

if Ω is not all of $\mathbb{R}^3$. Regarding $\mathbf{A}$ we assume that

$$A^{\alpha\beta}_{ij} = A^{\beta\alpha}_{ji}, \tag{4}$$

and that there is an open set $\mathcal{O}$ (in the domain of $\mathbf{A}$) with $0 \in \mathcal{O}$ on which the strong ellipticity condition holds, i.e. for some constant $C > 0$ we have

$$A^{\alpha\beta}_{ij}\varsigma^i\varsigma^j\eta_\alpha\eta_\beta \geq C|\varsigma|^2|\eta|^2 \tag{5}$$

satisfied on $\mathcal{O}$.

For the pure initial value problem (1),(2) with $\Omega = \mathbb{R}^3$, the energy method can be described roughly as follows. For appropriate functions w, one considers the family of linear hyperbolic systems

$$\ddot{u}^i = A_{ij}^{\alpha\beta}(\nabla w)\frac{\partial^2 u^j}{\partial x^\alpha \partial x^\beta} + f^i, \; x \in \Omega, \; t \in [0,T], \tag{6}$$

and shows that the mapping $\mathcal{T}$ which carries w into the solution u of the initial value problem (6),(2) has a fixed point if T is sufficiently small. To accomplish this, one needs an existence theorem for linear hyperbolic systems with variable coefficients and estimates for the solution u in terms of w, u_0, u_1, and f. The calculations which are used to derive such a priori estimates will also yield an existence theorem for the linear problem by means of a Galerkin argument. It follows from (4) that the system (6) has an energy integral. To obtain this energy integral, we multiply by $\dot{u}^i$ and integrate over $\mathbb{R}^3 \times [0,T]$. By virtue of (5) the energy has a certain positivity property. Differentiation of (6) with respect to time and/or space variables shows that each partial derivative of u satisfies a system with the same principal part as (6) and remainder terms involving lower order derivatives of u. Thus we can obtain "energy integrals" of higher order.

If the initial data satisfy

$$u_0 \in H^m(\Omega), \; u_1 \in H^{m-1}(\Omega), \; \nabla u_0(x) \in \mathcal{O} \;\; \forall x \in \Omega \tag{7}$$

for some positive integer m, and $\mathbf{A}$ and f are sufficiently smooth, then the energy estimates for derivatives through order m can be combined to produce an inequality of the form

$$\mathcal{E}_m\left(u(\cdot,t)\right) \leq E_m + \int_0^t \int_\Omega Q_m[w,u](x,s) \; dx \; ds, \; 0 \leq t \leq T, \tag{8}$$

where

$$\mathcal{E}_m\left(u(\cdot,t)\right) := \sum_{k=0}^m \|\frac{\partial^k}{\partial t^k}u(\cdot,t)\|_{m-k}^2, \tag{9}$$

E_m is a constant (depending only on $\|u_0\|_m$, $\|u_1\|_{m-1}$ and an appropriate norm of f), and $Q_m[w,u]$ is a polynomial in derivatives of w and u (with coefficients depending on ∇w). Since Sobolev spaces of sufficiently high order are Banach algebras, one can estimate $\int_\Omega Q_m[w,u](x,t) \; dx \; dt$ in terms of $\mathcal{E}_m\left(w(\cdot,t)\right)$ and $\mathcal{E}_m\left(u(\cdot,t)\right)$ provided m is large enough ($m \geq 4$), and then use (8) to show that the solution operator $\mathcal{T}$ maps an appropriate function space Z into itself if T is sufficiently small. One then equips Z with a complete metric and uses similar energy estimates to show that $\mathcal{T}$ is a contraction. More precisely, the space Z consists of functions for which $\mathcal{E}_m$ obeys a uniform bound and which satisfy some additional constraints resulting from the initial conditions and the requirement that $\nabla u \in \mathcal{O}$. It follows from a weak compactness argument that such a space is complete in a metric involving only first order derivatives. The calculations for the contraction argument are therefore simpler than those needed to show that $\mathcal{T}$ maps Z into itself. The contraction mapping theorem implies that $\mathcal{T}$ has a unique fixed point which is obviously the desired solution of (1),(2).

Certain modifications are needed to accommodate boundary conditions. Suppose now that Ω is a bounded domain with smooth boundary and let us seek a solution of the initial-boundary value problem (1)-(3). We now denote by $\mathcal{T}$ the mapping which carries w into the solution of the linear initial-boundary value problem (6),(2),(3). Although differentiation with respect to time and/or the spatial variables preserves the basic structure of the system (6), differentiation with respect to the spatial variables "destroys" the boundary conditions, i.e. we do not know anything about spatial derivatives of u at the boundary. Since integrations by parts are required to derive the energy integrals, and spatial derivatives of u do not necessarily vanish on $\partial\Omega$, the analogue of (8) contains boundary terms. Unfortunately, these boundary terms cannot be estimated in terms of $\mathcal{E}_m\left(u(\cdot,t)\right)$ and $\mathcal{E}_m\left(w(\cdot,t)\right)$. To overcome this difficulty, we differentiate (6) only with respect to time and note that by virtue of (3) time derivatives of u do indeed vanish on $\partial\Omega$, thus allowing us to obtain a "partial set" of energy integrals. If the initial data satisfy (7) and certain natural compatibility conditions, then these energy integrals yield an estimate of the form

$$\|\frac{\partial^m}{\partial t^m}u(\cdot,t)\|_0^2 + \|\frac{\partial^{m-1}}{\partial t^{m-1}}u(\cdot,t)\|_1^2 \leq E_m + \int_0^t \int_\Omega Q_m\,[w,u](x,s)\ dx\ ds \ \forall t \in [0,T],$$
(10)

where E_m and Q_m are as before. Bounds on higher order spatial derivatives are required to estimate Q_m. If, for fixed t, we regard (6),(3) as an elliptic boundary value problem, then standard regularity theory implies that $\|u(\cdot,t)\|_m$ can be estimated in terms of $\|\ddot{u}(\cdot,t)\|_{m-2}$, $\|f(\cdot,t)\|_{m-2}$, $\|u(\cdot,t)\|_{m-2}$ and $\|w(\cdot,t)\|_{m-1}$. By taking time derivatives of (6) and applying elliptic regularity theory to estimate spatial derivatives, we can obtain an inequality of the form (8). The contraction mapping theorem can then be used to show that $\mathcal{T}$ has a fixed point if T is sufficiently small.

Existence results for the Dirichlet initial-boundary value problem in nonlinear elasticity were proved by Kato [K8], Chen and von Wahl [C1] and Dafermos and Hrusa [D5]; for the initial-value problem on all of space we refer to Hughes, Kato and Marsden [H17]. We now quote the main result from [D5]. The following assumptions are needed.

(S1) Ω is a bounded domain in $\mathbb{R}^3$ with boundary of class $C^{m-1,1}$.

(S2) The elasticity tensor $\mathbf{A}$ is of class C^{m-1} on an open set $\mathcal{O}$.

(S3) For some $T > 0$, the forcing term f satisfies

$$f \in \bigcap_{k=0}^{m-2} C^{m-2-k}\left([0,T];H^k\left(\Omega\right)\right),$$
(11)_1

$$\frac{\partial^{m-1}f}{\partial t^{m-1}} \in L^1\left([0,T];L^2\left(\Omega\right)\right).$$
(11)_2

(S4) The initial data satisfy (7).

(E) The symmetry condition (4) and strong ellipticity condition (5) hold on $\mathcal{O}$.

For problems involving boundary conditions, it is imperative that the initial data be compatible with the boundary conditions. For $k = 2, ..., m$, let us denote by u_k the initial value of $\frac{\partial^k u}{\partial t^k}$ as computed recursively in terms of u_0, u_1 and f from equation (1). We shall say that the compatibility relations of order m hold if

$$u_k = 0 \text{ on } \partial\Omega, \ k = 0, ..., m - 1. \tag{12}$$

Failure of the compatibility relations should be interpreted as a singularity in the initial data at the boundary. Therefore, we also assume

(C) The initial data satisfy the compatibility relations of order m.

Theorem III.1:

Let m be an integer ≥ 4 and assume that (S1)-(S4), (E) and (C) hold. Then there exists $T' \in (0, T]$ such that the initial-boundary value problem (1)-(3) has a unique solution

$$u \in \bigcap_{k=0}^{m} C^{m-k}\left([0, T']; H^k(\Omega)\right). \tag{13}$$

Remarks III.2:

1. It follows from (13) and the Sobolev imbedding theorem that $u \in C^{m-2}(\bar{\Omega} \times [0, T'])$ and hence the differential equation is satisfied in the classical sense.
2. As long as $\|u\|_4 + \|\dot{u}\|_3$ remains bounded, ∇u remains in $\mathcal{O}$, and the forcing term retains the regularity (11), the solution can be continued in time and the regularity (13) is preserved. In fact, bounds on weaker norms suffice [C1].
3. The results of [D5] do not show continuous dependence on the data in the optimal norm, although a weaker type of continuous dependence is observed. Continuous dependence in the optimal norm was shown recently by Kato [K9].

The key ingredient in the proof of this theorem is Lemma III.3 below, which concerns the linear initial-boundary value problem

$$\ddot{u}^i = B_{ij}^{\alpha\beta}(x, t) \frac{\partial^2 u^j}{\partial x^\alpha \partial x^\beta} + g^i(x, t), \ x \in \Omega, \ t \in [0, T], \tag{14}_1$$

$$u|_{\partial\Omega} = 0, \tag{14}_2$$

$$u(x, 0) = u_0(x), \ \dot{u}(x, 0) = u_1(x), \tag{14}_3$$

where **B** satisfies

$$B_{ij}^{\alpha\beta} = B_{ji}^{\beta\alpha}, \tag{15}$$

and the (uniform) strong ellipticity condition

$$B_{ij}^{\alpha\beta}(x, t)\varsigma^i \varsigma^j \eta_\alpha \eta_\beta \geq C|\varsigma|^2 |\eta|^2 \ \forall x \in \Omega, \ \varsigma, \eta \in \mathbb{R}^3, \ t \in [0, T] \tag{16}$$

with $C > 0$. For our purposes, the main issue concerning (14) is the type of smoothness required of $\mathbf{B}$ (and the dependence of various norms of u on corresponding norms of $\mathbf{B}$). If $\mathbf{B}$ is of class C^{m-1}, then it is relatively straightforward to prove that the solution of (14) satisfies (13), provided that the data are regular enough. However, this result cannot be used in the iteration scheme.

For the purpose of stating an a priori estimate, we introduce two parameters K and L to measure the size of $\mathbf{B}$. We also introduce a parameter U which controls the data. We make the following assumptions:

(S2$'$) The coefficients $B_{ij}^{\alpha\beta}$ satisfy

$$\mathbf{B} \in \bigcap_{k=0}^{m-1} W^{m-1-k,\infty}\left([0,T];H^k\left(\Omega\right)\right) \tag{17}$$

and the norms of $\mathbf{B}$ in these spaces are bounded by L. The norm of $\mathbf{B}$ in $L^\infty\left([0,T];H^{m-2}\left(\Omega\right)\right)$ is bounded by K.

(E) The symmetry and strong ellipticity conditions (15) and (16) hold.

(S3$'$) The forcing function g satisfies

$$g \in \bigcap_{k=0}^{m-2} C^{m-2-k}\left([0,T];H^k\left(\Omega\right)\right), \quad \frac{\partial^{m-1}g}{\partial t^{m-1}} \in L^1\left([0,T];L^2\left(\Omega\right)\right), \tag{18}$$

and the norms in the indicated spaces are bounded by U.

(S4$'$) The initial data satisfy

$$u_0 \in H^m\left(\Omega\right), \quad u_1 \in H^{m-1}\left(\Omega\right). \tag{19}$$

As before, we denote by u_k the initial value of $\frac{\partial u^k}{\partial t^k}$ as computed from the equation for $k = 2, ..., m$. It is now easy to give an explicit recursion formula for the u_k, namely

$$u_k^i := \sum_{l=0}^{k-2} \binom{k-2}{l} \frac{\partial^l}{\partial t^l} B_{ij}^{\alpha\beta}\left(\cdot,0\right) \frac{\partial^2 u_{k-2-l}^j}{\partial x^\alpha \partial x^\beta} + \frac{\partial^{k-2}}{\partial t^{k-2}} g^i\left(\cdot,0\right). \tag{20}$$

Once again, we say that the compatibility relations of order m hold if (12) is satisfied. We now assume

(C$'$) The compatibility relations of order m hold and

$$\sum_{k=0}^{m} \|u_k\|_{m-k}^2 \le U^2. \tag{21}$$

Lemma III.3:

Let m be an integer ≥ 4 and assume that (S1), (S2$'$)-(S4$'$), (E) and (C$'$) hold. Then the initial-boundary value problem (14) has a unique solution

$$u \in \bigcap_{k=0}^{m} C^{m-k}\left([0,T];H^k\left(\Omega\right)\right). \tag{22}$$

Moreover, this solution obeys the a priori estimate

$$\mathcal{E}_m\left(u(\cdot,t)\right) \le \Gamma(U,K,L_0)\exp(T \cdot \Lambda(U,K,L,T)) \; \forall t \in [0,T], \qquad (23)_1$$

where Γ and Λ are functions that are bounded on bounded sets, and

$$L_0 := \sum_{k=0}^{m-2} \|\frac{\partial}{\partial t^k}\mathbf{B}(\cdot,0)\|_{m-1-k}. \qquad (23)_2$$

The proof is based on a Galerkin argument together with energy estimates for time derivatives and elliptic estimates for spatial derivatives. Rather than give a proof of this lemma, we refer to [D5], Theorem 3.1. In Section 4, we will actually give a proof of a similar lemma in the incompressible context. In [D5], a slightly stronger assumption on g is made, but it is easy to adapt the proof.

Remark III.4:

An analogous existence result holds for the pure initial value problem with $\Omega = \mathbb{R}^3$. However, we must modify (S2') because (17) implies $\mathbf{B}(x,t) \to 0$ as $|x| \to \infty$, which conflicts with the strong ellipticity conditions (16). The appropriate modification is:

S2'') There is a constant tensor $\bar{\mathbf{B}}$ such that

$$\mathbf{B} - \bar{\mathbf{B}} \in \bigcap_{k=0}^{m-1} W^{m-1-k,\infty}\left([0,T];H^k(\mathbb{R}^3)\right) \qquad (24)$$

and the norms of $\mathbf{B} - \bar{\mathbf{B}}$ in these spaces are bounded by L. The norm of $\mathbf{B} - \bar{\mathbf{B}}$ in $L^\infty\left([0,T];H^{m-2}(\Omega)\right)$ is bounded by K. The Euclidean norm of $\bar{\mathbf{B}}$ is bounded by $\min(L,K)$.

Also, the compatibility relations are redundant when $\Omega = \mathbb{R}^3$. Of course, the bound (21) is still assumed to hold.

3. Viscoelastic materials of integral type

In this section we discuss local existence for compressible viscoelastic materials of integral type with smooth memory functions. The constitutive law is assumed to be of the form (51) in Chapter I. This leads to an equation of motion of the form

$$\ddot{u}^i = \left\{ A_{ij}^{\alpha\beta}(\nabla u) + \int_{-\infty}^{t} L_{ij}^{\alpha\beta}(t-\tau, \nabla u(\cdot, t), \nabla u(\cdot, \tau))\, d\tau \right\} \frac{\partial^2 u^j}{\partial x^\alpha \partial x^\beta}$$

$$+ \int_{-\infty}^{t} M_{ij}^{\alpha\beta}(t-\tau, \nabla u(\cdot, t), \nabla u(\cdot, \tau)) \frac{\partial^2 u^j}{\partial x^\alpha \partial x^\beta}(\cdot, \tau)\, d\tau + f^i. \tag{25}$$

Here, as in Section 2, u represents displacement, $f = f(x,t)$ is a given body force and the reference density is assumed to be 1.

For purposes of illustration we begin with the integrodifferential equation

$$\ddot{u}^i = A_{ij}^{\alpha\beta}(\nabla u) \frac{\partial^2 u^j}{\partial x^\alpha \partial x^\beta}$$

$$+ \int_{-\infty}^{t} M_{ij}^{\alpha\beta}(t-\tau, \nabla u(\cdot, \tau)) \frac{\partial^2 u^j}{\partial x^\alpha \partial x^\beta}(\cdot, \tau)\, d\tau + f^i, \quad x \in \Omega,\ t \geq 0. \tag{26}$$

Equation (26) is not a realistic model for viscoelastic materials, if one allows large deformations or rotations, because the constitutive law leading to (26) is not frame-indifferent. We use this equation to illustrate the treatment of memory terms in hyperbolic problems. The same procedure applies to (25); this will be discussed in Theorem III.10 below.

In order to pose an initial-value problem for (26), the history of u prior to time $t = 0$, as well as the values of u and $\dot{u}$ at time $t = 0+$ must be prescribed. Without loss of generality, we may assume that $u(x,t) \equiv 0$ for $t < 0$; otherwise the term $\int_{-\infty}^{0}$ in (26) can be absorbed into the forcing term f provided that the given history of u is sufficiently smooth and well-behaved at $-\infty$. If the history vanishes, (26) reduces to

$$\ddot{u}^i = A_{ij}^{\alpha\beta}(\nabla u) \frac{\partial^2 u^j}{\partial x^\alpha \partial x^\beta}$$

$$+ \int_{0}^{t} M_{ij}^{\alpha\beta}(t-\tau, \nabla u(\cdot, \tau)) \frac{\partial^2 u^j}{\partial x^\alpha \partial x^\beta}(\cdot, \tau)\, d\tau + f^i, \quad x \in \Omega,\ t \geq 0. \tag{27$_1$}$$

We shall seek a solution that satisfies

$$u|_{\partial\Omega} = 0, \tag{27$_2$}$$

$$u(x,0) = u_0(x), \quad \dot{u}(x,0) = u_1(x). \tag{27$_3$}$$

Our assumptions on Ω, $\mathbf{A}$, u_0, u_1, and f are the same as in the elastic case.
(S1) Ω is a bounded domain in $\mathbb{R}^3$ with boundary of class $C^{m-1,1}$.
(S2) $\mathbf{A}$ is of class C^{m-1} on an open set $\mathcal{O}$.
(S3) The initial data satisfy (7).
(S4) For some $T' > 0$, the forcing term satisfies (11).

E) The symmetry condition (4) and the strong ellipticity condition (5) hold on $\mathcal{O}$.

Our only requirement of $\mathbf{M}$ will be that it is smooth on $[0, \infty) \times \mathcal{O}$, i.e.

(S5) $\mathbf{M}$ is of class C^{m-1} on $[0, \infty) \times \mathcal{O}$.

We denote by u_k the values of $\frac{\partial^k u}{\partial t^k}$ at time $t = 0+$ as computed from (27) for $k = 2, ..., m$. Once again we require

C) The compatibility conditions of order m, i.e. (12), hold.

Theorem III.5:

Let m be an integer ≥ 4 and assume that (S1)-(S5), (E) and (C) hold. Then there is a $T \in (0, T']$ such that the initial-boundary value problem (27) has a unique solution

$$u \in \bigcap_{k=0}^{m} C^{m-k} ([0, T]; H^k (\Omega)).\qquad(28)$$

Remarks III.6:

1. A similar result holds for the all-space problem with $\Omega = \mathbb{R}^3$.
2. The assumption on the temporal regularity of $\mathbf{M}$ can be considerably relaxed if the time derivatives of the history at $t = 0-$ are assumed to be equal to the u_k. Roughly speaking, $\mathbf{M}$ must have at least one time derivative which is integrable, uniformly in the second argument.
3. The smoothness assumption on $\mathbf{M}$ can be relaxed, but the most general assumption is not easy to state. If $\mathbf{M}$ is in convolution form, i.e.

$$M_{ij}^{\alpha \beta} (t - \tau, \mathbf{V}) = m_{i\gamma}^{k\beta} (t - \tau) E_{kj}^{\gamma \alpha} (\mathbf{V}),$$

then it suffices to assume that $\mathbf{E} \in C^{m-1}(\mathcal{O})$ and $\mathbf{m} \in W^{m-2,1}(0, T')$, i.e. derivatives of $\mathbf{m}$ through order $m - 2$ are integrable.
4. A similar theorem holds in any number of spatial dimensions. The only difference is in the minimum value of m. For a problem in n spatial dimensions, we must assume that $m \geq 3 + \left[\frac{n}{2}\right]$. In particular, we need $m \geq 3$ for $n = 1$, and $m \geq 4$ for $n = 2$ and $n = 3$.[1]
5. An existence theorem for the linearized problem was established by Dafermos [D2].

The proof of Theorem III.5 will be based on Lemma III.3 and the contraction mapping principle. We first observe that without loss of generality we may assume that the open set $\mathcal{O}$ is all of $\mathbb{R}^{3 \times 3}$ because we can always redefine $\mathbf{A}$ and $\mathbf{M}$ such that they agree with the given functions on a neighborhood of ∇u_0.

For $T, M > 0$ we denote by $Z(T, M)$ the set of all functions w which satisfy

$$w \in \bigcap_{k=0}^{m} W^{m-k, \infty} ([0, T]; H^k (\Omega)),\qquad(29)_1$$

---

[1] The case $n = 11$ (for which we need $m \geq 8$) might be of interest to physicists.

$$\frac{\partial^k w}{\partial t^k}(\cdot,0) = u_k, \quad k = 0, ..., m-1, \tag{29}$$

and

$$\sum_{k=0}^{m} \|w\|_{k,m-k}^2 \leq M^2, \tag{29}$$

where the u_k are as in Theorem III.5 and $\|w\|_{k,l}$ denotes the norm in $W^{k,\infty}([0,T]$, $H^l(\Omega))$. It follows from the trace theorem [L6] that $Z(T,M)$ is not empty if M is large.

For $w \in Z(T,M)$ we consider the linear problem

$$\ddot{u}^i = A_{ij}^{\alpha\beta}(\nabla w)\frac{\partial^2 u^j}{\partial x^\alpha \partial x^\beta} + f^i$$

$$+ \int_0^t M_{ij}^{\alpha\beta}(t-\tau, \nabla w(\cdot,\tau))\frac{\partial^2 w^j}{\partial x^\alpha \partial x^\beta}(\cdot,\tau)\, d\tau, \quad x \in \Omega, \ t \in [0,T], \tag{30}$$

$$u|_{\partial\Omega} = 0, \tag{30}$$

$$u(x,0) = u_0(x), \quad \dot{u}(x,0) = u_1(x). \tag{30}$$

For a given function $w \in Z(T,M)$, $(30)_1$ is a linear partial differential equation for u. If we identify $\mathbf{B}$ in (14) with $\mathbf{A}(\nabla w)$ and g in (14) with the last two terms in $(30)_1$, then it follows from Lemma III.3 that (30) has a unique solution which satisfies (22). We note that the compatibility relations for the linear problem are automatically satisfied. Indeed, the u_k in Lemma III.3 coincide with the u_k of Theorem III.5 by virtue of $(29)_2$. Let S denote the map which carries w into the unique solution u of (28).

The estimates required to prove Theorem III.5 make repeated use of the following well-known facts about Sobolev norms, which we record here for easy reference (we assume that (S1) holds).

Lemma III.7:

(i) Let $2 \leq k \leq m$. If ϕ lies in the ball of radius R in $H^k(\Omega)$ and G is a function of class C^k, then $G(\phi)$ belongs to $H^k(\Omega)$ and its norm is less than or equal to C, where C depends only on R and properties of G.

(ii) Suppose that $s_j \geq 0$, $j = 1, ..., k$ and that

$$r := \min_{1 \leq i \leq k} \min_{j_1 \leq ... \leq j_i} \left\{ s_{j_1} + ... + s_{j_i} - 2(i-1) \right\} \geq 0. \tag{31}$$

Then pointwise multiplication is a continuous mapping from $H^{s_1}(\Omega) \times ... \times H^{s_k}(\Omega)$ to $H^r(\Omega)$ (see [K6]).

Lemma III.8:

S maps $Z(T,M)$ into itself if M is sufficiently large and T is sufficiently small relative to M.

Proof:

For $w \in Z(T, M)$, let us set

$$_w B_{ij}^{\alpha \beta}(x, t) := A_{ij}^{\alpha \beta}(\nabla w(x, t)), \qquad (32)_1$$

and

$$_w g^i(x, t) := f^i(x, t) + \int_0^t M_{ij}^{\alpha \beta}(t - \tau, \nabla w(x, \tau)) \frac{\partial^2 w^j}{\partial x^\alpha \partial x^\beta}(x, \tau) \, d\tau, \qquad (32)_2$$

so that $(30)_1$ takes the form

$$\ddot{u}^i = {}_w B_{ij}^{\alpha \beta} \frac{\partial^2 u^j}{\partial x^\alpha \partial x^\beta} + {}_w g^i. \qquad (33)$$

We want to apply the a priori estimate of Lemma III.3 of Section 2 to show that S maps $Z(T, M)$ into itself for appropriate M and T. The crucial point is to obtain bounds for the various norms of $_w \mathbf{B}$ and $_w g$ in terms of M and T. Since the u_k and f remain fixed, we treat the various norms of these quantities as constants. For the same reason, the quantity L_0 in Lemma III.3 can be regarded as fixed. To simplify matters slightly we assume $T \leq 1$. Throughout this proof we use C to denote a generic positive constant which can be chosen independently of M and T, and we use $P(M)$ to denote a generic continuous function of M.

By standard Sobolev estimates for composite functions (see Lemma III.7,) above) we conclude that the norms of $_w \mathbf{B}$ in $W^{m-1-k,\infty}([0, T]; H^k(\Omega))$ ($k = , ..., m - 1$) are bounded by a continuous function of M, i.e. we may take $L = $ $^*(M)$ in Lemma III.3. To obtain a bound for $\|_w \mathbf{B}\|_{0,m-2}$ we write

$$_w \mathbf{B}(\cdot, t) = {}_w \mathbf{B}(\cdot, 0) + \int_0^t \frac{\partial}{\partial s} {}_w \mathbf{B}(\cdot, s) \, ds, \qquad (34)$$

from which it follows that

$$\|_w B_{ij}^{\alpha \beta}(\cdot, t)\|_{m-2} \leq \|A_{ij}^{\alpha \beta}(\nabla u_0)\|_{m-2} + \int_0^t \|C_{ijk}^{\alpha \beta \gamma}(\nabla w) \frac{\partial \dot{w}^k}{\partial x^\gamma}\|_{m-2} \, d\tau, \qquad (35)_1$$

where

$$C_{ijk}^{\alpha \beta \gamma}(\mathbf{V}) := \frac{\partial A_{ij}^{\alpha \beta}(\mathbf{V})}{\partial V_\gamma^k}. \qquad (35)_2$$

We can estimate the integral in $(35)_1$ by applying Lemma III.7 to the integrand. In this way we find

$$\|C_{ijk}^{\alpha \beta \gamma}(\nabla w) \frac{\partial \dot{w}^k}{\partial x^\gamma}\|_{m-2} \leq C \|\mathbf{C}\|_{m-2} \|\nabla \dot{w}\|_{m-2} \leq P(M). \qquad (36)$$

Hence $(35)_1$ yields

$$\|_w \mathbf{B}\|_{0,m-2} \leq C + T \cdot P(M). \qquad (37)$$

The key point to notice about (37) is that the right-hand side remains bounded as M gets large, provided T is reduced accordingly. This means that we can keep K fixed when using the a priori estimate (23).

Estimates for the appropriate norms of $_w g$ are obtained by a straightforward but tedious procedure. We first note that

$$\|_w g^i(\cdot,t)\|_{m-2} \le \|f^i(\cdot,t)\|_{m-2} + \int_0^t \|M_{ij}^{\alpha\beta}(t-\tau, \nabla w(\cdot,\tau))\frac{\partial^2 w^j}{\partial x^\alpha \partial x^\beta}(\cdot,\tau)\|_{m-2}\, d\tau$$

$$\le C + T \cdot P(M) \quad \forall t \in [0,T]. \tag{38}$$

Differentiating $(32)_2$ with respect to t we find

$$_w \dot{g}^i = \dot{f}^i + M_{ij}^{\alpha\beta}(0, \nabla w)\frac{\partial^2 w^j}{\partial x^\alpha \partial x^\beta}$$

$$+ \int_0^t \dot{M}_{ij}^{\alpha\beta}(t-\tau, \nabla w(\cdot,\tau))\frac{\partial^2 w^j}{\partial x^\alpha \partial x^\beta}(\cdot,\tau)\, d\tau, \tag{39}$$

where $\dot{M}$ denotes the derivative of M with respect to its first argument. The H^{m-3}-norm of the integral in (39) is clearly bounded by $T \cdot P(M)$. To estimate the second term on the right-hand side of (39), we write it as its initial value plus the integral of its time derivative and apply Lemma III.7 to the integrand. This yields

$$\|M_{ij}^{\alpha\beta}(0, \nabla w)\frac{\partial^2 w^j}{\partial x^\alpha \partial x^\beta}(\cdot,t)\|_{m-3} \le C + T \cdot P(M). \tag{40}$$

Thus we conclude that

$$\|_w \dot{g}^i(\cdot,t)\|_{m-3} \le C + T \cdot P(M) \quad \forall t \in [0,T]. \tag{41}$$

Application of a similar procedure to the higher order time derivatives of $_w g$ shows that we may take $U = C + T \cdot P(M)$ in Lemma III.3.

Putting all the above bounds together and appealing to the a priori estimate of Lemma III.3, we find that

$$\mathcal{E}_m(u(\cdot,t)) \le \Gamma^*(M,T)\exp(T \cdot \Lambda^*(M,T)) \quad \forall t \in [0,T], \tag{42}$$

where Λ^* is a function of M and T that is bounded on bounded sets, and there is a constant c_0 with the following property: for each $M > 0$ there exists $T_0 = T_0(M) > 0$ such that

$$\Gamma^*(M,T) \le c_0 \quad \forall T \in (0, T_0]. \tag{43}$$

It now follows easily that S maps $Z(T,M)$ into itself if M is large and T is sufficiently small relative to M.

We now equip $Z(T,M)$ with the metric d defined by

$$d(w, \overline{w}) := \left(\|w - \overline{w}\|_{0,2}^2 + \|w - \overline{w}\|_{1,1}^2 + \|w - \overline{w}\|_{2,0}^2\right)^{1/2}. \tag{44}$$

It is obvious that d is a metric. Because the functions in $Z(T,M)$ obey the uniform bound $(29)_3$, it follows from a weak-$*$ compactness argument that $Z(T,M)$ is complete under d.

Lemma III.9:

S is strictly contractive with respect to d if M is sufficiently large and T is sufficiently small relative to M.

Proof:

We use the same notation as in the proof of Lemma III.8. Moreover, we now fix M large enough so that for some $T_0 \in (0,1]$, S maps $Z(T,M)$ into itself for all $T \in (0,T_0]$.

Let $w, \overline{w} \in Z(T,M)$ be given and set $u := S(w), \overline{u} := S(\overline{w}), W := w - \overline{w}$, and $U := u - \overline{u}$. A simple calculation shows that

$$
\ddot{U}^i - A_{ij}^{\alpha\beta}(\nabla w)\frac{\partial^2 U^j}{\partial x^\alpha \partial x^\beta} = \left[A_{ij}^{\alpha\beta}(\nabla w) - A_{ij}^{\alpha\beta}(\nabla \overline{w})\right]\frac{\partial^2 \overline{u}^j}{\partial x^\alpha \partial x^\beta}
$$

$$
+ \int_0^t M_{ij}^{\alpha\beta}(t-\tau, \nabla w(\cdot,\tau))\frac{\partial^2 W^j}{\partial x^\alpha \partial x^\beta}(\cdot,\tau)\, d\tau
$$

$$
+ \int_0^t \left[M_{ij}^{\alpha\beta}(t-\tau,\nabla w(\cdot,\tau)) - M_{ij}^{\alpha\beta}(t-\tau,\nabla \overline{w}(\cdot,\tau))\right]\frac{\partial^2 \overline{w}^j}{\partial x^\alpha \partial x^\beta}(\cdot,\tau)\, d\tau, \qquad (45)_1
$$

$$
U\big|_{\partial\Omega} = 0, \qquad (45)_2
$$

$$
U(x,0) = \dot{U}(x,0) = 0. \qquad (45)_3
$$

We differentiate $(45)_1$ with respect to time, add and subtract appropriate terms, multiply the resulting equation by $\ddot{U}^i$ and integrate over space and time. After a number of integrations by parts we obtain

$$
\frac{1}{2}\int_\Omega |\ddot{U}(x,t)|^2\, dx + \frac{1}{2}\int_\Omega A_{ij}^{\alpha\beta}(\nabla w)\frac{\partial \dot{U}^i}{\partial x^\alpha}\frac{\partial \dot{U}^j}{\partial x^\beta}(x,t)\, dx
$$

$$
= \frac{1}{2}\int_0^t \int_\Omega C_{ijk}^{\alpha\beta\gamma}(\nabla w)\frac{\partial \dot{w}^k}{\partial x^\gamma}\frac{\partial \dot{U}^i}{\partial x^\alpha}\frac{\partial \dot{U}^j}{\partial x^\beta}(x,s)\, dx\, ds
$$

$$
- \int_0^t \int_\Omega C_{ijk}^{\alpha\beta\gamma}(\nabla w)\frac{\partial^2 w^k}{\partial x^\alpha \partial x^\gamma}\ddot{U}^i\frac{\partial \dot{U}^j}{\partial x^\beta}(x,s)\, dx\, ds
$$

$$
+ \int_0^t \int_\Omega \left[A_{ij}^{\alpha\beta}(\nabla w) - A_{ij}^{\alpha\beta}(\nabla \overline{w})\right]\frac{\partial^2 \ddot{\overline{u}}^j}{\partial x^\alpha \partial x^\beta}\ddot{U}^i(x,s)\, dx\, ds
$$

$$
+ \int_0^t \int_\Omega C_{ijk}^{\alpha\beta\gamma}(\nabla w)\frac{\partial \dot{w}^k}{\partial x^\gamma}\frac{\partial^2 U^j}{\partial x^\alpha \partial x^\beta}\ddot{U}^i(x,s)\, dx\, ds
$$

$$
+ \int_0^t \int_\Omega C_{ijk}^{\alpha\beta\gamma}(\nabla w)\frac{\partial \dot{W}^k}{\partial x^\gamma}\frac{\partial^2 \overline{w}^j}{\partial x^\alpha \partial x^\beta}\ddot{U}^i(x,s)\, dx\, ds
$$

93

$$+ \int_0^t \int_\Omega \left[C_{ijk}^{\alpha\beta\gamma}(\nabla w) - C_{ijk}^{\alpha\beta\gamma}(\nabla \overline{w}) \right] \frac{\partial \dot{w}^k}{\partial x^\gamma} \frac{\partial^2 \overline{u}^j}{\partial x^\alpha \partial x^\beta} \ddot{U}^i(x,s) \; dx \; ds$$

$$+ \int_0^t \int_\Omega M_{ij}^{\alpha\beta}(0, \nabla w) \frac{\partial^2 W^j}{\partial x^\alpha \partial x^\beta} \ddot{U}^i(x,s) \; dx \; ds$$

$$+ \int_0^t \int_\Omega \left[M_{ij}^{\alpha\beta}(0, \nabla w) - M_{ij}^{\alpha\beta}(0, \nabla \overline{w}) \right] \frac{\partial^2 \overline{w}^j}{\partial x^\alpha \partial x^\beta} \ddot{U}^i(x,s) \; dx \; ds$$

$$+ \int_0^t \int_0^s \int_\Omega \dot{M}_{ij}^{\alpha\beta}(s - \tau, \nabla w(x,\tau)) \frac{\partial^2 W^j}{\partial x^\alpha \partial x^\beta}(x,\tau) \ddot{U}^i(x,s) \; dx \; d\tau \; ds$$

$$+ \int_0^t \int_0^s \int_\Omega \left[\dot{M}_{ij}^{\alpha\beta}(s - \tau, \nabla w(x,\tau)) - \dot{M}_{ij}^{\alpha\beta}(s - \tau, \nabla \overline{w}(x,\tau)) \right]$$

$$\times \frac{\partial^2 \overline{w}^j}{\partial x^\alpha \partial x^\beta} \ddot{U}^i(x,s) \; dx \; d\tau \; ds, \quad x \in \Omega, \quad t \in [0,T]. \tag{46}$$

We claim that (46) implies an estimate of the form

$$\|\ddot{U}(\cdot,t)\|_0^2 + \|\dot{U}(\cdot,t)\|_1^2 \leq T \cdot P(M) \cdot (\|W\|_{1,1}^2 + \|W\|_{0,2}^2)$$

$$+ P(M) \int_0^t (\|\ddot{U}(\cdot,s)\|_0^2 + \|\dot{U}(\cdot,s)\|_1^2 + \|U(\cdot,s)\|_2^2) \; ds \quad \forall t \in [0,T]. \tag{47}$$

To see why this is so, we first observe that by virtue of Gårding's inequality

$$\int_\Omega A_{ij}^{\alpha\beta}(\nabla w) \frac{\partial \dot{U}^i}{\partial x^\alpha} \frac{\partial \dot{U}^j}{\partial x^\beta}(x,t) \; dx \geq \delta \|\dot{U}\cdot,t)\|_1^2 - P(M) \cdot \|\dot{U}(\cdot,t)\|_0^2 \quad \forall t \in [0,T], \tag{48}$$

where δ is a positive constant that can be chosen independently of M and T. Next, we observe that

$$\|\dot{U}(\cdot,t)\|_0^2 \leq \int_0^t \|\ddot{U}(\cdot,s)\|_0^2 \; ds \quad \forall t \in [0,T], \tag{49}$$

since $\dot{U}(\cdot,0) = 0$ and $T \leq 1$. Hence the left-hand side of (46) is bounded below by

$$\frac{1}{2}\|\ddot{U}(\cdot,t)\|_0^2 + \frac{\delta}{2}\|\dot{U}(\cdot,t)\|_1^2 - P(M) \int_0^t \|\ddot{U}(\cdot,s)\|_0^2 \; ds. \tag{50}$$

It remains to show that the right-hand side of (46) can be estimated by the right-hand side of (47). Most of the terms can be handled in more or less the same fashion; we give the details of several typical calculations.

To estimate the first term on the right-hand side of (46) we note that

$$\left| C_{ijk}^{\alpha\beta\gamma}(\nabla w) \frac{\partial \dot{w}^k}{\partial x^\gamma}(x,s) \right| \leq P(M), \quad \forall x \in \Omega, \; s \in [0,T] \tag{51}$$

94

by the Sobolev imbedding theorem, and hence

$$\int_0^t \int_\Omega \left| C_{ijk}^{\alpha\beta\gamma}(\nabla w) \frac{\partial \dot{w}^k}{\partial x^\gamma} \frac{\partial \dot{U}^i}{\partial x^\alpha} \frac{\partial \dot{U}^j}{\partial x^\beta}(x,s) \right| dx\, ds \le P(M) \int_0^t \|\dot{U}(\cdot,s)\|_1^2\, ds. \quad (52)$$

We shall now bound the third term on the right-hand side of (46). By the Cauchy-Schwarz inequality (and $|ab| \le \frac{1}{2}(a^2 + b^2)$) we have

$$\int_0^t \int_\Omega \left| \left[A_{ij}^{\alpha\beta}(\nabla w) - A_{ij}^{\alpha\beta}(\nabla \overline{w}) \right] \frac{\partial^2 \dot{\overline{u}}^j}{\partial x^\alpha \partial x^\beta} \ddot{U}^i(x,s) \right| dx\, ds$$

$$\le \frac{1}{2}\int_0^t \left\| \left[A_{ij}^{\alpha\beta}(\nabla w) - A_{ij}^{\alpha\beta}(\nabla \overline{w}) \right] \frac{\partial^2 \overline{u}^j}{\partial x^\alpha \partial x^\beta}(\cdot,s) \right\|_0^2 ds + \frac{1}{2}\int_0^t \|\ddot{U}(\cdot,s)\|_0^2\, ds. \quad (53)$$

To estimate the first term on the right side of (53) we apply Lemma III.7 to the integrand. This yields

$$\left\| \left[A_{ij}^{\alpha\beta}(\nabla w) - A_{ij}^{\alpha\beta}(\nabla \overline{w}) \right] \frac{\partial^2 \dot{\overline{u}}^j}{\partial x^\alpha \partial x^\beta} \right\|_0^2$$

$$\le C \cdot \|A(\nabla w) - A(\nabla \overline{w})\|_1^2 \cdot \|\nabla^2 \dot{\overline{u}}\|_1^2 \le P(M) \cdot \|W\|_2^2, \quad (54)$$

and therefore the left-hand side of (53) is bounded by

$$T \cdot P(M) \cdot \|W\|_{0,2}^2 + \frac{1}{2}\int_0^t \|\ddot{U}(\cdot,s)\|_0^2\, ds.$$

Applying similar calculations to the remaining terms in (46) we conclude that (47) holds.

In order to exploit the inequality (47) we must obtain a bound for $\|U(\cdot,t)\|_2$. For this purpose, we regard $(45)_1, (45)_2$ as an elliptic boundary value problem for fixed t. It follows from standard elliptic estimates that

$$\|U\|_2 \le P(M) \cdot (\|U\|_0 + \|\ddot{U}\|_0 + \|R\|_0) \quad (55)$$

where R denotes the right-hand side of $(45)_1$. It is easy to see that

$$\|R\|_0 \le P(M) \cdot \|W\|_1 + T \cdot P(M) \cdot \|W\|_2, \quad (56)$$

and since $\|W(\cdot,t)\|_1 \le T \cdot \|W\|_{1,1}$, we have

$$\|R(\cdot,t)\|_0 \le T \cdot P(M) \cdot (\|W\|_{1,1} + \|W\|_{0,2}) \quad \forall t \in [0,T]. \quad (57)$$

We next observe that

$$\|U(\cdot,t)\|_0 \le \int_0^t \|\dot{U}(\cdot,s)\|_0\, ds \le \int_0^t \|\ddot{U}(\cdot,s)\|_0\, ds. \quad (58)$$

95

Combining (47), (55), (57), and (58) we conclude that

$$\|U(\cdot,t)\|_2^2 + \|\dot{U}(\cdot,t)\|_1^2 + \|\ddot{U}(\cdot,t)\|_0^2 \leq T \cdot P(M) \cdot (\|W\|_{1,1}^2 + \|W\|_{0,2}^2)$$

$$+P(M) \int_0^t (\|U(\cdot,s)\|_0^2 + \|\dot{U}(\cdot,s)\|_1^2 + \|\ddot{U}(\cdot,s)\|_2^2) \; ds. \qquad (59)$$

Using (59) and Gronwall's inequality we find that

$$\|U(\cdot,t)\|_2^2 + \|\dot{U}(\cdot,t)\|_1^2 + \|\ddot{U}(\cdot,t)\|_0^2$$

$$\leq T \cdot P(M) \cdot \exp[T \cdot P(M)] \cdot (\|W\|_{0,2}^2 + \|W\|_{1,1}^2) \quad \forall t \in [0,T]. \qquad (60)$$

It follows easily from (60) that S is strictly contractive if T is sufficiently small. ∎

The preceding argument can be applied to equations governing the motions of much more general classes of viscoelastic materials. Roughly speaking, it is required that the instantaneous elasticity tensor should be strongly elliptic and possess "hyperelastic" symmetry, and that the dependence on the past history be suitably smooth. It is possible to consider a general functional dependence on the history (for the one-dimensional case see [H8]), however, it is difficult to state the appropriate hypotheses in this case. We now discuss the initial-boundary value problem associated with equation (25):

$$\ddot{u}^i = \left\{ A_{ij}^{\alpha\beta} + \int_{-\infty}^t L_{ij}^{\alpha\beta}\left(t-\tau, \nabla u(\cdot,t), \nabla u(\cdot,\tau)\right) \; d\tau \right\} \frac{\partial^2 u^j}{\partial x^\alpha \partial x^\beta}$$

$$+ \int_{-\infty}^t M_{ij}^{\alpha\beta}\left(t-\tau, \nabla u(\cdot,t), \nabla u(\cdot,\tau)\right) \frac{\partial^2 u^j}{\partial x^\alpha \partial x^\beta}(\cdot,\tau) \; d\tau + f^i, \qquad (61)_1$$

$$u\big|_{\partial\Omega} = 0. \qquad (61)_2$$

$$u(x,t) = \tilde{u}(x,t), \quad x \in \Omega, \; t < 0, \qquad (61)_3$$

$$u(x,0+) = u_0(x), \quad \dot{u}(x,0+) = u_1(x). \qquad (61)_4$$

To prove an existence theorem for (61) we use the same spaces $Z(T,M)$ as before. Of course the u_k are now computed from (61). For $w \in Z(T,M)$ let us set

$$_w\mathbf{B}(x,t) := \mathbf{A}(\nabla w(x,t)) + \int_{-\infty}^0 \mathbf{L}(t-\tau, \nabla w(x,t), \nabla\tilde{u}(x,\tau)) \; d\tau$$

$$+ \int_0^t \mathbf{L}(t-\tau, \nabla w(x,t), \nabla w(x,\tau)) \; d\tau \qquad (62)_1$$

and

$$_w g^i(x,t) := f^i(x,t) + \int_{-\infty}^0 M_{ij}^{\alpha\beta}(t-\tau, \nabla w(x,t), \nabla\tilde{u}(x,\tau)) \frac{\partial^2 \tilde{u}^j}{\partial x^\alpha \partial x^\beta}(x,\tau) \; d\tau$$

$$+ \int_0^t M_{ij}^{\alpha\beta} \left(t - \tau, \nabla w(x,t), \nabla w(x,\tau)\right) \frac{\partial^2 w^j}{\partial x^\alpha \partial x^\beta} (x,\tau) \, d\tau. \qquad (62)_2$$

As before, we denote by S the mapping which carries w into the solution u of

$$\ddot{u}^i = {}_w B_{ij}^{\alpha\beta} \frac{\partial^2 u^j}{\partial x^\alpha \partial x^\beta} + {}_w g^i, \quad x \in \Omega, \ t \in [0,T]. \qquad (63)_1$$

$$u|_{\partial\Omega} = 0, \qquad (63)_2$$

$$u(x,0) = u_0(x), \quad \dot{u}(x,0) = u_1(x). \qquad (63)_3$$

It is easy to see that a fixed point of S will be a solution of (61). We need to impose hypotheses on A, L, M, $\tilde{u}$, u_0, u_1 and f which will guarantee that (61) can be solved and that ${}_w B$ and ${}_w g$ obey estimates of the type used to prove Lemmas III.8 and III.9. It is clear that we may use the same assumptions on u_0, u_1 and f as before. The situation concerning A, L, M and $\tilde{u}$ is a bit more complicated, because there are numerous possible combinations of assumptions. One possibility is to impose hypotheses on the composite functions ${}_w B$ and ${}_w g$ rather than directly on A, L, M and $\tilde{u}$. This is not completely satisfactory because the meaning of such assumptions in applications is not very clear. We give one possible set of direct assumptions which guarantee an existence theorem of the same form as Theorem III.5. (S1)-(S4) and (C) remain unchanged. Instead of (E) we require

E*) The symmetry conditions

$$A_{ij}^{\alpha\beta} = A_{ji}^{\beta\alpha}, \quad L_{ij}^{\alpha\beta} = L_{ji}^{\beta\alpha}$$

and the strong ellipticity condition

$$\left(A_{ij}^{\alpha\beta}(\nabla u_0) + \int_{-\infty}^0 L_{ij}^{\alpha\beta}(-\tau, \nabla u_0, \nabla \tilde{u}(\cdot,\tau)) \, d\tau\right) \varsigma^i \varsigma^j \eta_\alpha \eta_\beta \geq \kappa |\varsigma|^2 |\eta|^2, \ \kappa > 0$$

hold.

Instead of (S5) we assume

S5*) L and M are of class C^{m-1} on $[0,\infty) \times \mathcal{O} \times \mathcal{O}$. Moreover, L and M and their derivatives through order $m-1$ are integrable on $[0,\infty)$ uniformly on bounded sets of $\mathcal{O} \times \mathcal{O}$.

We also need to impose some regularity on the given history. For simplicity we assume

S6*) $\tilde{u}$ belongs to

$$\bigcap_{k=0}^m C_b^k((-\infty,0]; H^{m-k}(\Omega))$$

and $\nabla \tilde{u}$ takes values in $\mathcal{O}$.

Theorem III.10:

Assume that (S1)-(S4), (S5*), (S6*), (E*) and (C) hold. Then there is a $T \in (0, T']$ such that the history value problem (61) has a unique solution

$$u \in \bigcap_{k=0}^{m} C^{m-k}([0, T]; H^k(\Omega)). \tag{64}$$

Of course, the solution is not smooth across $t = 0$ unless time derivatives as $t \to 0-$ computed from the history coincide with the u_k.

For future reference, we record a special case in one dimension. The problem we consider is:

$$u_{tt}(x, t) = \phi(u_x(x, t))_x - \int_0^t m(t - \tau)\psi(u_x(x, \tau))_x \, d\tau + f(x, t), \ x \in [0, 1], \ t > 0,$$

$$\tag{65}_1$$

$$u(0, t) = u(1, t) = 0, \ t > 0, \tag{65}_2$$

$$u(x, 0) = u_0(x), \ u_t(x, 0) = u_1(x). \tag{65}_3$$

Theorem III.11:

Assume that $\phi, \psi \in C^3(\mathbb{R})$, $m \in W_{loc}^{1,1}[0, \infty)$, $\phi' > 0$, $u_0 \in H^3(0, 1)$, $u_1 \in H^2(0, 1)$. Assume further that

$$f, f_t, f_x \in C([0, \infty); L^2(0, 1)), \quad f_{tt} \in L_{loc}^1([0, \infty); L^2(0, 1)), \tag{66}$$

and that the following compatibility conditions hold:

$$u_0(0) = u_0(1) = u_1(0) = u_1(1) = 0, \tag{67}_1$$

$$\phi'(u_0'(0))u_0''(0) + f(0, 0) = \phi'(u_0'(1))u_0''(1) + f(1, 0) = 0. \tag{67}_2$$

Then the initial-boundary value problem (65) has a unique solution u on a maximal time interval $[0, T_0)$ with

$$u \in \bigcap_{k=0}^{3} C^k([0, T_0); H^{3-k}(0, 1)). \tag{68}$$

Moreover, if

$$\sup_{t \in [0, T_0)} \left\{ \|u(\cdot, t)\|_3 + \|u_t(\cdot, t)\|_2 \right\} < \infty, \tag{69}$$

then $T_0 = \infty$.

Remarks III.12:

1. That (69) implies $T_0 = \infty$ follows from a standard argument; indeed, as long as $\|u\|_3 + \|u_t\|_2$ remains bounded, a stepping argument can be used to continue the solution. One can in fact show that pointwise bounds on u_{xx} and u_{xt} are sufficient to continue the solution.

2. A similar theorem (with the obvious change regarding the compatibility conditions) can be established for Neumann boundary data or for the Cauchy problem on the whole real line. In the latter case, it is sufficient to assume $u_0' \in H^2(\mathbb{R})$ rather than $u_0 \in H^3(\mathbb{R})$.

3. Although the three-dimensional analogue of (65) is not frame-indifferent, it is possible to construct three-dimensional constitutive models which are frame-indifferent and lead to (65) in a certain class of one-dimensional motions. Moreover, in a purely one-dimensional theory, there is no analogue of frame-indifference.

The equations considered in this chapter are essentially hyperbolic, with the memory term playing only the role of a perturbation. One therefore expects a finite propagation speed (cf. also Section II.3). In fact, the energy method can be used to prove such a result (for an abstract approach based on semigroup methods, see Desch, Grimmer and Zeman [D8]). For purposes of illustration, we shall stick to a simple one-dimensional model problem. More general equations such as (61) can be treated in an analogous fashion, but the calculations required to give a detailed proof are rather involved. The problem we shall discuss is:

$$\ddot{u} = \phi(u_x)_x - \int_0^t m(t - \tau)\psi(u_x(\cdot, \tau))_x \, d\tau, \tag{70}_1$$

$$u(x, 0) = u_0(x), \quad \dot{u}(x, 0) = u_1(x), \quad x \in \mathbb{R}. \tag{70}_2$$

Theorem III.13:

Assume that ϕ, ψ and m are as in Theorem III.11. Moreover, assume that u_0', $u_1 \in H^2(\mathbb{R})$, that $u_0(x) = u_1(x) = 0$ for $|x| \leq M_0$ and that a solution of (70) exists up to $T = M_0/\sqrt{\phi'(0)}$. Then we have $u = 0$ for $|x| \leq M_0 - t\sqrt{\phi'(0)}$.

Proof:
Without loss of generality, we assume that $\phi(0) = \psi(0) = 0$. We define

$$\Phi(p) := \int_0^p \phi(q) \, dq, \tag{71}$$

and we set $c := \sqrt{\phi'(0)}$. Let γ be any positive number greater than c. For each $M \in (0, M_0]$, we multiply $(70)_1$ by $\dot{u}$ and integrate over the triangle $-M \leq x \leq M$, $0 \leq t \leq (M - |x|)/\gamma$. Several integrations by parts yield

$$\int_{-M}^{M} \left\{ \frac{1}{2}\dot{u}^2 + \Phi(u_x) + \frac{1}{\gamma}\phi(u_x)\dot{u} \right\}(x, \frac{M - |x|}{\gamma}) \, dx$$

99

$$= \int_{-M}^{M} (u_x + \frac{1}{\gamma}\dot{u})(x, \frac{M-|x|}{\gamma}) \int_{0}^{\frac{M-|x|}{\gamma}} m(\frac{M-|x|}{\gamma} - \tau)\psi(u_x(x,\tau))\, d\tau \, dx$$

$$- \int_{-M}^{M} \int_{0}^{\frac{M-|x|}{\gamma}} m(0)u_x(x,t)\psi(u_x(x,t)) + u_x(x,t) \int_{0}^{t} m'(t-\tau)\psi(u_x(x,\tau))\, d\tau \, dt \, dx.$$

$$(72)$$

We now set

$$E(M) := \int_{-M}^{M} \{\dot{u}^2 + u_x^2\}(x, \frac{M-|x|}{\gamma})\, dx. \qquad (73)$$

Noting that in a neighborhood of 0 we have

$$\phi(u_x) \sim c^2 u_x, \quad \Phi(u_x) \sim \frac{1}{2}c^2 u_x^2, \qquad (74)$$

we see that as long as u_x stays small (this is guaranteed for small M), the left-hand side of (72) has a lower bound of the form $k_1 E(M)$ with $k_1 > 0$. Using Cauchy-Schwarz and elementary substitutions ($t = \frac{\sigma M - |x|}{\gamma}$, $\tau = \frac{\sigma' M - |x|}{\gamma}$), we find that on the right-hand side of (72), the first term has an upper bound of the form

$$k_2 M \sqrt{E(M)} \int_{0}^{1} \sqrt{E(\sigma' M)}\, d\sigma',$$

the second term has an upper bound of the form

$$k_3 M \int_{0}^{1} E(\sigma M)\, d\sigma,$$

and the third term has an upper bound of the form

$$k_4 M^2 \int_{0}^{1} \sqrt{E(\sigma M)} \int_{0}^{\sigma} \sqrt{E(\sigma' M)}\, d\sigma' \, d\sigma.$$

Hence we have an inequality of the form

$$k_1 E(M) \leq k_2 M \sqrt{E(M)} \int_{0}^{1} \sqrt{E(\sigma' M)}\, d\sigma' + k_3 M \int_{0}^{1} E(\sigma M)\, d\sigma$$

$$+ k_4 M^2 \int_{0}^{1} \sqrt{E(\sigma M)} \int_{0}^{1} \sqrt{E(\sigma' M)}\, d\sigma' \, d\sigma. \qquad (75)$$

Since (75) holds for all $M \in (0, M_0]$, it follows that $E(M) = 0$.

4. Incompressible nonlinear elasticity

The energy method can also be applied to problems involving motions of incompressible matèrials. However, the incompressibility constraint introduces serious technical complications in the analysis. In this section, we treat the equations of incompressible nonlinear elasticity. Viscoelastic materials with incompressibility can be handled in a similar fashion, treating the memory term as a perturbation (as in Section 3), but the calculations are very involved. In order to focus on the main ideas concerning incompressibility, we shall therefore restrict our attention to elasticity. We follow the presentation of [H15]. Of course, in contrast to the problems considered earlier in this chapter, the speed of propagation is not finite in the incompressible case.

We consider a homogeneous body with reference configuration Ω and unit reference density. By x we denote material coordinates, and by $y(x,t)$ we denote the spatial position at time t of the particle with reference position x. We use the notation $F_\alpha^i = \partial y^i / \partial x^\alpha$ and we write $\partial x^\alpha / \partial y^i$ for the components of the inverse matrix $\mathbf{F}^{-1}$. The equation of motion is obtained by adding a pressure term to equation (19) of Chapter I:

$$\ddot{y}^i = -\frac{\partial p}{\partial x^\alpha}\frac{\partial x^\alpha}{\partial y^i} + \frac{\partial^2 W(\mathbf{F})}{\partial F_\alpha^i \partial F_\beta^j}\frac{\partial^2 y^j}{\partial x^\alpha \partial x^\beta} + f^i. \tag{76}_1$$

Here $W = W(\mathbf{F})$ is the stored energy function, $p = p(x,t)$ is an unknown pressure, and $f = f(x,t)$ is a prescribed body force. The motion must satisfy the incompressibility constraint

$$\det \mathbf{F} = 1. \tag{76}_2$$

We seek a solution to $(76)_1$ and $(76)_2$ subject to the initial conditions

$$y(x,0) = y_0(x), \quad \dot{y}(x,0) = y_1(x), \tag{76}_3$$

and the Dirichlet boundary condition

$$y(x,t) = x \text{ for } t > 0, \; x \in \partial\Omega. \tag{76}_4$$

In order to make the pressure unique, we shall normalize it by

$$\int_\Omega p(x,t)\, dx = 0. \tag{76}_5$$

We make the following smoothness assumptions:
(S1) Ω is a bounded domain in $\mathbb{R}^3$ with a boundary of class $C^{3,1}$.
(S2) The stored energy function W is of class[2] C^5.

---

[2] We always think of W as being defined even for $\det \mathbf{F} \neq 1$. This can always be achieved by extension.

(S3) The initial data satisfy $y_0 \in H^4(\Omega)$, $y_1 \in H^3(\Omega)$.

(S4) For some $T > 0$, we have

$$f \in \bigcap_{k=0}^{2} C^k([0,T]; H^{2-k}(\Omega)),$$

and the third order time derivative of f lies in $L^1([0,T]; L^2(\Omega))$.

Moreover, we assume that the elastic energy satisfies the strong ellipticity condition.

(E) There is a continuous function $\kappa : (0, \infty) \to (0, \infty)$ such that, with

$$A_{ij}^{\alpha\beta}(\mathbf{F}) := \frac{\partial^2 W}{\partial F_\alpha^i \partial F_\beta^j}, \tag{77}$$

we have

$$A_{ij}^{\alpha\beta}(\mathbf{F})\varsigma^i \varsigma^j \eta_\alpha \eta_\beta \geq \kappa(|\mathbf{F}|)|\varsigma|^2 |\eta|^2 \tag{78}$$

for all $\varsigma, \eta \in \mathbb{R}^3$ and all $\mathbf{F}$ with $\det \mathbf{F} = 1$.

The quantity $\mathbf{A}$ has the same physical significance as in Section 2, except that it is now considered as a function of the deformation gradient $\mathbf{F}$ rather than the displacement gradient ∇u. We have chosen to formulate the equations in terms of deformations rather than displacements because the components of $\mathbf{F}^{-1}$ appear in the equations.

Finally, we must assume compatibility of the initial data with the boundary and incompressibility conditions. The following conditions are required:

(C1) $y_0 = x$ and $y_1 = 0$ on $\partial\Omega$.

(C2) $\det \nabla y_0 = 1$, and y_1 satisfies the following equation, which is obtained by differentiating the incompressibility condition $(76)_2$ with respect to time:

$$\frac{\partial y_1^i}{\partial x^\alpha} \frac{\partial x^\alpha}{\partial y_0^i} = 0. \tag{79}$$

(C3) The initial values of $\ddot{y}$ and $\partial^3 y/\partial t^3$, henceforth denoted by y_2 and y_3, vanish on $\partial\Omega$.

The last condition requires some explanation. The initial values of $\ddot{y}$ and $\partial^3 y/\partial t^3$ have to be determined from the differential equation. However, in order to find $\ddot{y}$ from the equation, we must first find the pressure at time $t = 0$. The procedure for this is as follows. First we differentiate $(76)_2$ twice with respect to time to obtain

$$\frac{\partial \ddot{y}^i}{\partial x^\alpha} \frac{\partial x^\alpha}{\partial y^i} - \frac{\partial \dot{y}^i}{\partial x^\alpha} \frac{\partial x^\alpha}{\partial y^j} \frac{\partial \dot{y}^j}{\partial x^\beta} \frac{\partial x^\beta}{\partial y^i} = 0, \tag{80}$$

from which we compute the initial value of

$$\frac{\partial \ddot{y}^i}{\partial x^\alpha} \frac{\partial x^\alpha}{\partial y^i}. \tag{81}$$

We then apply the "divergence operator" $(\partial x^\gamma / \partial y^i)(\partial / \partial x^\gamma)$ to the equation of motion $(76)_1$ to obtain an elliptic differential equation for p, having the form

$$\frac{\partial x^\gamma}{\partial y^i} \frac{\partial}{\partial x^\gamma} \Big[\frac{\partial x^\alpha}{\partial y^i} \frac{\partial p}{\partial x^\alpha} \Big] = g, \tag{82}_1$$

where g is known at time $t = 0$ and lies in $H^1(\Omega)$. On the boundary $\partial\Omega$, we multiply equation $(76)_1$ by $(\partial x^\gamma / \partial y^i) n_\gamma$, where n is the outer unit normal. Since $(76)_4$ implies that $\ddot{y}$ must vanish on $\partial\Omega$, the appropriate boundary condition for p has the form

$$\frac{\partial x^\gamma}{\partial y^i} n_\gamma \Big[\frac{\partial x^\alpha}{\partial y^i} \frac{\partial p}{\partial x^\alpha} \Big] = h, \tag{82}_2$$

where h is known at time $t = 0$ and lies in $H^{3/2}(\partial\Omega)$. Equations (82) form a Neumann-type boundary value problem for p (in fact, they are the Neumann problem if we transform to Eulerian coordinates). This problem has a unique solution $p \in H^3(\Omega)$ subject to the constraint $(76)_5$, provided the right-hand sides satisfy the condition

$$\int_\Omega g \; dx = \int_{\partial\Omega} h \; dS. \tag{83}$$

To verify that (83) holds we must show that the integral over Ω of the second term on the left-hand side of (80) is zero. If we transform to Eulerian coordinates, with v denoting to the velocity field, we obtain for the integral in question

$$\int_\Omega (\nabla_y v) : (\nabla_y v)^T \; dy = -\int_\Omega (v \cdot \nabla_y)(\mathrm{div}_y \; v) \; dy = 0. \tag{84}$$

After having solved for p at time $t = 0$, henceforth denoted p_0, we can insert it into $(76)_1$ and compute $\ddot{y}$ at time $t = 0$. Differentiating $(76)_1$ with respect to time we obtain, in an analogous fashion, a Neumann problem for $\dot{p}$ which we must solve in order to find $\partial^3 y / \partial t^3$ at $t = 0$. Condition (C3) means that the so computed initial data for $\ddot{y}$ and $\partial^3 y / \partial t^3$ must vanish on $\partial\Omega$.

Theorem III.14:

Let the hypotheses (S1)-(S4), (E) and (C1)-(C3) be satisfied. Then there is a $T' \in (0, T]$ such that the problem (76) has a unique solution (y, p) on $\Omega \times [0, T']$ with the regularity

$$y \in \bigcap_{k=0}^{4} C^{4-k}([0, T']; H^k(\Omega)), \quad p \in \bigcap_{k=2}^{4} C^{4-k}([0, T']; H^{k-1}(\Omega)).$$

Remarks III.15:

1. It follows from the Sobolev imbedding theorem that the solution satisfies $y \in C^2(\bar{\Omega} \times [0, T'])$, $p \in C^1(\bar{\Omega} \times [0, T'])$, and it is therefore a classical solution of the differential equations.

2. The result can be modified to accommodate Dirichlet boundary data other than $y = x$ provided appropriate smoothness and compatibility conditions are satisfied.

3. It follows from $(76)_2$ and $(76)_4$ that the mappings $x \to y(x, t)$ are actually globally invertible, see Ball [B2] and Meisters and Olech [M12].

4. If more regularity of Ω, W and the data is assumed, and the appropriate additional compatibility conditions are satisfied, then the solution also has higher regularity.

5. The initial value problem on all of space has been studied by Schochet [S2] (following ideas of Klainerman and Majda [K13],[K14]) and by Ebin and Saxton [E1]. In addition to proving existence, Schochet [S2] shows that the solution to the incompressible problem is the limit of a sequence of solutions of compressible problems.

6. In Section V.8, we shall discuss an existence result for incompressible K-BKZ materials on all of space.

The solution of (76) will be obtained by a contraction argument which involves solving a sequence of linear initial-boundary value problems. There are, of course, numerous ways to formulate (76) as a fixed point problem; for technical reasons the precise form of the fixed point problem is a very delicate issue. Since the constraint $\det \mathbf{F} = 1$ is not in quasilinear form, we shall work with a differentiated version of (76). As noted in Section 1, we differentiate with respect to time because of the boundary conditions. We shall actually differentiate $(76)_1$ twice and $(76)_2$ three times. This yields a linear initial-boundary value problem for $\ddot{y}$ and $\ddot{p}$. Bounds for y, $\dot{y}$, p and $\dot{p}$ will be obtained via elliptic estimates.

We now describe the basic iteration scheme. Assuming that (y, p) is a sufficiently smooth solution of (76), we introduce the notations $v := \dot{y}$, $z := \ddot{y} - \lambda(y - x)$, $q := \dot{p}$, and $\phi := \ddot{p}$, where λ is a positive constant to be chosen later. The original system $(76)_1$, $(76)_2$ can now be written as follows:

$$z^i = -\frac{\partial p}{\partial x^\alpha} \frac{\partial x^\alpha}{\partial y^i} + \frac{\partial^2 W(\nabla y)}{\partial F_\alpha^i \partial F_\beta^j} \frac{\partial^2 y^j}{\partial x^\alpha \partial x^\beta} - \lambda(y^i - x^i) + f^i, \tag{85$_1$}$$

$$\det(\nabla y) = 1. \tag{85$_2$}$$

After differentiating (85) once with respect to time, we obtain

$$\dot{z}^i = -\frac{\partial q}{\partial x^\alpha} \frac{\partial x^\alpha}{\partial y^i} + \frac{\partial^2 W(\nabla y)}{\partial F_\alpha^i \partial F_\beta^j} \frac{\partial^2 v^j}{\partial x^\alpha \partial x^\beta}$$

$$+ \frac{\partial p}{\partial x^\alpha} \frac{\partial x^\alpha}{\partial y^k} \frac{\partial v^k}{\partial x^\beta} \frac{\partial x^\beta}{\partial y^i} + \frac{\partial^3 W(\nabla y)}{\partial F_\alpha^i \partial F_\beta^j \partial F_\gamma^k} \frac{\partial v^k}{\partial x^\gamma} \frac{\partial^2 y^j}{\partial x^\alpha \partial x^\beta} - \lambda v^i + \dot{f}^i, \tag{86$_1$}$$

$$\frac{\partial v^i}{\partial x^\alpha} \frac{\partial x^\alpha}{\partial y^i} = 0. \tag{86$_2$}$$

Differentiating once more with respect to time, we obtain

$$\ddot{z}^i + \frac{\partial \phi}{\partial x^\alpha}\frac{\partial x^\alpha}{\partial y^i} - \frac{\partial^2 W(\nabla y)}{\partial F_\alpha^i \partial F_\beta^j}\frac{\partial^2 z^j}{\partial x^\alpha \partial x^\beta}$$

$$= \lambda \frac{\partial^2 W(\nabla y)}{\partial F_\alpha^i \partial F_\beta^j}\frac{\partial^2 y^j}{\partial x^\alpha \partial x^\beta} + 2\frac{\partial q}{\partial x^\alpha}\frac{\partial x^\alpha}{\partial y^k}\frac{\partial v^k}{\partial x^\beta}\frac{\partial x^\beta}{\partial y^i}$$

$$-2\frac{\partial p}{\partial x^\alpha}\frac{\partial x^\alpha}{\partial y^l}\frac{\partial v^l}{\partial x^\gamma}\frac{\partial x^\gamma}{\partial y^k}\frac{\partial v^k}{\partial x^\beta}\frac{\partial x^\beta}{\partial y^i} + \frac{\partial p}{\partial x^\alpha}\frac{\partial x^\alpha}{\partial y^k}\left[\frac{\partial z^k}{\partial x^\beta} + \lambda\frac{\partial y^k}{\partial x^\beta} - \lambda\delta_\beta^k\right]\frac{\partial x^\beta}{\partial y^i}$$

$$+2\frac{\partial^3 W(\nabla y)}{\partial F_\alpha^i \partial F_\beta^j \partial F_\gamma^k}\frac{\partial v^k}{\partial x^\gamma}\frac{\partial^2 v^j}{\partial x^\alpha \partial x^\beta} + \frac{\partial^3 W(\nabla y)}{\partial F_\alpha^i \partial F_\beta^j \partial F_\gamma^k}\left[\frac{\partial z^k}{\partial x^\gamma} + \lambda\frac{\partial y^k}{\partial x^\gamma} - \lambda\delta_\gamma^k\right]\frac{\partial^2 y^j}{\partial x^\alpha \partial x^\beta}$$

$$+\frac{\partial^4 W(\nabla y)}{\partial F_\alpha^i \partial F_\beta^j \partial F_\gamma^k \partial F_\delta^l}\frac{\partial v^k}{\partial x^\gamma}\frac{\partial v^l}{\partial x^\delta}\frac{\partial^2 y^j}{\partial x^\alpha \partial x^\beta} - \lambda z^i - \lambda^2 y^i + \lambda^2 x^i + \ddot{f}^i, \tag{$87)_1$}$$

$$\frac{\partial z^i}{\partial x^\alpha}\frac{\partial x^\alpha}{\partial y^i} + \lambda\left[\frac{\partial y^i}{\partial x^\alpha} - \delta_\alpha^i\right]\frac{\partial x^\alpha}{\partial y^i} - \frac{\partial v^i}{\partial x^\alpha}\frac{\partial x^\alpha}{\partial y^j}\frac{\partial v^j}{\partial x^\beta}\frac{\partial x^\beta}{\partial y^i} = 0. \tag{$87)_2$}$$

In the derivation of (86) and (87) we have used the identity

$$\frac{\partial}{\partial t}\mathbf{F}^{-1} = -\mathbf{F}^{-1}\frac{\partial \mathbf{F}}{\partial t}\mathbf{F}^{-1} \tag{88}$$

to differentiate terms involving $\partial x^\alpha / \partial y^i$.

Instead of using $(87)_2$ directly, we shall differentiate this equation with respect to time and transform it to a more convenient form. This transformation will exploit the identity

$$\frac{\partial}{\partial x^\alpha}\frac{\partial x^\alpha}{\partial y^i} = 0, \tag{89}$$

which is obtained as follows. Employing (88) with $\partial/\partial t$ replaced by $\partial/\partial x^\alpha$, we find

$$\frac{\partial}{\partial x^\alpha}\frac{\partial x^\alpha}{\partial y^i} = -\frac{\partial x^\alpha}{\partial y^j}\frac{\partial^2 y^j}{\partial x^\alpha \partial x^\beta}\frac{\partial x^\beta}{\partial y^i} = -\mathrm{tr}\left(\mathbf{F}^{-1}\frac{\partial \mathbf{F}}{\partial x^\beta}\right)\frac{\partial x^\beta}{\partial y^i}, \tag{90}$$

which yields (89) because $(76)_2$ implies that

$$\mathrm{tr}\,\mathbf{F}^{-1}\frac{\partial \mathbf{F}}{\partial x^\beta} = 0. \tag{91}$$

Using (89), we can rewrite the first two terms in $(87)_2$ as

$$\frac{\partial}{\partial x^\alpha}\left[z^i\frac{\partial x^\alpha}{\partial y^i} + \lambda(y^i - x^i)\frac{\partial x^\alpha}{\partial y^i}\right]. \tag{92}$$

After differentiation with respect to time, this expression becomes

$$\frac{\partial}{\partial x^\alpha}\left[\dot{z}^i\frac{\partial x^\alpha}{\partial y^i} + \lambda v^i\frac{\partial x^\alpha}{\partial y^i} - [z^i + \lambda(y^i - x^i)]\frac{\partial x^\alpha}{\partial y^j}\frac{\partial v^j}{\partial x^\beta}\frac{\partial x^\beta}{\partial y^i}\right]. \tag{93}$$

105

It is convenient to transform the last term in $(87)_2$ to Eulerian coordinates; in these coordinates the term is simply equal to

$$(\nabla_y v) : (\nabla_y v)^T = \text{div}_y \left[(v \cdot \nabla_y) v \right] \tag{94}$$

(∇_y and $'$ denote Eulerian space and time derivatives, while $\frac{d}{dt} =' +(v \cdot \nabla_y$ denotes the material time derivative). We set $w := (v \cdot \nabla_y) v$ and take the material time derivative to obtain

$$\frac{d}{dt} \text{div}_y \, w = \text{div}_y \left[w' + (v \cdot \nabla_y) w - (w \cdot \nabla_y) v \right]. \tag{95}$$

To proceed further, we note the identity

$$\text{div}_y \left[(v \cdot \nabla_y) w \right] = \text{div}_y \left[v \, \text{div}_y \, w + (w \cdot \nabla_y) v \right]. \tag{96}$$

Combining (95) and (96) we find that

$$\frac{d}{dt} \text{div}_y \, w = \text{div}_y \left[w' + v \, \text{div}_y \, w \right] = \text{div}_y \left[(v' \cdot \nabla_y) v + (v \cdot \nabla_y) v' + v \, \text{div}_y \, w \right]$$

$$= \text{div}_y \left[2(v' \cdot \nabla_y) v + v \, \text{div}_y \, w \right]$$

$$= \text{div}_y \left[2(\frac{dv}{dt} \cdot \nabla_y) v - 2 \left[((v \cdot \nabla_y) v) \cdot \nabla_y \right] v + v \left[(\nabla_y v) : (\nabla_y v)^T \right] \right]. \tag{97}$$

By transforming (97) back to Lagrangian coordinates and combining the result with (93), we finally obtain

$$\frac{\partial x^\alpha}{\partial y^i} \frac{\partial}{\partial x^\alpha} \left\{ \dot{z}^i + \lambda v^i - 3 \left[z^j + \lambda (y^j - x^j) \right] \frac{\partial v^i}{\partial x^\beta} \frac{\partial x^\beta}{\partial y^j} \right.$$

$$\left. + 2 v^k \frac{\partial x^\beta}{\partial y^k} \frac{\partial v^j}{\partial x^\beta} \frac{\partial x^\gamma}{\partial y^j} \frac{\partial v^i}{\partial x^\gamma} - v^i \frac{\partial v^j}{\partial x^\beta} \frac{\partial x^\beta}{\partial y^k} \frac{\partial v^k}{\partial x^\gamma} \frac{\partial x^\gamma}{\partial y^j} \right\} = 0 \tag{98}$$

in place of $(87)_2$.

We now describe the basic iteration scheme. Given functions $y_{(n)}$, $z_{(n)}$, $p_{(n)}$ and $\phi_{(n)}$ on $\Omega \times [0, T]$, we determine $y_{(n+1)}$, $z_{(n+1)}$, $p_{(n+1)}$ and $\phi_{(n+1)}$ as follows. First, we set $v_{(n)} := \dot{y}_{(n)}$ and $q_{(n)} := \dot{p}_{(n)}$. For technical reasons we also need to introduce a function $\tilde{v}_{(n)}$ which is another approximation of v possessing better regularity properties than $v_{(n)}$. We set

$$\tilde{v}_{(n)} = \Pi \left(v_{(n)}, y_1(x) + \int_0^t z_{(n)}(x, \tau) + \lambda (y_{(n)}(x, \tau) - x) \, d\tau \right), \tag{99}$$

where Π is the projection operator from an appropriate product $X_1 \times X_2$ of Hilbert spaces onto the diagonal. The spaces X_1 and X_2 will be chosen so that

106

he functions in X_1 have more spatial regularity than those in X_2, while the unctions in X_2 have more temporal regularity.

We then determine $z_{(n+1)}$ and $\phi_{(n+1)}$ by solving a pair of equations related o $(87)_1$ and (98). To describe this pair of equations, it is convenient to denote he expression on the right-hand side of $(87)_1$ by

$$G(y, \nabla y, \nabla^2 y, v, \nabla v, \nabla^2 v, z, \nabla z, \ddot{f}, x).$$

Ve replace equation $(87)_1$ with

$$\ddot{z}^i_{(n+1)} = -\frac{\partial \phi_{(n+1)}}{\partial x^\alpha} \frac{\partial x^\alpha}{\partial y^i_{(n)}} + \frac{\partial^2 W(\nabla y_{(n)})}{\partial F^i_\alpha \partial F^j_\beta} \frac{\partial^2 z^j_{(n+1)}}{\partial x^\alpha \partial x^\beta}$$

$$+ G^i\left(y_{(n)}, \nabla y_{(n)}, \nabla^2 y_{(n)}, v_{(n)}, \nabla v_{(n)}, \nabla^2 v_{(n)}, z_{(n)}, \nabla z_{(n)}, \ddot{f}, x\right). \qquad (100)_1$$

n the analogue of (98) we shall use $\tilde{v}_{(n)}$ rather than $v_{(n)}$. More precisely, we eplace (98) with

$$\frac{\partial x^\alpha}{\partial y^i_{(n)}} \frac{\partial}{\partial x^\alpha} \left\{ \ddot{z}^i_{(n+1)} + \lambda \tilde{v}^i_{(n)} - 3\left[z^j_{(n)} + \lambda(y^j_{(n)} - x^j)\right] \right\} \frac{\partial \tilde{v}^i_{(n)}}{\partial x^\beta} \frac{\partial x^\beta}{\partial y^j_{(n)}}$$

$$+ 2\tilde{v}^k_{(n)} \frac{\partial x^\beta}{\partial y^k_{(n)}} \frac{\partial \tilde{v}^j_{(n)}}{\partial x^\beta} \frac{\partial x^\gamma}{\partial y^j_{(n)}} \frac{\partial \tilde{v}^i_{(n)}}{\partial x^\gamma} - \tilde{v}^i_{(n)} \frac{\partial \tilde{v}^j_{(n)}}{\partial x^\beta} \frac{\partial x^\beta}{\partial y^k_{(n)}} \frac{\partial \tilde{v}^k_{(n)}}{\partial x^\gamma} \frac{\partial x^\gamma}{\partial y^j_{(n)}} \right\}$$

$$=: \frac{\partial x^\alpha}{\partial y^i_{(n)}} \frac{\partial}{\partial x^\alpha} \left\{ \ddot{z}^i_{(n+1)} - H^i\left(\tilde{v}, \nabla \tilde{v}, y, \nabla y, z, x\right) \right\} = 0. \qquad (100)_2$$

f we impose the initial conditions

$$z_{(n+1)}(x, 0) = z_0(x) := y_2(x) - \lambda(y_0(x) - x),$$

$$\dot{z}_{(n+1)}(x, 0) = z_1(x) := y_3(x) - \lambda y_1(x), \qquad (100)_3$$

he boundary conditions

$$z_{(n+1)} = 0 \text{ for } x \in \partial \Omega, \ t \geq 0, \qquad (100)_4$$

und the normalization

$$\int_\Omega \phi_{(n+1)}(x, t) \, dx = 0, \qquad (100)_5$$

he problem (100) can be solved uniquely for $z_{(n+1)}$ and $\phi_{(n+1)}$.

Finally, we obtain $y_{(n+1)}$ and $p_{(n+1)}$ by solving a nonlinear elliptic system. 'or each $t \in [0, T']$, we consider the boundary value problem

$$-\frac{\partial p_{(n+1)}}{\partial x^\alpha} \frac{\partial x^\alpha}{\partial y^i_{(n+1)}} + \frac{\partial^2 W(\nabla y_{(n+1)})}{\partial F^i_\alpha \partial F^j_\beta} \frac{\partial^2 y^j_{(n+1)}}{\partial x^\alpha \partial x^\beta} - \lambda(y^i_{(n+1)} - x^i) = z^i_{(n+1)} - f^i,$$

$$(101)_1$$

107

$$\det \nabla y_{(n+1)} = 1, \tag{101}_2$$

$$\int_\Omega p_{(n+1)} = 0, \tag{101}_3$$

$$y_{(n+1)}(x,t) = x \text{ for } x \in \partial\Omega. \tag{101}_4$$

Using the implicit function theorem and standard results for linear elliptic equations, we can solve this boundary value problem for $y_{(n+1)}(\cdot,t)$ and $p_{(n+1)}(\cdot,t)$ on a sufficiently small time interval. By formally differentiating (101) with respect to time, we can obtain estimates for $\dot{y}$, $\ddot{y}$, $\dot{p}$ and $\ddot{p}$. The formal differentiation with respect to time can be justified by taking appropriate difference quotients and passing to the limit.

To implement the procedure outlined above, we must solve linear initial-boundary value problems of the form

$$\ddot{z}^i = -\chi_i^\alpha \frac{\partial \phi}{\partial x^\alpha} + A_{ij}^{\alpha\beta} \frac{\partial^2 z^j}{\partial x^\alpha \partial x^\beta} + G^i, \tag{102}_1$$

$$\chi_i^\alpha \left(\frac{\partial \dot{z}^i}{\partial x^\alpha} - \frac{\partial H^i}{\partial x^\alpha} \right) = 0, \tag{102}_2$$

$$z = 0 \text{ on } \partial\Omega, \tag{102}_3$$

$$z(x,0) = z_0(x), \ \dot{z}(x,0) = z_1(x), \tag{102}_4$$

$$\int_\Omega \phi(x,t) \, dx = 0, \tag{102}_5$$

where G^i, H^i and the coefficients χ_i^α, $A_{ij}^{\alpha\beta}$ are given.

The solution of (102) is based on the Galerkin method together with elliptic estimates. It is important to keep track of the quantities on which the constants in the estimates depend. In the estimates derived below, we always regard Ω, W, the forcing term f and the initial data as given once and for all. We shall therefore not explicitly point out any dependence on these quantities. Since we shall obtain a solution with ∇y uniformly close to ∇y_0, we assume without loss of generality that the function κ in the strong ellipticity condition has a lower bound $\underline{\kappa} > 0$. Therefore we shall not point out any dependence on the constant of ellipticity.

We shall have to solve linear elliptic systems of the form

$$-\chi_i^\alpha \frac{\partial \phi}{\partial x^\alpha} + A_{ij}^{\alpha\beta} \frac{\partial^2 z^j}{\partial x^\alpha \partial x^\beta} + B_{ij}^\alpha \frac{\partial z^j}{\partial x^\alpha} - \lambda z^i = R^i, \tag{103}_1$$

$$\chi_i^\alpha \frac{\partial z^i}{\partial x^\alpha} = S, \tag{103}_2$$

$$z = 0 \text{ on } \partial\Omega, \tag{103}_3$$

$$\int_\Omega \phi \, dx = 0. \tag{103}_4$$

108

For our purposes, it is particularly important to understand how the solution (z, ϕ) depends on the norms of the coefficients χ_i^α, $A_{ij}^{\alpha\beta}$, B_{ij}^α. For convenience we introduce a parameter K which controls the size of the coefficients.

We make the following assumptions:

1*) The coefficients χ_i^α and $A_{ij}^{\alpha\beta}$ are in $H^2(\Omega)$, and B_{ij}^α is in $H^1(\Omega)$. Moreover, we have $\|\chi\|_2 + \|A\|_2 + \|B\|_1 \leq K$.

2*) For some given integer k with $-1 \leq k \leq 1$, we have $R \in H^k(\Omega)$, $S \in H^{k+1}(\Omega)$.

(I) The coefficients χ_i^α are the components of F^{-1}, where F is the gradient of a globally invertible mapping $y(x)$ with $\det F = 1$.

E*) We have $A_{ij}^{\alpha\beta} = A_{ji}^{\beta\alpha}$ and the following strong ellipticity condition is satisfied:

$$A_{ij}^{\alpha\beta} \varsigma^i \varsigma^j \eta_\alpha \eta_\beta \geq \underline{\kappa} |\varsigma|^2 |\eta|^2 \quad \forall \varsigma, \eta \in \mathbb{R}^3.$$

C*) $\int_\Omega S \, dx = 0$.

The following lemma holds:

Lemma III.16:

Let assumptions (S1), (S1), (S2*), (I), (E*) and (C*) hold. If $\lambda > 0$ is chosen sufficiently large relative to K, then the problem (103) has a unique solution. We have $z \in H^{k+2}(\Omega)$, $\phi \in H^{k+1}(\Omega)$ and an estimate of the form*

$$\|z\|_{k+2} + \|\phi\|_{k+1} \leq C(\|R\|_k + \|S\|_{k+1}) \tag{104}$$

holds. The constant C depends solely on K and λ.

The proof employs standard techniques in elliptic theory and we omit details. One first obtains the existence of a unique weak solution by using Gårding's inequality and a standard variational argument. Higher regularity is obtained by using the fact that the system (103) is elliptic in the sense of Agmon, Douglis and Nirenberg [A3]. The regularity results as stated in [A3] apply only to solutions which a priori satisfy $z \in H^2(\Omega)$, $p \in H^1(\Omega)$ (as stated in [A3], the coefficients would also be required to be of class C^2, but it is clear from the proofs that this is not actually required, since $H^2(\Omega)$ is an algebra in three spatial dimensions). In the compressible case, H^2-regularity of H^1-solutions is well known. In the incompressible case, regularity of weak solutions is proved by Giaquinta and Modica [G2] for traction boundary conditions. As they remark, their techniques can also be applied with Dirichlet conditions.

For future use, we note that if R, S and the coefficients (regarded as taking values in the appropriate Sobolev spaces) are continuous (or bounded measurable) functions of a parameter t, then this property is inherited by the solution.

We also have to consider the solution of the nonlinear system (85) (with the boundary condition (76)$_4$ and the normalization (76)$_5$) for given z and f. At the initial time $t = 0$ we have the solution $y = y_0$, $p = p_0$ for $z = z_0$ (the initial value of z) and $f = f_0(x) = f(x, 0)$. By using Lemma III.16, the implicit function

109

theorem and the regularity theory of Agmon, Douglas and Nirenberg [A3], we obtain the following result.

Lemma III.17:

Let $\lambda > 0$ be chosen sufficiently large (relative to the data for the original problem). If z and f lie in $H^1(\Omega)$ and $\|z - z_0\|_1$, $\|f - f_0\|_1$ are sufficiently small, then (85), (76)$_4$, (76)$_5$ has a solution $(y, p) \in H^3(\Omega) \times H^2(\Omega)$. Within a neighborhood of (y_0, p_0) in $H^3(\Omega) \times H^2(\Omega)$ this solution is unique and it depends smoothly on $z \in H^1(\Omega)$ and $f \in H^1(\Omega)$. If moreover $z \in H^2(\Omega)$, $f \in H^2(\Omega)$, then $y \in H^4(\Omega)$, $p \in H^3(\Omega)$ and we have an estimate of the form

$$\|y\|_4 + \|p\|_3 \leq C(1 + \|z\|_2 + \|f\|_2), \tag{105}$$

where the constant C depends only on λ, $\|z - z_0\|_1$ and $\|f - f_0\|_1$.

We now consider problem (102). As before, it is imperative to understand how certain norms of the solution depend on corresponding norms of the coefficients. We use two parameters K and L to measure the sizes of χ_i^α and $A_{ij}^{\alpha\beta}$. In addition, it is convenient to introduce a parameter U which measures the sizes of G and H. The initial values of $\mathbf{A}$ and χ will remain fixed in our iteration scheme, and we therefore do not point out dependence on this quantity in our estimates. We make the following assumptions:

(S1′) The coefficients χ_i^α and $A_{ij}^{\alpha\beta}$ lie in $W^{2,\infty}([0,T]; H^1(\Omega)) \cap W^{1,\infty}([0,T]; H^2(\Omega))$ $\cap L^\infty([0,T]; H^3(\Omega))$ with norms bounded by L. Their norms in $L^\infty([0,T]; H^2(\Omega))$ are bounded by K.

(S2′) $G \in C([0,T]; L^2(\Omega))$, $\dot{G} \in L^1([0,T]; L^2(\Omega))$ and the norms are bounded by U.

(S3′) $H \in W^{2,1}([0,T]; L^2(\Omega)) \cap L^1([0,T]; H^2(\Omega))$ and the norms are bounded by U.

(S4′) $z_0 \in H^2(\Omega)$, $z_1 \in H^1(\Omega)$.

(E′) We have $A_{ij}^{\alpha\beta} = A_{ji}^{\beta\alpha}$ and the following strong ellipticity condition holds:

$$A_{ij}^{\alpha\beta} \varsigma^i \varsigma^j \eta_\alpha \eta_\beta \geq \underline{\kappa} |\varsigma|^2 |\eta|^2 \quad \forall \varsigma, \eta \in \mathbb{R}^3.$$

(C1′) z_0 and z_1 vanish on $\partial\Omega$.

(C2′) z_1 satisfies (102)$_2$.

(C3′) H vanishes on $\partial\Omega$.

(I) The coefficients χ_i^α are the components of $\mathbf{F}^{-1}(t)$, where $\mathbf{F}(t)$ is the gradient of a globally invertible mapping $y(x,t)$ with $\det \mathbf{F} = 1$. We note that

$$\frac{\partial}{\partial x^\alpha} \chi_i^\alpha = 0$$

by virtue of (I).

Lemma III.18:

Let assumptions (S1), (S1')-(S4'), (E), (C1')-(C3) and (I) hold. Then the initial-boundary value problem (102) has a unique solution (z, ϕ) defined on $\Omega \times [0, T]$ with the regularity

$$z \in \bigcap_{k=0}^{2} C^k([0, T]; H^{2-k}(\Omega)), \quad \phi \in C([0, T]; H^1(\Omega)).$$

Moreover, the solution obeys the a priori estimate

$$E[z, \phi](t) \leq \Gamma(U, K, T, T \cdot L) + \Lambda(U, K, T, L) \int_0^t E[z, \phi](s) \, ds \ \forall t \in [0, T], \quad (106)_1$$

and hence

$$E[z, \phi](T) \leq \Gamma(U, K, T, T \cdot L) \exp(T \cdot \Lambda(U, K, T, L)), \quad (106)_2$$

where

$$E[z, \phi](t) := \max_{s \in [0, t]} \left(\sum_{k=0}^{2} \left\| \frac{\partial^k z}{\partial t^k}(\cdot, s) \right\|_{2-k}^2 + \|\phi(\cdot, s)\|_1^2 \right), \quad (106)_3$$

and Γ and Λ are functions which are bounded on bounded sets.

It is important to note that Γ depends on L only through the combination $T \cdot L$, and therefore Γ can be controlled for large L by making T small.

Proof: To establish the existence and uniqueness of a solution, we assume without loss of generality that $H = 0$. (Reduction to the case $H = 0$ is accomplished by subtracting from z the reference function $\int_0^t H(\cdot, \tau) \, d\tau$.) In order to solve (102), we differentiate with respect to time and then apply a Galerkin method to the differentiated equation. After differentiation, $(102)_1$ and $(102)_2$ become

$$\frac{\partial^3 z^i}{\partial t^3} = -\chi_i^\alpha \frac{\partial \dot{\phi}}{\partial x^\alpha} + A_{ij}^{\alpha\beta} \frac{\partial^2 \dot{z}^j}{\partial x^\alpha \partial x^\beta}$$

$$-\dot{\chi}_i^\alpha \frac{\partial \phi}{\partial x^\alpha} + \dot{A}_{ij}^{\alpha\beta} \frac{\partial^2 z^j}{\partial x^\alpha \partial x^\beta} + \dot{G}^i, \quad (107)_1$$

$$\chi_i^\alpha \frac{\partial \ddot{z}^i}{\partial x^\alpha} + \dot{\chi}_i^\alpha \frac{\partial \dot{z}^i}{\partial x^\alpha} = 0. \quad (107)_2$$

Because of (I) the latter equation can be rewritten as

$$\frac{\partial}{\partial x^\alpha} \left(\chi_i^\alpha \ddot{z}^i + \dot{\chi}_i^\alpha \dot{z}^i \right) = 0. \quad (107)_2^*$$

Recalling that F_α^i is the matrix inverse to χ_i^α, we can write $(107)_2$ in the form

$$\chi_i^\alpha \frac{\partial}{\partial x^\alpha} \left(\ddot{z}^i - \dot{F}_\beta^i \chi_j^\beta \dot{z}^j \right) = 0. \quad (107)_2^{**}$$

111

In order to solve (107), we must prescribe initial data for $\tilde{z}$, which need to be determined from $(102)_1$. We must therefore find ϕ at time $t = 0$. For this purpose, we apply the operator $\chi_i^\gamma \frac{\partial}{\partial x^\gamma}$ to $(102)_1$, and we multiply by $\chi_i^\gamma n_\gamma$ on the boundary $\partial\Omega$. To handle the term involving $\ddot{z}^i$, we use $(107)_2^{**}$ and the fact that $\ddot{z}^i$ should vanish on the boundary. The initial values of z and $\dot{z}$ are known. We therefore obtain a problem of the following form for $\phi(\cdot, 0)$:

$$\chi_i^\gamma \frac{\partial}{\partial x^\gamma}\left[\chi_i^\alpha \frac{\partial\phi}{\partial x^\alpha}\right] = \chi_i^\gamma \frac{\partial h^i}{\partial x^\gamma}, \tag{108$_1$}$$

$$\chi_i^\gamma n_\gamma \left[\chi_i^\alpha \frac{\partial\phi}{\partial x^\alpha}\right] = \chi_i^\gamma n_\gamma h^i \text{ on } \partial\Omega. \tag{108$_2$}$$

When transformed to Eulerian coordinates, this is simply the Neumann problem. Subject to the constraint $(102)_5$, there is a unique weak solution $\phi \in H^1(\Omega)$ for every $h \in L^2(\Omega)$ (this is known as the Hodge projection theorem). We therefore have a unique initial datum $z_2 \in L^2(\Omega)$ for $\tilde{z}$ which is consistent with $(102)_4$ and $(102)_3$.

Instead of considering $(107)_1$ directly, we shall work with the equation

$$\frac{\partial^3 z^i}{\partial t^3} = -\chi_i^\alpha \frac{\partial\dot{\phi}}{\partial x^\alpha} + A_{ij}^{\alpha\beta} \frac{\partial^2 \dot{z}^j}{\partial x^\alpha \partial x^\beta}$$

$$-\dot{\chi}_i^\alpha \frac{\partial\tilde{\phi}}{\partial x^\alpha} + \dot{A}_{ij}^{\alpha\beta} \frac{\partial^2 \tilde{z}^j}{\partial x^\alpha \partial x^\beta} + \dot{G}^i, \tag{109}$$

where $\tilde{\phi}$ and $\tilde{z}$ are expressed in terms of $\ddot{z} - \lambda z$ by the following elliptic problem obtained from (102):

$$\ddot{z}^i - \lambda z^i = -\chi_i^\alpha \frac{\partial\tilde{\phi}}{\partial x^\alpha} + A_{ij}^{\alpha\beta} \frac{\partial^2 \tilde{z}^j}{\partial x^\alpha \partial x^\beta} - \lambda \tilde{z}^i + G^i, \tag{110$_1$}$$

$$\chi_i^\alpha(x,t) \frac{\partial \tilde{z}^i(x,t)}{\partial x^\alpha} = \int_0^t \dot{\chi}_i^\alpha(x,\tau) \frac{\partial \tilde{z}^i(x,\tau)}{\partial x^\alpha} d\tau + \chi_i^\alpha(x,0) \frac{\partial z_0^i(x)}{\partial x^\alpha}. \tag{110$_2$}$$

The latter equation has been obtained by integrating $(102)_2$ (with $H = 0$) with respect to time. Of course we shall eventually show that $\tilde{z} = z$ for the solution we construct, and hence (109) is equivalent to $(107)_1$, but the approximations used for $\tilde{z}$ and z will actually differ. It follows from the elliptic estimates discussed above that (110) (with the boundary condition $\tilde{z} = 0$ and the normalization $\int_\Omega \tilde{\phi}(x)\, dx = 0$) can be solved for $\tilde{z}$ and $\tilde{\phi}$ if λ is chosen appropriately. Moreover, an estimate of the form

$$\|\tilde{z}\|_{L^\infty([0,T];H^2(\Omega))} + \|\tilde{\phi}\|_{L^\infty([0,T];H^1(\Omega))}$$

$$\leq C\left(\|\ddot{z} - \lambda z\|_{L^\infty([0,T];L^2(\Omega))} + \|G\|_{L^\infty([0,T];L^2(\Omega))} + \|z_0\|_{H^2(\Omega)}\right) \tag{111}$$

holds.

Let now $\tilde{V}$ be the space of all divergence-free vector fields in $H_0^1(\Omega)$ and let $\{\xi_1, \xi_2, \xi_3, ...\}$ be a basis for $\tilde{V}$. Moreover, let $V(t) = \mathbf{F}(t)\tilde{V} = \{w \in H_0^1(\Omega) \mid \mathbf{F}^{-1}(t)w \in \tilde{V}\}$, and let $\eta_\kappa(t) = \mathbf{F}(t)\xi_\kappa$. Let $\tilde{V}_n = \mathrm{span}\,\{\xi_1, \xi_2, ..., \xi_n\}$ and $V_n(t) = \mathbf{F}(t)\tilde{V}_n$. We seek an approximate solution of the form

$$\dot{z}_n(t) = \sum_{\kappa=1}^{n} a_{n\kappa}(t)\eta_\kappa(t), \tag{112}$$

(i.e., $\dot{z} \in V_n(t)$), which satisfies the following approximate version of (109)

$$\left\langle F_\gamma^i \omega^\gamma,\; \frac{\partial^3(z_n)^i}{\partial t^3} - A_{ij}^{\alpha\beta}\frac{\partial^2(\dot{z}_n)^j}{\partial x^\alpha \partial x^\beta} + \dot{\chi}_i^\alpha \frac{\partial \tilde{\phi}_n}{\partial x^\alpha} - \dot{A}_{ij}^{\alpha\beta}\frac{\partial^2(\tilde{z}_n)^j}{\partial x^\alpha \partial x^\beta} - \dot{G}^i \right\rangle = 0 \tag{113}$$

for every $\omega \in \tilde{V}_n$. Here $\tilde{\phi}_n$ and $\tilde{z}_n$ are determined from (110) with z on the left side of $(110)_1$ replaced by z_n, and $\langle \cdot, \cdot \rangle$ denotes the inner product in $L^2(\Omega)$. Finally, we define initial data by

$$z_n(x,0) = z_{n0}(x) := z_0(x), \quad \dot{z}_n(x,0) = z_{n1}(x) := P_n^1 z_1(x),$$

$$\ddot{z}_n(x,0) = z_{n2}(x) := \dot{\mathbf{F}}(x,0)\mathbf{F}^{-1}(x,0)z_{n1}(x)$$

$$+ P_n^0\left(z_2(x) - \dot{\mathbf{F}}(x,0)\mathbf{F}^{-1}(x,0)z_{n1}(x)\right). \tag{114}$$

Here P_n^1 is the projection in $V(0)$ onto the subspace $V_n(0)$ and P_n^0 is the projection in $L^2(\Omega)$ onto $V_n(0)$.

Equation (113) is a linear system of ordinary integro-differential equations for the coefficients $a_{n\kappa}(t)$, which can be solved by standard methods. Hence a solution to (113), (114) always exists. To obtain an energy estimate for this solution, we set

$$\omega = \mathbf{F}^{-1}\left(\ddot{z}_n - \dot{\mathbf{F}}\mathbf{F}^{-1}\dot{z}_n\right) = \sum_{\kappa=1}^{n} \dot{a}_{n\kappa}(t)\xi_\kappa \tag{115}$$

and integrate from 0 to t. This yields

$$\frac{1}{2}\int_\Omega \left\{ |\ddot{z}_n(x,t)|^2 + A_{ij}^{\alpha\beta}\frac{\partial(\dot{z}_n)^i}{\partial x^\alpha}\frac{\partial(\dot{z}_n)^j}{\partial x^\beta}\right\}\, dx$$

$$= \frac{1}{2}\|z_{n2}\|^2_{L^2(\Omega)} + \langle \dot{\mathbf{F}}(t)\mathbf{F}^{-1}(t)\dot{z}_n(t), \ddot{z}_n(t)\rangle - \langle \dot{\mathbf{F}}(0)\mathbf{F}^{-1}(0)z_{n1}, z_{n2}\rangle$$

$$- \int_0^t \langle \ddot{\mathbf{F}}(\tau)\mathbf{F}^{-1}(\tau)\dot{z}_n(\tau) - \dot{\mathbf{F}}(\tau)\mathbf{F}^{-1}(\tau)\dot{\mathbf{F}}(\tau)\mathbf{F}^{-1}(\tau)\dot{z}_n(\tau)$$

$$+ \dot{\mathbf{F}}(\tau)\mathbf{F}^{-1}(\tau)\ddot{z}_n(\tau), \ddot{z}_n(\tau)\rangle\, d\tau$$

113

$$+\frac{1}{2}\langle\frac{\partial(z_{n1})^i}{\partial x^\alpha}, A_{ij}^{\alpha\beta}(\cdot,0)\frac{\partial(z_{n1})^j}{\partial x^\beta}\rangle + \frac{1}{2}\int_0^t\langle\frac{\partial(\dot z_n)^i}{\partial x^\alpha}, \dot A_{ij}^{\alpha\beta}\frac{\partial(\dot z_n)^j}{\partial x^\beta}\rangle\,d\tau$$

$$-\int_0^t\langle(\ddot z_n)^i, \left(\frac{\partial}{\partial x^\alpha}A_{ij}^{\alpha\beta}\right)\frac{\partial(\dot z_n)^j}{\partial x^\beta}\rangle\,d\tau$$

$$+\int_0^t\langle\frac{\partial}{\partial x^\alpha}\left(A_{ij}^{\alpha\beta}\dot F_\gamma^i\chi_k^\gamma(\dot z_n)^k\right), \frac{\partial(\dot z_n)^j}{\partial x^\beta}\rangle\,d\tau$$

$$+\int_0^t\langle(\ddot z_n)^i - \dot F_\gamma^i\chi_k^\gamma(\dot z_n)^k, -\dot\chi_i^\alpha\frac{\partial\tilde\phi_n}{\partial x^\alpha} + \dot A_{ij}^{\alpha\beta}\frac{\partial^2(\tilde z_n)^j}{\partial x^\alpha\partial x^\beta} + \dot G^i\rangle\,d\tau. \tag{116}$$

It follows from (116) and a standard argument that we have uniform bounds on the norms of $\ddot z_n$ in $L^\infty([0,T];L^2(\Omega))$ and $\dot z_n$ in $L^\infty([0,T];H^1(\Omega))$. After passing to a subsequence, we may therefore assume weak-$*$ convergence of z_n and $\dot z_n$ in $L^\infty([0,T];H^1(\Omega))$ and of $\ddot z_n$ in $L^\infty([0,T];L^2(\Omega))$. An argument along the lines of Chapter III, §1 in [T2] can be used to show that the limit z is in fact a solution of (109) and $(102)_2$. Obviously, $\ddot z \in L^\infty([0,T];L^2(\Omega))$ and $\dot z \in L^\infty([0,T];H^1(\Omega))$. Moreover, it can be shown that $\ddot z$ is weakly continuous into $L^2(\Omega)$ (and in that sense assumes the given initial data), that ϕ is in $L^\infty([0,T];L^2(\Omega))$, and that

$$\frac{\partial^3 z^i}{\partial t^3} + \chi_i^\alpha\frac{\partial\dot\phi}{\partial x^\alpha}$$

lies in $L^\infty([0,T];H^{-1}(\Omega))$. The latter need not be true for each of the two terms separately. The reason for this is that the Hodge decomposition cannot be applied in H^{-1}. In this respect the initial-boundary value problem differs from the problem on all of space.

To see that $\ddot z$ is weakly continuous, we first note that $\langle\omega, F^T\ddot z\rangle$ is continuous for every solenoidal test function ω. Therefore, $PF^T\ddot z$ is weakly continuous, where P denotes the Hodge projection in $L^2(\Omega)$. If we put $w := F^{-1}\ddot z$, then it follows from $(107)_2$ that

$$w - F^{-1}\dot F F^{-1}\dot z = P(w - F^{-1}\dot F F^{-1}\dot z), \tag{117}$$

and hence $w - Pw$ is continuous in $L^2(\Omega)$ as a function of time. As a consequence, $PF^T F Pw$ is weakly continuous, and since the operator $PF^T F P$ is invertible on $PL^2(\Omega)$, Pw is weakly continuous. It therefore follows that w and hence also $\ddot z$ is weakly continuous from $[0,T]$ to $L^2(\Omega)$.

In order to show that we actually obtain a solution of (102), we have to prove that $\tilde\phi = \phi$ and $\tilde z = z$. By integrating (109) with respect to time, we find

$$\ddot z^i = -\chi_i^\alpha\frac{\partial\phi}{\partial x^\alpha} + A_{ij}^{\alpha\beta}\frac{\partial^2 z^j}{\partial x^\alpha\partial x^\beta} + G^i$$

$$+\int_0^t -\dot\chi_i^\alpha\left(\frac{\partial\tilde\phi}{\partial x^\alpha} - \frac{\partial\phi}{\partial x^\alpha}\right) + \dot A_{ij}^{\alpha\beta}\left(\frac{\partial^2\tilde z^j}{\partial x^\alpha\partial x^\beta} - \frac{\partial^2 z^j}{\partial x^\alpha\partial x^\beta}\right)\,d\tau. \tag{118}$$

In view of (110), we can rewrite (118) as

$$-\chi_i^\alpha \left(\frac{\partial \phi}{\partial x^\alpha} - \frac{\partial \tilde{\phi}}{\partial x^\alpha}\right) + A_{ij}^{\alpha\beta} \left(\frac{\partial^2 z^j}{\partial x^\alpha \partial x^\beta} - \frac{\partial^2 \tilde{z}^j}{\partial x^\alpha \partial x^\beta}\right) - \lambda(z^i - \tilde{z}^i)$$

$$+ \int_0^t -\dot{\chi}_i^\alpha \left(\frac{\partial \tilde{\phi}}{\partial x^\alpha} - \frac{\partial \phi}{\partial x^\alpha}\right) + \dot{A}_{ij}^{\alpha\beta} \left(\frac{\partial^2 \tilde{z}^j}{\partial x^\alpha \partial x^\beta} - \frac{\partial^2 z^j}{\partial x^\alpha \partial x^\beta}\right) \, d\tau = 0. \tag{119}$$

Similarly, by integrating $(102)_2$ and using $(110)_2$ we obtain

$$\chi_i^\alpha \left(\frac{\partial \tilde{z}^i}{\partial x^\alpha} - \frac{\partial z^i}{\partial x^\alpha}\right) = \int_0^t \dot{\chi}_i^\alpha \left(\frac{\partial \tilde{z}^i}{\partial x^\alpha} - \frac{\partial z^i}{\partial x^\alpha}\right) \, d\tau. \tag{120}$$

It follows from the elliptic uniqueness result established in the previous section and a straightforward perturbation argument that $\tilde{z} = z$ and $\tilde{\phi} = \phi$.

It remains to be shown that $\ddot{z}$ is actually strongly continuous into $L^2(\Omega)$ and that $\dot{z}$ is strongly continuous into $H^1(\Omega)$. For this purpose, we first rewrite (109) in a different form by first multiplying the equation by $\mathbf{F}^T$ (this transforms the ϕ-term into $\nabla\dot{\phi}$) and then substituting $\dot{z} = \mathbf{F}\tilde{w}$ (this transforms the incompressibility condition into div $\tilde{w} = 0$). We can then apply the technique on pp. 276-279 of [L6], which employs mollifiers with respect to time. This is why it is important to write the incompressibility constraint in a form which persists under such mollification.

Finally, to obtain the a priori estimate (106), we proceed along the lines of the derivation of (3.15) in [D5]. The basic idea is that by virtue of the weak-* lower semicontinuity of L^∞-type norms, bounds for the approximate solutions are inherited by the actual solution. See Section 3 of [D5] for the details in a similar situation. Of course, in the derivation of (106), we must account for the fact that we have subtracted the reference function $\int_0^t H(\cdot, \tau) \, d\tau$ from z. ∎

The proof that the iteration converges is based on a contraction argument similar to that given in Section 3. We omit the details and merely state the analogues of Lemmas III.8 and III.9. The iteration involves the quadruplet of functions (y, z, p, ϕ). Let $T > 0$ and $M > 0$ be given. We denote by $\|y\|_{k,l}$ the norm of y in $W^{k,\infty}([0,T]; H^l(\Omega))$. By $Z(T, M)$ we denote the set of all (y, z, p, ϕ) which satisfy the following conditions:

$$\|y\|_{0,4}^2 + \|y\|_{1,3}^2 + \|y\|_{2,2}^2 \le M^2, \tag{121$_1$}$$

$$\|p\|_{0,3}^2 + \|p\|_{1,2}^2 + \|p\|_{2,1}^2 \le M^2, \tag{121$_2$}$$

$$\|z\|_{0,2}^2 + \|z\|_{1,1}^2 + \|z\|_{2,0}^2 \le M^2, \tag{121$_3$}$$

$$\|\phi\|_{0,1} \le M, \tag{121$_4$}$$

$$z(x,0) = z_0(x), \tag{122$_1$}$$

$$y(x,0) = y_0(x), \tag{122$_2$}$$

115

$$\det \nabla y = 1, \tag{123}$$

$$y = x \text{ on } \partial\Omega, \tag{124}_1$$

$$z = 0 \text{ on } \partial\Omega, \tag{124}_2$$

Clearly the set $Z(T, M)$ contains $(y_0, z_0, 0, 0)$ (provided M is chosen large enough), and hence it is not empty. If $(y, z, p, \phi) \in Z(T, M)$, we have

$$v = \dot{y} \in L^\infty\left([0, T]; H^3(\Omega) \cap H_0^1(\Omega)\right) \cap W^{1,\infty}\left([0, T]; H^2(\Omega) \cap H_0^1(\Omega)\right)$$

$$\subset L^2([0, T]; H^3(\Omega) \cap H_0^1(\Omega)) \cap H^1([0, T]; H^2(\Omega) \cap H_0^1(\Omega)) =: X_1. \tag{125}_1$$

On the other hand we have

$$y_1(x) + \int_0^t z(x, \tau) + \lambda(y(x, \tau) - x)\ d\tau$$

$$\in W^{1,\infty}\left([0, T]; H^2(\Omega) \cap H_0^1(\Omega)\right) \cap W^{2,\infty}\left([0, T]; H_0^1(\Omega)\right)$$

$$\subset H^1([0, T]; H^2(\Omega) \cap H_0^1(\Omega)) \cap H^2([0, T]; H_0^1(\Omega)) =: X_2. \tag{125}_2$$

We now choose Π in (99) to be the orthogonal projection of the Hilbert space $X_1 \times X_2$ onto the diagonal $X_1 \cap X_2$. Let Σ be the mapping that takes $(y_{(n)}, z_{(n)}, p_{(n)}, \phi_{(n)})$ to $(y_{(n+1)}, z_{(n+1)}, p_{(n+1)}, \phi_{(n+1)})$ by the procedure outlined below Theorem III.14. It follows from the results above that Σ is well defined if M is large enough relative to the initial data, T is small enough relative to M, and λ is sufficiently large. The reason why T must be small relative to M is to ensure that $\|z_{n+1}(\cdot, t) - z_0\|_1$ remains small enough so that Lemma III.17 can be used to solve (101).

On $Z(T, M)$, we define the following pseudometric:

$$d((y, z, p, \phi), (\hat{y}, \hat{z}, \hat{p}, \hat{\phi})) = \|y - \hat{y}\|_{0,3} + \|y - \hat{y}\|_{1,2} + \|p - \hat{p}\|_{0,2} + \|p - \hat{p}\|_{1,1}$$

$$+ \|z - \hat{z}\|_{0,1} + \|z - \hat{z}\|_{1,0}. \tag{126}$$

The next two lemmas guarantee convergence of the iteration.

Lemma III.19:

If M is chosen sufficiently large and T sufficiently small relative to M, then Σ maps the set $Z(T, M)$ into itself.

Lemma III.20:

If M is chosen sufficiently large, and T sufficiently small relative to M, then the mapping $\Sigma\colon Z(T, M) \to Z(T, M)$ is a contraction with respect to the pseudometric d.

We note that the parameter λ appearing in the iteration can be controlled for M large and T sufficiently small relative to M.

It follows immediately from Lemmas III.19 and III.20 that $y_{(n)}$, $z_{(n)}$ and $p_{(n)}$ converge in the sense described by d. We can then conclude directly from $(100)_1$, $(100)_5$ that $\phi_{(n)}$ converges in the sense of distributions and consequently in the weak-$*$ topology of $L^\infty([0,T]; H^1(\Omega))$. Let (y, z, p, ϕ) be the limit of $(y_{(n)}, z_{(n)}, p_{(n)}, \phi_{(n)})$ and let $\tilde{v}$ be the limit of $\tilde{v}_{(n)}$. It is obvious that (y, z, p, ϕ) satisfies (100) and (101) (with the indices n and $n+1$ left out). To verify that we actually have a solution of the original problem (76), it remains to be checked that $\ddot{y} = z + \lambda(y - x)$ (which implies that $\tilde{v} = v$ and $\tilde{p} = \phi$. For this purpose, we first note that by differentiating $(101)_1$ twice with respect to time we obtain $(100)_1$ with z replaced by $\ddot{y} - \lambda(y - x)$ and ϕ replaced by $\ddot{p}$ on the right-hand side. Next we integrate $(100)_2$ with respect to time. This yields

$$\frac{\partial x^\alpha}{\partial y^i} \frac{\partial z^i}{\partial x^\alpha} = \frac{\partial x^\alpha}{\partial y_0^i} \frac{\partial z_0^i}{\partial x^\alpha}$$

$$+ \int_0^t -\frac{\partial x^\alpha}{\partial y^j} \frac{\partial v^j}{\partial x^\beta} \frac{\partial x^\beta}{\partial y^i} \frac{\partial z^i}{\partial x^\alpha} + \frac{\partial x^\alpha}{\partial y^i} \frac{\partial}{\partial x^\alpha} H^i(\tilde{v}, \nabla \tilde{v}, y, \nabla y, z, x) \, d\tau. \qquad (127)$$

By differentiating $(101)_2$ twice with respect to time, we obtain the same equation with z replaced by $\ddot{y} - \lambda(y - x)$ and $\tilde{v}$ replaced by v. The next lemma yields an estimate for $\tilde{v} - v$ in terms of $z - \ddot{y} + \lambda(y - x)$.

Lemma III.21:

Let X_1, X_2 and Π be defined as in (125) above. Then there is a constant C such that for every $v_1 \in X_1$ and $v_2 \in X_2$ we have

$$\|v_1 - \Pi(v_1, v_2)\|_{X_1} + \|v_2 - \Pi(v_1, v_2)\|_{X_2} \leq C \|v_1 - v_2\|_{H^1([0,T]; H^2(\Omega) \cap H_0^1(\Omega))}. \qquad (128)$$

For the proofs of Lemmas III.19-21, we refer to [H15]. Lemma III.21 yields

$$\|\tilde{v} - v\|_{L^2([0,T]; H^3(\Omega))} \leq C \|\ddot{y} - z - \lambda(y - x)\|_{L^2([0,T]; H^2(\Omega))}, \qquad (129)$$

and in conjunction with the uniqueness statement in Lemma III.16 and a standard perturbation argument, this implies that $z = \ddot{y} - \lambda(y - x)$ and $\phi = \ddot{p}$ as claimed.

Finally, the uniqueness of the solution of (76) follows from a straightforward energy estimate.

5. A model problem with a singular memory function

As remarked earlier, there are theoretical and experimental indications that certain viscoelastic materials may be modelled by equations with singular memory functions. In this section we study the one-dimensional model problem

$$u_{tt}(x,t) = \phi(u_x(x,t))_x - \int_{-\infty}^{t} m(t-\tau)\psi(u_x(x,\tau))_x \, d\tau + f(x,t), \quad x \in [0,1], \; t \geq 0,$$

$$\tag{$130)_1$}$$

$$u(0,t) = u(1,t) = 0, \tag{$130)_2$}$$

$$u(x,t) = \tilde{u}(x,t), \; x \in [0,1], \; t \leq 0, \tag{$130)_3$}$$

$$u(x,0+) = \tilde{u}(x,0-), \; u_t(x,0+) = \tilde{u}_t(x,0-) \tag{$130)_4$}$$

under hypotheses which permit m to have an integrable singularity at 0. We follow the presentation of [H13]. We shall also discuss a related problem involving a stronger singularity in the memory function [H16]. In contrast to Section 3, the results presented here depend in a more essential way on the special form of the equations considered.

To simplify the exposition, we assume that the given history $\tilde{u}$ satisfies equation $(130)_1$ and the boundary conditions $(130)_2$ for $t \leq 0$ and that $(130)_4$ holds. This guarantees that all of the appropriate compatibility conditions are automatically satisfied. For problems with singular kernels, stronger compatibility conditions are needed than those required in cases with bounded smooth kernels, in order to avoid singularities in the solution as $t \to 0+$ which reflect that of the kernel. There is no serious loss of generality in assuming that $\tilde{u}$ satisfies the equation of motion for $t \leq 0$ because if the relevant compatibility conditions hold at $t = 0$ we can define f on $[0,1] \times (-\infty, 0]$ in such a way that the modified f is smooth across $t = 0$ and $\tilde{u}$ satisfies $(130)_1$ for $t \leq 0$.

The iteration scheme of Section 3 is insensitive to sign conditions on m and ψ. It is therefore clear that this scheme will not be useful if m is singular. Indeed, we have seen in Section II.3 that singular kernels lead to smoothing of solutions, at least in the linear case. Reversing the sign of m would create an instantaneous loss of regularity (like with the backward heat equation) and local existence would not hold (except possibly in a space of analytic functions). We must therefore use an iteration which involves the memory term in a more essential way. This will lead us to the study of linear integrodifferential equations with variable coefficients and singular kernels.

We assume that m is integrable at infinity and introduce a new kernel a through the equation

$$a(s) := \int_{s}^{\infty} m(\tau) \, d\tau, \; s > 0. \tag{131}$$

We make the following assumptions:
(s1) $\phi, \psi \in C^3(\mathbb{R})$.

118

2) For some $T > 0$, we have

$$f, \, f_x, \, f_t \in L^\infty\left((-\infty, T]; L^2(0,1)\right) \cap L^2\left((-\infty, T]; L^2(0,1)\right),$$

$$f_{tt} \in L^2\left((-\infty, T]; L^2(0,1)\right). \tag{132}$$

3) All derivatives of $\tilde{u}$ up to third order lie in $L^\infty\left((-\infty, T]; L^2(0,1)\right)$ $\cap L^2\left((-\infty, T]; L^2(0,1)\right)$. Moreover, $\tilde{u}_{xxtt}$ lies in $L^2\left((-\infty, T]; L^2(0,1)\right)$.

e) $\phi' > 0$, $\psi' > 0$.

a) The kernel a satisfies a, $a' \in L^1(0,\infty)$, $a \geq 0$, $a' = -m \leq 0$, $a'' \geq 0$, and the measure a'' has a nontrivial absolutely continuous component.

c) The history $\tilde{u}$ satisfies the equation of motion $(130)_1$ and the boundary conditions $(130)_2$.

Theorem III.22:

Assume that (s1)-(s3), (e), (a) and (c) hold. Then there exists $T' \in (0, T]$ such that the initial-history value problem (130) has a unique solution u such that all derivatives of u through order three belong to $L^\infty\left((-\infty, T']; L^2(0,1)\right)$. Moreover, the solution can be continued as long as the norms of these derivatives remain finite and the forcing function satisfies (132).

Remarks III.23:

1. A solution u with the regularity given by Theorem III.22 is of class C^2 and is therefore a classical solution.

2. A similar result holds for Neumann boundary conditions, mixed boundary conditions or the initial value problem on all of $\mathbb{R}$.

3. It is shown in [H13] that if the kernel a satisfies the "(A2)-condition"

$$\sup_I \left(\frac{1}{|I|} \int_I (1 + \omega^2 \, \mathrm{Re}\, \hat{a}(i\omega)) \, d\omega\right) \left(\frac{1}{|I|} \int_I \frac{1}{1 + \omega^2 \, \mathrm{Re}\, \hat{a}(i\omega)} \, d\omega\right) < \infty, \tag{133}$$

where the sup is taken over all intervals $I \subset \mathbb{R}$, and in addition $f_x \in C([0,T]; L^2(0,1))$, then the third order derivatives of u are in $C([0,T']; L^2(0,1))$. Here $\hat{a}$ denotes the Laplace transform of a. We note that (133) holds if $\widehat{a'}(i\omega) \sim \omega^{-\delta}$ as $\omega \to \infty$ with $\delta \in (0,1]$; in particular, this is the case (with $\delta = 1$) if a' is smooth on $[0, \infty)$.

4. The above theorem is of course applicable to bounded smooth kernels. The results of Section II.3 suggest that singular kernels should lead to improved regularity of the solution. In fact, one of the terms appearing in the energy estimates indicates a slight improvement in regularity if $a'(0) = -\infty$. However, comparison with the linear case indicates that this regularity is not optimal.

5. We note that the proof given below can be simplified considerably if it is assumed that $\phi' - a(0)\psi' > 0$.

The proof of Theorem III.22 is based on the following iteration scheme. Fc functions w in an appropriate class, we replace $(130)_1$ with the linear integrodi ferential equation

$$u_{tt}(x,t) = \phi'(w_x(x,t))u_{xx}(x,t) + \int_{-\infty}^{t} a'(t-\tau)\psi'(w_x(x,\tau))u_{xx}(x,\tau)\, d\tau$$

$$+f(x,t), \quad x \in [0,1], \ t \in [0,T]. \tag{134}$$

We want to show that the mapping S which carries w into the solution u of (134 $(130)_2$-$(130)_4$ has a fixed point. The first step is to prove an existence theorem for the linear history value problem

$$u_{tt}(x,t) = \alpha(x,t)u_{xx}(x,t) + \int_{-\infty}^{t} a'(t-\tau)\beta(x,\tau)u_{xx}(x,\tau)\, d\tau + f(x,t),$$

$$x \in [0,1], \ t \in [0,T], \tag{135}$$

$$u(0,t) = u(1,t) = 0, \tag{135}$$

$$u(x,t) = \tilde{u}(x,t), \ t \le 0, \tag{135}$$

$$u(x,0+) = \tilde{u}(x,0-), \ u_t(x,0+) = \tilde{u}_t(x,0-). \tag{135}$$

We now make the following assumptions:
(s1′) All derivatives of α and β through second order lie in $L^\infty\left((-\infty,T]; L^2(0,1)\right)$
(e′) $\alpha(x,t) \ge \underline{\alpha} > 0$, $\beta(x,t) \ge \underline{\beta} \ge 0$ for all x and t.
(c′) The history $\tilde{u}$ satisfies $(135)_1$, $(135)_2$.

The general form of the a priori estimate for solutions to (135) is quite com plicated. In our iteration scheme, we will change only the coefficients α and and these will be held fixed on $(-\infty,0]$. In the statement of our a priori estimate we therefore regard f, $\tilde{u}$ and a as fixed once and for all, and we also regard α an β on $(-\infty,0]$ and the constants $\underline{\alpha}$ and $\underline{\beta}$ as fixed. Let us set

$$\Gamma := \text{ess} - \sup_{s \in [0,T]} \int_0^1 \left(\alpha^2 + \alpha_x^2 + \alpha_t^2 + \alpha_{xx}^2 + \alpha_{xt}^2 + \alpha_{tt}^2 \right.$$

$$\left. +\beta^2 + \beta_x^2 + \beta_t^2 + \beta_{xx}^2 + \beta_{xt}^2 + \beta_{tt}^2 \right)(x,s)\, dx, \tag{136}$$

and

$$\mathcal{E}[u](t) := \text{ess} - \sup_{s \in [0,t]} \int_0^1 \left(u_{xxx}^2 + u_{xxt}^2 + u_{xtt}^2 + u_{ttt}^2 \right)(x,s)\, dx. \tag{137}$$

120

Lemma III.24:

Let (s1′), (s2), (s3), (e′), (a) and (c′) hold. Then (135) has a unique solution with all derivatives through order three belonging to $L^\infty((-\infty, T]; L^2(0,1))$, and the following a priori estimate holds:

$$\mathcal{E}[u](T) \leq C(1 + \Gamma \cdot T)\exp\big(C(1+\Gamma)(T + T^3)\big). \tag{138}$$

For the proof of Lemma III.24 it will be convenient to regularize the kernel. For each $\delta > 0$, we define $a_\delta : [0, \infty) \to \mathbb{R}$ by

$$a_\delta(s) = \int_{-\delta}^{\delta} J_\delta(\tau) a(s + \delta - \tau)\, d\tau \tag{139}$$

here J_δ is a standard mollifier with support contained in $(-\frac{\delta}{2}, \frac{\delta}{2})$. It follows om (a) and (139) that

$$a_\delta \in C^\infty[0,\infty), \; a_\delta \geq 0, \; a_\delta' \leq 0, \; a_\delta'' \geq 0, \; a_\delta, \; a_\delta' \in L^1(0,\infty),$$

$$\|a_\delta\|_{L^1} \leq \|a\|_{L^1}, \; \|a_\delta'\|_{L^1} \leq \|a'\|_{L^1} = a(0). \tag{140}$$

Moreover, $a_\delta \to a$ pointwise (and in L^1) as $\delta \to 0$. The use of regularized kernels not essential, but it appears to be the easiest way of constructing approximate problems which are known to have solutions.

For $T \in \mathbb{R}$, $h > 0$, $t \in (-\infty, T]$ and $\Phi \in L^2((-\infty, T]; L^2(0,1))$, we employ the notations

$$\Delta_h u(\cdot, t) := u(\cdot, t) - u(\cdot, t - h), \tag{141}$$

nd

$$Q(\Phi, t, a_\delta) := \int_{-\infty}^{t} \int_{0}^{1} \Phi(x,s) \int_{-\infty}^{s} a_\delta(s - \tau)\Phi(x,\tau)\, d\tau\, dx\, ds. \tag{142}$$

The following result is proved in [H13].

Lemma III.25:

Assume that (a) holds and let $\varepsilon > 0$ be given. Then, there are constants $C(\varepsilon), \delta_0(\varepsilon) > 0$ with the following property: If $\delta \in (0, \delta_0(\varepsilon)]$, $T \in \mathbb{R}$, $\Phi \in L^2((-\infty, T]; L^2(0,1))$, and $\lim_{h\downarrow 0}\frac{1}{h^2}Q(\Delta_h \Phi, t, a_\delta)$ exists for a.e. $t \in (-\infty, T]$, then

$$\lim_{h\downarrow 0}\frac{1}{h^2}Q(\Delta_h \Phi, t, a_\delta) \geq \big(\frac{1}{2}a_\delta(0) - \varepsilon\big)\|\Phi(t)\|^2 - C(\varepsilon)\int_{-\infty}^{t}\|\Phi(s)\|^2\, ds$$

$$\text{a.e. } t \in (-\infty, T], \tag{143}$$

where $\|\cdot\|$ denotes the norm in $L^2(0,1)$.

121

Proof of Lemma III.24:

Consider the approximating problems

$$u_{tt}^{(\delta)}(x,t) = \alpha(x,t)u_{xx}^{(\delta)}(x,t) + \int_{-\infty}^{t} a_{\delta}'(t-\tau)\beta(x,\tau)u_{xx}^{(\delta)}(x,\tau)\ d\tau + f^{(\delta)}(x,t),$$

$$x \in [0,1],\ t \in (-\infty, T], \tag{144}$$

$$u^{(\delta)}(0,t) = u^{(\delta)}(1,t) = 0,\ t \in (-\infty, T], \tag{144}$$

$$u^{(\delta)}(x,t) = \tilde{u}(x,t),\ x \in [0,1],\ t \in (-\infty, 0], \tag{144}$$

$$u^{(\delta)}(x,0+) = \tilde{u}(x,0-),\ u_t^{(\delta)}(x,0+) = \tilde{u}_t(x,0-), \tag{144}$$

for $\delta > 0$, where a_{δ} is defined by (139) and $f^{(\delta)}$ approximates f in such way that $\tilde{u}$ satisfies equation $(144)_1$ for $t \leq 0$, and $f^{(\delta)}, f_x^{(\delta)}, f_t^{(\delta)} \to f, f_x$, in $L^{\infty}((-\infty,T];L^2(0,1)) \cap L^2((-\infty,T];L^2(0,1))$, $f_{tt}^{(\delta)} \to f_{tt}$ in $L^2((-\infty,T$ $L^2(0,1))$ as $\delta \downarrow 0$. [The existence of such an approximation to f follows from our assumptions on f and $\tilde{u}$ and a straightforward extension theorem. It is her that the assumption $\tilde{u}_{xxtt} \in L^2((-\infty,0];L^2(0,1))$ is used.] It follows from stan dard theory for equations with regular kernels that for each $\delta > 0$, (144) ha a unique solution $u^{(\delta)}$ with $u^{(\delta)}, u_x^{(\delta)}, u_t^{(\delta)}, u_{xx}^{(\delta)}, u_{xt}^{(\delta)}, u_{tt}^{(\delta)}, u_{xxx}^{(\delta)}, u_{xxt}^{(\delta)}, u_{xtt}^{(\delta)}, u_{ttt}^{(\delta)}$ $L^{\infty}((-\infty,T];L^2(0,1))$.

Our objective is to show that $u^{(\delta)}$ obeys certain a priori bounds, uniforml in δ, that imply the existence of a sequence $\{u^{(\delta_n)}\}_{n=1}^{\infty}$ which converges to solution as $\delta_n \downarrow 0$. In order to simplify the notation, we suppress the superscript on $u^{(\delta)}$, and $f^{(\delta)}$. For the purpose of deriving such bounds, we set

$$V := \mathrm{ess} - \sup_{s \in (-\infty,0]} \int_0^1 \{\tilde{u}_x^2 + \tilde{u}_{xx}^2 + \tilde{u}_{xt}^2 + \tilde{u}_{xxx}^2 + \tilde{u}_{xxt}^2 + \tilde{u}_{xtt}^2 + \tilde{u}_{ttt}^2\}(x,s)\ dx$$

$$+ \int_{-\infty}^{0} \int_0^1 \{\tilde{u}_{xx}^2 + \tilde{u}_{xxx}^2 + \tilde{u}_{xxt}^2 + \tilde{u}_{xtt}^2 + \tilde{u}_{ttt}^2\}(x,s)\ dx\ ds, \tag{145}$$

$$F := \mathrm{ess} - \sup_{s \in (-\infty,T]} \int_0^1 \{f_x^2 + f_t^2\}(x,s)\ dx$$

$$+ \int_{-\infty}^{T} \int_0^1 \{f_x^2 + f_t^2 + f_{tt}^2\}(x,s)\ dx\ ds \tag{146}$$

and

$$\Gamma_+ := 2\Gamma,$$

$$\Gamma_- := 2\,\mathrm{ess} - \sup_{s \in (-\infty,0]} \int_0^1 \{\,\alpha^2 + \alpha_x^2 + \alpha_t^2 + \alpha_{xx}^2 + \alpha_{xt}^2 + \alpha_{tt}^2$$

$$+ \beta^2 + \beta_x^2 + \beta_t^2 + \beta_{xx}^2 + \beta_{xt}^2 + \beta_{tt}^2\,\}(x,s)\ dx, \tag{147}$$

122

nd we observe that there exists a constant $\underline{\lambda} > 0$ such that

$$\frac{\alpha(x,t)}{\beta(x,t)} \geq \underline{\lambda} \quad \forall x \in [0,1], \ t \in (-\infty, T], \tag{148}$$

y virtue of $(s1')$ and (e').

An integration by parts in $(144)_1$ yields

$$u_{tt} = \gamma^{(\delta)} u_{xx} + \int_{-\infty}^{t} a_\delta(t-\tau)[\beta u_{xx}]_t(x,\tau) \, d\tau + f, \tag{149}$$

here

$$\gamma^{(\delta)}(x,t) := \alpha(x,t) - a_\delta(0)\beta(x,t). \tag{150}$$

'e apply the backward difference operator Δ_h (in the time variable) to (149),
hus obtaining

$$\Delta_h u_{tt} = \Delta_h[\gamma^{(\delta)} u_{xx}] + \int_{-\infty}^{t} a_\delta(t-\tau)\Delta_h[(\beta u_{xx})_t](x,\tau) \, d\tau + \Delta_h f. \tag{151}$$

hen, we multiply (151) by $\Delta_h[(\beta u_{xx})_t]$ and integrate over $[0,1] \times (-\infty, t]$, $t \in$
$, T]$. After several integrations by parts, we divide by h^2 and let $h \downarrow 0$. The
utcome of this tedious, but straightforward, computation is

$$\frac{1}{2} \int_0^1 \left\{ \beta\gamma^{(\delta)} u_{xxt}^2 + \beta u_{xtt}^2 \right\}(x,t) \, dx + \lim_{h \downarrow 0} \frac{1}{h^2} Q(\Delta_h[(\beta u_{xx})_t], t, a_\delta)$$

$$+ \int_0^1 \left\{ \beta\gamma_t^{(\delta)} u_{xx} u_{xxt} + \beta f_t u_{xxt} \right\}(x,t) \, dx$$

$$= \int_{-\infty}^{t} \int_0^1 \left\{ \frac{3}{2}\beta\gamma_t^{(\delta)} u_{xxt}^2 - \frac{3}{2}\beta_t \gamma^{(\delta)} u_{xxt}^2 + \frac{1}{2}\beta_t u_{xtt}^2 \right.$$

$$- \beta_x u_{xtt} u_{ttt} + \beta\gamma_{tt}^{(\delta)} u_{xx} u_{xxt} + 2\beta_t u_{xxt} u_{ttt} + \beta_{tt} u_{xx} u_{ttt}$$

$$- \beta_{tt}\gamma^{(\delta)} u_{xx} u_{xxt} - \beta_t \gamma_t^{(\delta)} u_{xx} u_{xxt} - \beta_{tt}\gamma_t^{(\delta)} u_{xx}^2$$

$$\left. + \beta f_{tt} u_{xxt} - \beta_t f_t u_{xxt} - \beta_{tt} f_t u_{xx} \right\}(x,s) \, dx \, ds$$

$$a.e. \ t \in (-\infty, T], \tag{152}$$

here Q is defined by (142). [We note that $u_t, u_{tt}, \Delta_h u, \Delta_h u_t$, and $\Delta_h u_{tt}$ all
anish at $x = 0,1$ by virtue of $(144)_2$. All of the spatial integrations by parts
sed in the derivation of (152) were carried out in such a way that the boundary
rms (at $x = 0,1$) vanish.]

It is not a priori evident that $\lim_{h \downarrow 0} \dfrac{1}{h^2} Q(\Delta_h[(\beta u_{xx})_t], t, a_\delta)$ exists for a.e. $t \in$
$-\infty, T]$. However, all of the other limits involved in the derivation of (152) exist
r a.e. $t \in (-\infty, T]$, and consequently so does the limit in question.

123

Using Lemma III.25 (with ε sufficiently small relative to $\underline{\lambda}$), and the algebra inequality $|AB| \leq \eta A^2 + \frac{1}{4\eta}B^2 \; \forall \eta > 0$, we find that the left-hand side of (152) bounded from below by

$$\int_0^1 \left\{ \frac{1}{4}\underline{\lambda}\beta^2\, u_{xxt}^2 + \frac{1}{2}\underline{\beta}u_{xtt}^2 \right\}(x,t)\; dx$$

$$-C\int_0^1 \left\{ (\alpha_t^2 + \beta_t^2)u_{xx}^2 + f_t^2 \right\}(x,t)\; dx - C\int_{-\infty}^t \int_0^1 \left\{ \beta^2 u_{xxt}^2 + \beta_t^2 u_{xx}^2 \right\}(x,s)\; dx\; ds$$

$$\forall t \in (-\infty, t], \quad \delta \in (-\infty, \delta_0], \tag{15.}$$

where C is a positive constant (which depends on $\underline{\lambda}$ and $\underline{\beta}$).

Differentiating $(144)_1$ with respect to t and x, and splitting the convolutic integrals, we obtain

$$u_{ttt} = \alpha u_{xxt} + \alpha_t u_{xx} + f_t + \int_{-\infty}^0 a_\delta'(t-\tau)[\beta\tilde{u}_{xxt} + \beta_t\tilde{u}_{xx}](x,\tau)\; d\tau$$

$$+ \int_0^t a_\delta'(t-\tau)[\beta u_{xxt} + \beta_t u_{xx}](x,\tau)\; d\tau, \tag{154}$$

$$\alpha u_{xxx} + \int_0^t a_\delta'(t-\tau)[\beta u_{xxx}](x,\tau)\; d\tau = u_{xtt} - \alpha_x u_{xx} - f_x$$

$$- \int_{-\infty}^0 a_\delta'(t-\tau)[\beta\tilde{u}_{xxx} + \beta_x\tilde{u}_{xx}](x,\tau)\; d\tau$$

$$- \int_0^t a_\delta'(t-\tau)[\beta_x u_{xx}](x,\tau)\; d\tau. \tag{154}$$

It follows easily from $(154)_1$ that

$$\int_0^1 u_{ttt}^2(x,t)\; dx \leq 5\int_0^1 \left\{ \alpha^2 u_{xxt}^2 + \alpha_t^2 u_{xx}^2 + f_t^2 \right\}(x,t)\; dx$$

$$+ 10a(0)^2 \; \text{ess} - \sup_{s \in [0,t]} \int_0^2 \left\{ \beta^2 u_{xxt}^2 + \beta_t^2 u_{xx}^2 \right\}(x,s)\; dx$$

$$+ 10a(0)^2 \; \text{ess} - \sup_{s \in (-\infty,0]} \int_0^1 \left\{ \beta^2 \tilde{u}_{xxt}^2 + \beta_t^2 \tilde{u}_{xx}^2 \right\}(x,s)\; dx$$

$$\text{a.e. } t \in [0,T]. \tag{155}$$

Using Gronwall's inequality in $(154)_2$, we obtain, after a straightforward computation,

$$\int_0^1 [\alpha u_{xxx}^2]^2(x,t)\; dx$$

$$\leq 8\exp[2a(0)\underline{\lambda}^{-1}] \, \text{ess} - \sup_{s \in [0,t]} \int_0^1 \left\{ u_{xtt}^2 + \alpha_x^2 u_{xx}^2 + f_x^2 + a(0)^2 \beta_x^2 u_{xx}^2 \right\}(x,s)$$

$$+ 4a(0)^2 \exp[2a(0)\underline{\lambda}^{-1}] \, \text{ess} - \sup_{s \in (-\infty,0]} \int_0^1 \left\{ \beta^2 \tilde{u}_{xxx}^2 + \beta_x^2 \tilde{u}_{xx}^2 \right\}(x,s)\; dx$$

$$\text{a.e. } t \in [0,T]. \tag{156}$$

124

Combining (152), (155), (156), and recalling the lower bound (153), we conclude that there exists a positive constant K such that

$$E[u](t) \le K\{F + (1 + \Gamma_- + \Gamma_+ T)V\} + K \cdot (1 + \Gamma_+) \cdot (1 + T^2) \int_0^t E[u](s)\, ds$$

$$\forall t \in [0, T],\ \delta \in (0, \delta_0]. \tag{157}$$

The constant K depends on $\underline{\alpha}, \beta, \lambda$, and a, but is independent of $F, V, \Gamma_-, \Gamma_+, T$, nd δ.) Gronwall's inequality and (156) yield

$$E[u](T) \le K\{F + (1 + \Gamma_- + \Gamma_+ T)V\} \exp[K \cdot (1 + \Gamma_+) \cdot (T + T^3)] \tag{158}$$

or all $\delta \in (0, \delta_0]$.

To assist the reader in following the derivation of (157), we show the detailed stimation of a few typical terms. By the Sobolev embedding theorem, $\beta_x^2(x, t) \le \Gamma_-$ for all $x \in [0, 1], t \in (-\infty, 0]$, and $\beta_x^2(x, t) \le \Gamma_+$ for all $x \in [0, 1], t \in [0, T]$. Therefore,

$$\left| \int_{-\infty}^t \int_0^1 \beta_x u_{xtt} u_{ttt}(x, s)\, dx\, ds \right| \le \frac{1}{2} \int_{-\infty}^t \int_0^1 \{\beta_x^2 u_{xtt}^2 + u_{ttt}^2\}(x, s)\, dx\, ds$$

$$= \frac{1}{2} \int_{-\infty}^0 \int_0^1 \{\beta_x^2 \tilde{u}_{xtt}^2 + \tilde{u}_{ttt}^2\}(x, s)\, dx\, ds$$

$$+ \frac{1}{2} \int_0^t \int_0^1 \{\beta_x^2 u_{xtt}^2 + u_{ttt}^2\}(x, s)\, dx\, ds$$

$$\le \frac{1}{2}(\Gamma_- + 1)V + \frac{1}{2}(\Gamma_+ + 1) \int_0^t E[u](s)\, ds\ \forall t \in [0, T]. \tag{159}$$

Next, we observe that

$$\frac{1}{2} \max_{\xi \in [0, 1]} \tilde{u}_{xx}^2(\xi, s) \le \int_0^1 \{\tilde{u}_{xx}^2 + \tilde{u}_{xxx}^2\}(x, s)\, dx\ \forall s \in (-\infty, 0], \tag{160}$$

nd consequently

$$\frac{1}{2} \int_{-\infty}^0 \max_{\xi \in [0, 1]} \tilde{u}_{xx}^2(\xi, s)\, ds \le V. \tag{161}$$

n addition, we note that

$$u_{xx}(x, t) = \tilde{u}_{xx}(x, 0) + \int_0^t u_{xxt}(x, s)\, ds\ \forall x \in [0, 1],\ t \in [0, T], \tag{162}$$

rom which we easily deduce the estimates

$$\int_0^1 u_{xx}^2(x, t)\, dx \le 2 \int_0^1 \tilde{u}_{xx}^2(x, 0)\, dx + 2t \int_0^t \int_0^1 u_{xxt}^2(x, s)\, dx\, ds$$

$$\le 2V + 2T^2\, E[u](t)\ \forall t \in [0, T], \tag{163}$$

125

and

$$\frac{1}{2}\max_{x\in[0,1]} u_{xx}^2(x,t) \le 2V + (1+2T^2)E[u](t) \quad \forall t \in [0,T]. \tag{164}$$

Using (160) and (164), we find

$$\left| \int_{-\infty}^{t} \int_{0}^{1} \beta_{tt} u_{xx} u_{ttt}(x,s) \ dx \ ds \right| \le \frac{1}{2} \int_{-\infty}^{t} \int_{0}^{1} \{\beta_{tt}^2 u_{xx}^2 + u_{ttt}^2\}(x,s) \ dx \ ds$$

$$\le \frac{1}{2} \int_{-\infty}^{t} \max_{\xi\in[0,1]} u_{xx}^2(\xi,s) \int_{0}^{1} \beta_{tt}^2(x,s) \ dx \ ds + \frac{1}{2} \int_{-\infty}^{t} \int_{0}^{1} u_{ttt}^2(x,s) \ dx \ ds$$

$$\le \frac{1}{2} \int_{-\infty}^{0} \max_{\xi\in[0,1]} \tilde{u}_{xx}^2(\xi,s) \int_{0}^{1} \beta_{tt}^2(x,s) \ dx \ ds + \frac{1}{2} \int_{-\infty}^{0} \int_{0}^{1} \tilde{u}_{ttt}^2(x,s) \ dx \ ds$$

$$+ \frac{1}{2} \int_{0}^{t} \max_{\xi\in[0,1]} u_{xx}^2(\xi,s) \int_{0}^{1} \beta_{tt}^2(x,s) \ dx \ ds + \frac{1}{2} \int_{0}^{t} \int_{0}^{1} u_{ttt}^2(x,s) \ dx \ ds$$

$$\le \frac{1}{2}(\Gamma_- + 1)V + \Gamma_+ VT + \frac{1}{2}\Gamma_+ \cdot (1+2T^2) \int_{0}^{t} E[u](s) \ ds + \frac{1}{2} \int_{0}^{t} E[u](s) \ ds$$

$$\forall t \in [0,T]. \tag{165}$$

The other terms can all be handled in a similar manner.

We conclude from (158) that $u_{xxx}^{(\delta)}, u_{xxt}^{(\delta)}, u_{xtt}^{(\delta)}$, and $u_{ttt}^{(\delta)}$ are bounded in $L^\infty([0,T]; L^2(0,1))$ independently of $\delta \in (0,\delta_0]$. It follows from (163) (and similar inequalities for the other derivatives) that $u_{xx}^{(\delta)}, u_{xt}^{(\delta)}, u_{tt}^{(\delta)}, u_{x}^{(\delta)}, u_{t}^{(\delta)}$ and $u^{(\delta)}$ are also bounded in $L^\infty([0,T]; L^2(0,1))$ independently of $\delta \in (0,\delta_0]$. Therefore there exists a function $u: [0,1] \times (-\infty, T] \to \mathbb{R}$, with $u = \tilde{u}$ on $[0,1] \times (-\infty,0]$ and a sequence $\{\delta_n\}_{n=1}^{\infty}$, with $\delta_n \downarrow 0$ as $n \to \infty$, such that

$$u^{(\delta_n)}, u_x^{(\delta_n)}, u_t^{(\delta_n)}, u_{xx}^{(\delta_n)}, u_{xt}^{(\delta_n)}, u_{tt}^{(\delta_n)}, u_{xxx}^{(\delta_n)}, u_{xxt}^{(\delta_n)}, u_{xtt}^{(\delta_n)}, u_{ttt}^{(\delta_n)}$$

$$\to u, u_x, u_t, \text{ etc. weakly star in } L^\infty((-\infty, T]; L^2(0,1)) \tag{166}$$

as $n \to \infty$. Standard embedding theorems and (166) imply

$$u^{(\delta_n)}, u_x^{(\delta_n)}, u_t^{(\delta_n)}, u_{xx}^{(\delta_n)}, u_{xt}^{(\delta_n)}, u_{tt}^{(\delta_n)} \to u, u_x, u_t, u_{xx}, u_{xt}, u_{tt}$$

$$\text{uniformly on } [0,1] \times (-\infty, T] \tag{167}$$

as $n \to \infty$. It thus follows easily that u satisfies (135). The uniqueness claim follows by a standard argument. Finally, we note that the a priori estimate (158) also holds for the "exact" solution u. ∎

Proof of Theorem III.22:

For each $M, T > 0$, let $Z(T,M)$ denote the set of all functions $w: [0,1] \times (-\infty, T] \to \mathbb{R}$ such that

$$w, w_x, w_t, w_{xx}, w_{xt}, w_{tt}, w_{xxx}, w_{xxt}, w_{xtt}, w_{ttt} \in L^\infty((-\infty, T]; L^2(0,1)), \tag{168}$$

$$w(0,t) = w(1,t) = 0 \ \forall t \in (-\infty, T], \tag{168$_2$}$$

$$w(x,t) = \tilde{u}(x,t) \ \forall x \in [0,1], \ t \in (-\infty, 0], \tag{168$_3$}$$

and

$$\text{ess} - \sup_{t \in [0,T]} \int_0^1 \left\{ w_{xxx}^2 + w_{xxt}^2 + w_{xtt}^2 + w_{ttt}^2 \right\}(x,t) \ dx \leq M. \tag{168$_4$}$$

We note that $Z(T,M)$ is nonempty for M sufficiently large. Henceforth, we always make this assumption.

It follows from (e) that $\inf_{\xi \in \mathbb{R}} [\phi'(\xi)/\psi'(\xi)] \geq 0$. We temporarily make the stronger assumption

$$\underline{\phi} := \inf_{\xi \in \mathbb{R}} \phi'(\xi) > 0, \ \underline{\psi} := \inf_{\xi \in \mathbb{R}} \psi'(\xi) > 0, \ \underline{\nu} := \inf_{\xi \in \mathbb{R}} \frac{\phi'(\xi)}{\psi'(\xi)} > 0, \tag{169}$$

which will be removed later. Identifying α with $\phi'(w_x)$ and β with $\psi'(w_x)$, it follows immediately from Lemma III.24 that for $w \in Z(T,M)$ the history value problem

$$u_{tt}(x,t) = \phi'(w_x)u_{xx}(x,t) + \int_{-\infty}^t a'(t-\tau)\psi'(w_x)u_{xx}(x,\tau) \ d\tau + f(x,t),$$

$$x \in [0,1], \ t \in (-\infty, T], \tag{170}$$

has a unique solution u with $u, u_x, u_t, u_{xx}, u_{xt}, u_{tt}, u_{xxx}, u_{xxt}, u_{xtt}, u_{ttt} \in L^\infty((-\infty, T]; L^2(0,1))$. Moreover, the corresponding α, β, and λ can be chosen independently of M and T.

Let S denote the mapping which carries w into the solution of (170).

Our goal is to show that, for appropriately chosen M and T, S has a unique fixed point in $Z(T,M)$ which is obviously a solution of (130). For this purpose, we employ the contraction mapping principle and the complete metric d given by

$$d(w,\bar{w})^2 := \max_{t \in [0,T]} \left(\int_0^1 \left\{ (w_{xx} - \bar{w}_{xx})^2 + (w_{xt} - \bar{w}_{xt})^2 + (w_{tt} - \bar{w}_{tt})^2 \right\}(x,t) \ dx \right). \tag{171}$$

Observe that for $w \in Z(T,M)$ we have

$$w_{xx}(x,t) = \tilde{u}_{xx}(x,0) + \int_0^t w_{xxt}(x,s) \ ds \ \forall x \in [0,1], \ t \in [0,T]. \tag{172}$$

Therefore,

$$\int_0^1 w_{xx}^2(x,t) \ dx \leq 2 \int_0^1 \tilde{u}_{xx}^2(x,0) \ dx + 2t \int_0^t \int_0^1 w_{xxt}^2(x,s) \ dx \ ds$$

$$\leq 2V + 2Mt^2 \ \forall t \in [0,T], \tag{173}$$

127

where V is defined by (145), and so clearly

$$\sup_{t \in [0,T]} \int_0^1 w_{xx}^2(x,t) \, dx \le 2V + 2MT^2 \quad \forall w \in Z(T,M). \tag{174}$$

Similarly, the following inequalities hold for all $w \in Z(T,M)$

$$\sup_{t \in [0,T]} \int_0^1 w_{xt}^2(x,t) \, dx \le 2V + 2MT^2, \tag{175}_1$$

$$\sup_{\substack{x \in [0,1] \\ t \in [0,T]}} w_{xx}^2(x,t) \le 4V + 2(1 + 2T^2)M, \tag{175}_2$$

$$\sup_{\substack{x \in [0,1] \\ t \in [0,T]}} w_{xt}^2(x,t) \le 4V + 2(1 + 2T^2)M, \tag{175}_3$$

$$\sup_{\substack{x \in [0,1] \\ t \in [0,T]}} w_x^2(x,t) \le 2V(1 + 4T^2) + (4T^2 + 8T^4)M. \tag{175}_4$$

The a priori estimate (138) and the above inequalities show that S maps $Z(M,T)$ into itself provided that T is sufficiently small relative to M. From now on, we assume that T is small enough so that S maps $Z(T,M)$ into itself.

To show that S is a contraction, let $M, T > 0$ and $w, \bar{w} \in Z(M,T)$ be given, and set $u := Sw$, $\bar{u} := S\bar{w}$, $W := w - \bar{w}$, $U := u - \bar{u}$. A simple computation shows that U satisfies

$$U_{tt} = \phi'(w_x)U_{xx} + \int_0^t a'(t - \tau)\psi'(w_x)U_{xx}(x,\tau) \, d\tau + [\phi'(w_x) - \phi'(\bar{w}_x)]\bar{u}_{xx}$$

$$+ \int_0^t a'(t - \tau)[\psi'(w_x) - \psi'(\bar{w}_x)]\bar{u}_{xx}(x,\tau) \, d\tau, \quad \forall x \in [0,1], \ t \in [0,T], \tag{176}_1$$

$$U(0,t) = U(1,t) = 0, \quad \forall t \in [0,T], \tag{176}_2$$

$$U(x,t) = 0 \quad \forall x \in [0,1], \ t \in (-\infty, 0]. \tag{176}_3$$

Integrating the first convolution term in $(176)_1$ by parts, we obtain

$$U_{tt} = \chi'(w_x)U_{xx} + \int_0^t a(t - \tau)[\psi'(w_x)U_{xx}]_t(x,\tau) \, d\tau + [\phi'(w_x) - \phi'(\bar{w}_x)]\bar{u}_{xx}$$

$$+ \int_0^t a'(t - \tau)[\psi'(w_x) - \psi'(\bar{w}_x)]\bar{u}_{xx}(x,\tau) \, d\tau, \tag{177}$$

where

$$\chi(\xi) := \phi(\xi) - a(0)\psi(\xi) \quad \forall \xi \in \mathbb{R}. \tag{178}$$

We multiply (177) by $[\psi'(w_x)U_{xx}]_t$ and integrate over $[0,1] \times [0,t], t \in [0,T]$, performing various integrations by parts and exploiting $(176)_2, (176)_3$. This yields

$$\frac{1}{2} \int_0^1 \{\psi'(w_x)\chi'(w_x)U_{xx}^2 + \psi'(w_x)U_{xt}^2\}(x,t)\ dx + Q([\psi'(w_x)U_{xx}]_t, t, a)$$

$$= -\int_0^1 [\phi'(w_x) - \phi'(\bar{w}_x)]\psi'(w_x)\bar{u}_{xx}U_{xx}(x,t)\ dx$$

$$-\int_0^1 \psi'(w_x)U_{xx}(x,t) \int_0^t a'(t-\tau)[\psi'(w_x) - \psi'(\bar{w}_x)]\bar{u}_{xx}(x,\tau)\ d\tau\ dx$$

$$+\int_0^t \int_0^1 \{\frac{1}{2}\psi''(w_x)w_{xt}U_{xt}^2 - \psi''(w_x)w_{xx}U_{xt}U_{tt} + \psi''(w_x)w_{xt}U_{xx}U_{tt}$$

$$+\frac{1}{2}[\chi''(w_x)\psi'(w_x) - \chi'(w_x)\psi''(w_x)]w_{xt}U_{xx}^2$$

$$+[\phi'(w_x) - \phi'(\bar{w}_x)]\psi'(w_x)\bar{u}_{xxt}U_{xx}$$

$$+[\phi''(w_x) - \phi''(\bar{w}_x)]\psi'(w_x)\bar{u}_{xx}U_{xx}W_{xt}\}(x,s)\ dx\ ds$$

$$+\int_0^t \int_0^1 \psi'(w_x)U_{xx}(x,s) \int_0^s a'(s-\tau)\{[\psi'(w_x) - \psi'(\bar{w}_x)]\bar{u}_{xxt}$$

$$+[\psi''(w_x) - \psi''(\bar{w}_x)]\bar{u}_{xx}W_{xt}\}(x,\tau)\ d\tau\ dx\ ds \quad \forall t \in [0,T]. \quad (179)$$

Using Lemma III.25 with ε sufficiently small, we see that the left-hand side of (179) is bounded from below by

$$\int_0^1 \{\frac{1}{4}\underline{\nu}\underline{\psi}^2 U_{xx}^2 + \frac{1}{2}\underline{\psi}U_{xt}^2\}(x,t)\ dx$$

$$- C\int_{-\infty}^t \int_0^1 [\psi'(w_x)U_{xx}^2](x,s)\ dx\ ds\ \forall t \in [0,T], \quad (180)$$

where C is a constant that can be chosen independently of M and T.

It follows from $(176)_1$ that

$$\int_0^1 U_{tt}^2(x,t)dt \le 3\int_0^1 \{\phi'(w_x)^2 U_{xx}^2 + [\phi'(w_x) - \phi'(\bar{w}_x)]^2 \bar{u}_{xx}^2\}(x,t)\ dx$$

$$+ 6a(0)^2 \max_{s \in [0,t]} \int_0^1 \{\psi'(w_x)^2 U_{xx}^2 + [\psi'(w_x) - \psi'(\bar{w}_x)]^2 \bar{u}_{xx}^2(x,s)\}\ dx$$

$$\forall t \in [0,T]. \quad (181)$$

We combine (179) and (181) and proceed as in the derivation of (157). Exploiting the fact that $W \equiv 0$ on $[0,1] \times (-\infty, 0]$, we obtain (after a rather long computation) an estimate of the form

$$d(Sw, S\bar{w}) \le P(M,T) \exp(T \cdot R(M,T))d(w, \bar{w})\ \forall w, \bar{w} \in Z(T,M) \quad (182)$$

for every $M, T > 0$, where $P, R : [0,\infty) \times [0,\infty) \to [0,\infty)$ are continuous functions with $P(M,0) = 0\ \forall M > 0$.

The derivation of (182) from (179) and (181) is in much the same spirit as the derivation of (157). We show the detailed estimation of the first term on the right-hand side of (179). For each $\eta > 0$, we have

$$|\int_0^1 [\phi'(w_x) - \phi'(\bar{w}_x)]\psi'(w_x)\bar{u}_{xx}U_{xx}(x,t)\ dx| \leq \eta \int_0^1 U_{xx}^2(x,t)\ dx$$

$$+ (4\eta)^{-1} \int_0^1 [\phi'(w_x) - \phi'(\bar{w}_x)]^2 \psi'(w_x)^2 \bar{u}_{xx}^2(x,t)\ dx\ \forall t \in [0,T]. \quad (183)$$

If we choose η sufficiently small, the first integral on the right-hand side of (183) can be absorbed by the first integral in (179). To estimate the last integral we first observe that by the mean value theorem

$$[\phi'(w_x) - \phi'(\bar{w}_x)]^2(x,t) \leq \Phi(M,T)W_x^2(x,t)\ \forall x \in [0,1],\ t \in [0,T], \quad (184)$$

where $\Phi(M,T) := \max \phi''(\xi)^2$ and the max is taken over all ξ with $\xi^2 \leq 2V(1 + 4T^2) + (4T^2 + 8T^4)M$. Using the fact that $W \equiv 0$ on $[0,1] \times (-\infty, 0]$, the type of argument used to derive (174) yields

$$W_x^2(x,t) \leq 4M(T^2 + T^4)\ \forall x \in [0,1],\ t \in [0,T]. \quad (185)$$

Next, we set $\Psi(M,T) := \max \psi'(\xi)^2$ where the max is taken over all ξ with $\xi^2 \leq 2V(1 + 4T^2) + (4T^2 + 8T^4)M$. Then, using (184), (185), and the fact that $\bar{u} \in Z(T,M)$ we find

$$\int_0^1 [\phi'(w_x) - \phi'(\bar{w}_x)]^2 \psi'(w_x)^2 \bar{u}_{xx}^2(x,t)\ dx$$

$$\leq 8M(T^2 + T^4)\Phi(M,T)\Psi(M,T)(V + MT^2)\ \forall t \in [0,T]. \quad (186)$$

The remaining steps in the derivation of (182) can be carried out in a similar fashion.

The contraction mapping principle and (182) imply that S has a unique fixed point $u \in Z(T,M)$ for a sufficiently small choice of $T > 0$. It is obvious that u satisfies (130) on $[0,1] \times (-\infty, T]$. The uniqueness statement in Theorem III.22 is immediate. The continuation claim of Theorem III.22 follows from a standard argument. (See, e.g. [D4].)

It is easy to remove the extraneous assumption (169). To do so, we construct functions $\tilde{\phi}, \tilde{\psi} \in C^3(\mathbb{R})$ which satisfy

$$\tilde{\phi}(\xi) = \phi(\xi),\ \tilde{\psi}(\xi) = \psi(\xi)\quad \forall \xi \in \left[-2\sqrt{V}, 2\sqrt{V}\right], \quad (187)_1$$

$$\inf_{\xi \in \mathbb{R}} \tilde{\phi}'(\xi) > 0,\ \inf_{\xi \in \mathbb{R}} \tilde{\psi}'(\xi) > 0,\ \sup_{\xi \in \mathbb{R}} \tilde{\psi}'(\xi) < \infty, \quad (187)_2$$

and we consider equation $(130)_1$ with ϕ and ψ replaced by $\tilde{\phi}$ and $\tilde{\psi}$, respectively. The preceding argument shows that the modified history value problem has a unique solution u on $(-\infty, T]$ for some $T > 0$. The Sobolev embedding theorem implies that

$$\sup_{\substack{x \in [0,1] \\ t \in (-\infty, 0]}} \tilde{u}_x^2(x,t) \leq V. \tag{188}$$

By virtue of $(187)_1$, (188), and the continuity properties of u_x, u is a solution of the original problem on some smaller interval $(-\infty, \bar{T}]$ with $\bar{T} > 0$. The additional properties of u as a solution of the original problem all follow easily. ∎

It is possible to treat somewhat more general equations with singular kernels by a similar approach. Consider the equation

$$u_{tt}(x,t) = \int_{-\infty}^t m(t-\tau) h(u_x(x,t), u_x(x,\tau))_x \, d\tau + f(x,t). \tag{189}$$

The iteration scheme for (189) involves the solution of linear equations of the form

$$u_{tt}(x,t) = \alpha(x,t) u_{xx}(x,t) - \int_{-\infty}^t m(t-\tau)\beta(x,t,\tau) u_{xx}(x,\tau) \, d\tau + f(x,t). \tag{190}$$

The key idea is to split β as follows:

$$\beta(x,t,\tau) = \beta(x,\tau,\tau) + \{\beta(x,t,\tau) - \beta(x,\tau,\tau)\}. \tag{191}$$

The second term on the right vanishes as $\tau \to t$ and cancels the singularity in m. The resulting term in (190) can be treated as a perturbation. The term resulting from $\beta(x,\tau,\tau)$ is dealt with as above.

We now briefly discuss a model with a stronger singularity in the kernel. Equation $(130)_1$ is not meaningful unless m is integrable (and hence a defined by (131) is finite) at 0. However, we can rewrite it in the form

$$u_{tt}(x,t) = \chi(u_x(x,t))_x + \int_{-\infty}^t a(t-\tau)\psi(u_x)_{xt}(x,\tau) \, d\tau + f(x,t) \tag{192}$$

provided the history is well-behaved at $-\infty$. In this form, the equation has meaning if a has an integrable singularity at 0. We now give an existence theorem for (192), which permits a to have such an integrable singularity. On the other hand, we now require stronger monotonicity assumptions on a. We treat the initial-value problem

$$u_{tt}(x,t) = \chi(u_x(x,t))_x + \int_0^t a(t-\tau)\psi(u_x)_{xt}(x,\tau) \, d\tau + f(x,t), \; x \in [0,1], \; t \geq 0,$$

$$\tag{193}_1$$

$$u(0,t) = u(1,t) = 0, \; t \geq 0, \tag{193}_2$$

131

$$u(x,0) = u_0(x), \; u_t(x,0) = u_1(x), \; x \in [0,1]. \tag{193$_3$}$$

Our assumptions on f will permit us to absorb the integral from $-\infty$ to 0 in (192) into the forcing term for smooth enough histories. In order to ensure that the original history value problem is equivalent to the initial-value problem (193), we must require that the limit as $t \to 0-$ of the given history is equal to u_0. The assumptions formulated below are what is required to solve (193). The reduction of the history value problem to (193) would actually require a little more smoothness on ψ and more smoothness of the history than is obtained for the solution.

Concerning the kernel a we now assume

(a)* $a \in L^1(0,\infty)$, $a \geq 0$, $a' \leq 0$, $a'' \geq 0$, $a''' \leq 0$,

and we make the following smoothness assumptions

(s1)* $\chi, \psi \in C^3(\mathbb{R})$,

(s2)* $u_0 \in H^3(0,1)$, $u_1 \in H^2(0,1)$,

(s3)* $f, f_x, f_{xx} \in L^2_{loc}([0,\infty); L^2(0,1))$.

In order to obtain a smooth solution of (193), we need the data to be compatible with the boundary conditions; we make the following rather strong compatibility assumptions:

(c)* $u_0(0) = u_0(1) = u_1(0) = u_1(1) = 0$, $u_0''(0) = u_0''(1) = 0$, $f(0,t) = f(1,t) = 0$ a.e. $t \geq 0$,

which guarantee that the data admit smooth, spatially periodic, odd extensions (of period 2). Finally, in order to ensure the evolutionarity of equation (193)$_1$, we require

(e)* $\psi'(\xi) > 0$, $\chi'(\xi) + a(0+)\psi'(\xi) > 0 \; \forall \xi \in \mathbb{R}$.

Observe that if $a(0+) = \infty$, then (e)* imposes no restrictions on χ'. In this regard, a kernel which is infinite at 0 has the same effect as a Newtonian viscosity.

Theorem III.26:

Assume that (a)*, (s1)*-(s3)*, (c)* and (e)* hold. Then the initial value problem (193) has a unique solution u on a maximal time interval $[0, T_0)$, $T_0 > 0$, with

$$u, u_x, u_t, u_{xx}, u_{xt}, u_{xxx}, u_{xxt} \in L^\infty_{loc}([0,T_0); L^2(0,1)), \tag{194$_1$}$$

$$u_{tt}, u_{xtt} \in L^2_{loc}([0,T_0); L^2(0,1)). \tag{194$_2$}$$

Moreover, if

$$\mathrm{ess} - \sup_{t \in [0,T_0]} \int_0^1 \{u_{xxx}^2 + u_{xxt}^2\}(x,t) \; dx + \int_0^{T_0} \int_0^1 u_{xtt}^2(x,t) \; dx \; dt < \infty, \tag{195}$$

then $T_0 = \infty$.

Remarks III.27:

1. Standard embedding theorems and (194) imply that u, u_x, u_t, u_{xx} and u_{xt} are continuous on $[0,1] \times [0,T_0)$. However, we do not obtain a classical solution from this theorem.

2. A similar theorem holds for the initial-value problem on all of $\mathbb{R}$.
3. The question of optimal regularity of the solution appears to be rather delicate. If more regularity of the data is assumed, and a is smooth on $(0, \infty)$, then it is easy to prove that the solution has additional smoothness for $t > 0$. Smoothness at $t = 0$ would be obtained in the original history value problem, provided the history satisfies appropriate compatibility conditions. However, when converting to the initial-value problem (193), cancelling singularities in the time derivatives of f and the memory term appear even if all terms in (192) are smooth (note that (s3)* requires no temporal regularity of f).

The proof is similar in spirit to that of Theorem III.22. Since we are now working with an initial value problem, we replace Q in (142) by

$$Q(\Phi, t, a_\delta) = \int_0^t \int_0^1 \Phi(x, s) \int_0^s a_\delta(s - \tau)\Phi(x, \tau) \, d\tau \, dx \, ds. \qquad (196)$$

To derive the analogue of equation (152), we apply a spatial difference operator rather than a temporal difference operator. The energy estimate that is obtained has the form

$$\frac{1}{2} \int_0^1 \beta u_{xxt}^2(x, t) + \beta \gamma^{(\delta)} u_{xxx}^2(x, t) \, dx$$

$$+ \lim_{h \to 0} \frac{1}{h^2} Q(D_h[(\beta u_{xx})_t], t, a_\delta) = \ldots \qquad (197)$$

Here β and $\gamma^{(\delta)}$ have the same meaning as before; D_h denotes a spatial difference operator (for the definition of D_h near the boundary note that we have extended the problem periodically in space). The left-hand side provides bounds for u_{xxx} and u_{xxt} in $L^\infty([0, T]; L^2(0, 1))$ by means of a variant of Lemma III.25. To estimate the right-hand side we need a bound for u_{xtt} in $L^2([0, T]; L^2(0, 1))$. Such a bound can be obtained by applying D_h to the linearized equation of motion and using the following inequality of Staffans [S13] with $\Phi = D_h[(\beta u_{xx})_t]$.

Lemma III.28:

Assume that (a)* holds, and let $A := 5\|a\|_{L^1}$. Then, for every $\delta, T > 0$ and $\Phi \in L^2([0, T]; L^2(0, 1))$, we have

$$\int_0^t \left\| \int_0^s a_\delta(s - \tau)\Phi(\tau) \, d\tau \right\|^2 ds \leq A \cdot Q(\Phi, t, a_\delta) \ \forall t \in [0, T]. \qquad (198)$$

The contraction argument is carried out in a different function space for technical reasons. See [H16] for details.

IV Global existence results

1. Introduction

In the preceding chapter we established results which guarantee that certain initial value problems have smooth solutions on (possibly small) time intervals. We now study the problem of extending these solutions globally in time; we also discuss the related problem of asymptotic behavior as $t \to \infty$.

It is well known that a solution x of the ordinary differential equation $\dot{x} = f(x,t)$ can be continued for as long as $|x(t)|$ remains finite. To continue a solution of a partial differential equation, one generally needs bounds on various derivatives of the solution. The precise form of the bounds that are needed, and the methods which can be used to obtain these bounds, depend on the mathematical type of the equation, the nature of the nonlinear terms, etc.

For parabolic problems involving viscosity of the Newtonian type, the implicit function theorem can be used to establish global existence theorems for small data. Such results are not always optimal. For the Newtonian fluid itself with Dirichlet boundary conditions, global existence of solutions for large data is known in the two-dimensional case, but remains an open problem in three dimensions. Oskolkov (see e.g. [O4],[O5]) has considered a number of model equations for viscoelastic fluids with a Newtonian contribution to the viscosity and obtained global existence for large data in three dimensions. However, while retaining the nonlinear terms present in the original Navier-Stokes system, he neglects a number of others, including those required to make the constitutive law frame-indifferent. In one space dimension, a number of authors have proved global existence for large data for equations of the form $u_{tt} = \phi(u_x)_x + (\eta(u_x)u_{xt})_x$. See Greenberg, MacCamy and Mizel [G10], Greenberg [G11], MacCamy [M2], Greenberg and MacCamy [G12], Kanel'[K1], Dafermos [D1], Andrews [A4], Andrews and Ball [A5], Pego [P2] and Antman and Malek-Madani [A7],[A8]. Engler [E2] has considered one-dimensional shearing motions of a K-BKZ fluid with a Newtonian viscosity; his result will be discussed in Section 6.

For viscoelastic materials of integral type with smooth memory functions, the equations of motion are essentially hyperbolic. The analysis of Section II.4 shows that solutions of such equations will generally develop singularities in finite time if the data are large. The study of acceleration waves (see Section II.4) suggests that globally defined smooth solutions exist if the data are small. We shall see that this is indeed the case. In contrast with the local existence results discussed in the previous chapter, the global theorems depend crucially on the sign of the memory terms. In fact, the essential requirement is that the rest state be linearly stable.

The implicit function theorem (at least in its standard form) is not applicable to nonlinear hyperbolic problems, because the required coercive estimates for the linearized problem do not hold. Therefore, a different approach is needed. Nishida [N4] used an argument based on Riemann invariants to show that the damped quasilinear wave equation $u_{tt} + u_t = \phi(u_x)_x$ has a unique globally defined solution if the initial data are small. Subsequently, Matsumura [M9] devised an argument based on energy estimates to obtain similar results. The energy approach has the advantage that it works in more than one spatial dimension and for more general types of boundary value problems. Both the energy method and Riemann invariants have been used to obtain small-data global existence theorems in viscoelasticity. These results will be discussed in Sections 4 and 5.

For viscoelastic materials of integral type with singular kernels, the equations of motion have properties between the parabolic and hyperbolic cases. So far, global existence of classical solutions to nonlinear problems with singular kernels is known only for certain problems with small data. Londen [L10] and Engler [E3] have established global existence of weak solutions for certain specific model problems with singular kernels.

The chapter is organized as follows: In Section 2, we use the implicit function theorem to prove a small-data global existence theorem for a model parabolic problem. Then, in Section 3, we use the frictionally damped wave equation $u_{tt} + u_t = \phi(u_x)_x$ to illustrate the energy approach of Matsumura. Section 4 is devoted to viscoelastic materials of integral type in one space dimension; the analysis is based on energy estimates in conjunction with properties of strongly positive definite kernels. In Section 5, we prove a global existence theorem for spatially periodic three-dimensional motions of incompressible K-BKZ fluids. The proof is an adaptation of the argument of Kim [K12]. Finally, in Section 6, we present the results of Engler on large-data global existence of shearing flows of a viscoelastic fluid with a Newtonian viscosity and a K-BKZ type memory term.

2. Global existence for a parabolic problem

In this section, we demonstrate through a simple example how to obtain global existence results for problems of parabolic type (with small data) by using the implicit function theorem. We consider the initial-boundary value problem

$$u_{tt} = (\eta(u_x)u_{xt})_x + \phi(u_x)_x - \int_0^t m(t-\tau)\psi(u_x(x,\tau))_x \ d\tau + f(x,t), \ x \in [0,1], \ t \geq 0,$$

$$\tag{1}_1$$

$$u(0,t) = u(1,t) = 0, \ t \geq 0,$$

$$\tag{1}_2$$

$$u(x,0) = u_0(x), \ u_t(x,0) = u_1(x), \ x \in [0,1],$$

$$\tag{1}_3$$

under the assumptions

(S1) $\phi, \psi, \eta \in C^2(\mathbb{R})$,

(S2) $f \in L^2((0,1) \times (0,\infty))$,

(S3) $u_0 \in H^2(0,1), \ u_1 \in H^1(0,1)$,

(M) $m \in L^1(0,\infty), \ m > 0, \ m' \leq 0$,

(E) $\eta(0) > 0, \ \phi'(0) > 0, \ \psi'(0) > 0, \ \phi'(0) - \psi'(0) \int_0^\infty m(s) \ ds > 0$,

(C) $u_0(0) = u_0(1) = u_1(0) = u_1(1) = 0$.

Theorem IV.1:

Assume that (S1)-(S3), (M), (E) and (C) hold and that $\|f\|_{L^2((0,1)\times(0,\infty))}$, $\|u_0\|_{H^2}, \ \|u_1\|_{H^1}$ are sufficiently small. Then (1) has a unique solution $u \in H^2([0,\infty); L^2(0,1)) \cap H^1([0,\infty); H^2(0,1))$.

Remark IV.2:

1. Observe that our assumptions permit the kernel m to have an integrable singularity at zero.

2. The technique used to prove Theorem IV.1 can be generalized to treat more complicated equations, including problems in several space dimensions. It is also possible to obtain solutions with more regularity and to treat other boundary conditions. We note that if fluids are considered, one cannot expect u to decay to 0. Rather, one expects that u decays to a constant function. Therefore, different function spaces must be used (cf. e.g. [R4]).

Proof:

We set $X := H^2([0,\infty); L^2(0,1)) \cap H^1([0,\infty); H^2(0,1) \cap H_0^1(0,1))$ and $Y := L^2((0,1) \times (0,\infty)) \times (H^2(0,1) \cap H_0^1(0,1)) \times H_0^1(0,1)$, and for $u \in X$ we consider the mapping

$$u \rightarrow F(u) := \left\{ u_{tt} - (\eta(u_x)u_{xt})_x - \phi(u_x)_x + \int_0^t m(t-\tau)\psi(u_x(\cdot,\tau))_x \ d\tau, \right.$$

$$\left. u(\cdot,0), u_t(\cdot,0) \right\}.$$

$$\tag{2}$$

Observe that a function $u \in X$ is a solution of (1) if and only if it satisfies

$$F(u) = (f, u_0, u_1).$$

$$\tag{3}$$

We solve equation (3) by using the implicit function theorem in Banach space. It follows from Theorem 2.1 in Vol. II, p. 9 of [L6] that $u \mapsto u_t(\cdot, 0)$ is a bounded linear mapping from X to $H_0^1(0,1)$. Using the assumed smoothness of ϕ, ψ, and η, and basic properties of Sobolev spaces, we conclude that F is a C^1 mapping from X to Y. Moreover, we note that $F(0) = (0,0,0)$.

To apply the implicit function theorem, it remains to verify that the linearization of F about zero, i.e. the mapping

$$u \to \left\{ u_{tt} - \eta(0)u_{xxt} - \phi'(0)u_{xx} + \int_0^t m(t-\tau)\psi'(0)u_{xx}(\cdot, \tau)\, d\tau, u(\cdot, 0), u_t(\cdot, 0) \right\} \quad (4)$$

is an isomorphism of X onto Y. We therefore consider the problem

$$u_{tt} - \eta(0)u_{xxt} - \phi'(0)u_{xx} + \int_0^t m(t-\tau)\psi'(0)u_{xx}(\cdot, \tau)\, d\tau = f(x,t), \quad (5)_1$$

$$u(x,0) = u_0(x), \quad u_t(x,0) = u_1(x), \quad (5)_2$$

with $(f, u_0, u_1) \in Y$.

To solve (5) we use Fourier series in the spatial variable and the Laplace transform with respect to time, i.e. we set

$$u(x,t) = \sum_{k=1}^{\infty} u_k(t)\sin(k\pi x), \quad (6)_1$$

and

$$\hat{u}_k(\lambda) = \int_0^{\infty} e^{-\lambda t} u_k(t)\, dt. \quad (6)_2$$

Analogous notation is used to represent f, and we denote by u_{k0} and u_{k1} the Fourier coefficients of the initial data. After transforming (5) we obtain

$$\lambda^2 \hat{u}_k(\lambda) - \lambda u_{k0} - u_{k1} + \eta(0)\pi^2 k^2 \lambda \hat{u}_k(\lambda) - \eta(0)\pi^2 k^2 u_{k0}$$

$$+ \big(\phi'(0) - \hat{m}(\lambda)\psi'(0)\big)\pi^2 k^2 \hat{u}_k(\lambda) = \hat{f}_k(\lambda), \quad (7)$$

from which we conclude that

$$\hat{u}_k(i\omega) = \frac{\hat{f}_k(i\omega) + \eta(0)\pi^2 k^2 u_{k0} + u_{k1} + i\omega u_{k0}}{-\omega^2 + \pi^2 k^2 i\omega\eta(0) + \pi^2 k^2 \big(\phi'(0) - \hat{m}(i\omega)\psi'(0)\big)} \quad \forall k \in \mathbb{N},\ \omega \in \mathbb{R}. \quad (8)$$

We denote the denominator in (8) by $N(k, \omega)$. It follows from our assumptions on m that

$$\mathrm{Re}\,\big(\phi'(0) - \psi'(0)\hat{m}(i\omega)\big) > 0 \ \forall \omega \in \mathbb{R}, \quad (9)_1$$

$$\lim_{\omega \to \pm\infty} \hat{m}(i\omega) = 0, \quad (9)_2$$

$$\mathrm{sgn}\,\mathrm{Im}\,\hat{m}(i\omega) = -\mathrm{sgn}\,\omega. \quad (9)_3$$

137

Using (9), it is easy to show that there is a constant $\gamma > 0$ such that

$$|N(k,\omega)| \geq \gamma(k^2 + \omega^2 + k^2|\omega|) \quad \forall k \in \mathbb{N}, \ \omega \in \mathbb{R}, \tag{10}$$

and hence we conclude there is a constant C_1 such that

$$|k^2 \hat{u}_k(i\omega)| \leq C_1\left(|\hat{f}_k(i\omega)| + \frac{k^2|u_{k0}|}{1+|\omega|} + \frac{|u_{k1}|}{1+|\omega|}\right) \quad \forall k \in \mathbb{N}, \ \omega \in \mathbb{R}. \tag{11}$$

The norm of (f, u_0, u_1) in Y (up to equivalence) is given by

$$\|(f, u_0, u_1)\|_Y^2 := \sum_{k=1}^{\infty}\left\{\frac{1}{\pi}\int_{-\infty}^{\infty} |\hat{f}_k(i\omega)|^2 \, d\omega\right.$$

$$\left. + \pi^4 k^4 |u_{k0}|^2 + \pi^2 k^2 |u_{k1}|^2 \right\}. \tag{12}$$

Using (11) and elementary properties of Fourier series and the Laplace transform, we find that $u \in L^2([0,\infty); H^2(0,1) \cap H_0^1(0,1))$ and there is a constant C_2 such that

$$\|u\|_{L^2([0,\infty);H^2 \cap H_0^1)} \leq C_2 \|(f, u_0, u_1)\|_Y. \tag{13}$$

It remains to be shown that u_{tt} and u_{xxt} are also square integrable on $(0,1) \times (0,\infty)$. The Laplace transforms of their Fourier components are given by

$$-\omega^2 \hat{u}_k(i\omega) - u_{k1} - i\omega u_{k0}$$

$$= \frac{1}{N(k,\omega)}\left\{-\omega^2 \hat{f}_k(i\omega) - \eta(0)i\omega\pi^2 k^2 u_{k1} - i\omega\pi^2 k^2 \left(\phi'(0) - \hat{m}(i\omega)\psi'(0)\right)u_{k0}\right.$$

$$\left. - \pi^2 k^2 \left(\phi'(0) - \hat{m}(0)\psi'(0)\right)u_{k1}\right\}, \tag{14}$$

and

$$i\omega\pi^2 k^2 \hat{u}_k(i\omega) - \pi^2 k^2 u_{k0}$$

$$= \frac{i\omega\pi^2 k^2 \hat{f}_k(i\omega) + i\omega\pi^2 k^2 u_{k1} - \pi^4 k^4 \left(\phi'(0) - \hat{m}(i\omega)\psi'(0)\right)u_{k0}}{N(k,\omega)}, \tag{14}$$

respectively. This leads to the estimate

$$\left| -\omega^2 \hat{u}_k(i\omega) - u_{k1} - i\omega u_{k0}\right| + \left|i\omega\pi^2 k^2 \hat{u}_k(i\omega) - \pi^2 k^2 u_{k0}\right|$$

$$\leq C_3\left(|\hat{f}_k(i\omega)| + \frac{k^2|u_{k0}|}{1+|\omega|} + \frac{k^2|u_{k1}|}{k^2+|\omega|}\right), \quad \forall k \in \mathbb{N}, \ \omega \in \mathbb{R}, \tag{14}$$

for some constant C_3. Observing that

$$\sum_{k=1}^{\infty}\int_{-\infty}^{\infty} \frac{k^4|u_{k1}|^2}{(k^2+|\omega|)^2} \, d\omega = 2\sum_{k=1}^{\infty} k^2|u_{k1}|^2,$$

138

e conclude that $u_{tt}, u_{xxt} \in L^2([0,\infty); L^2(0,1))$ and that

$$\|u_{tt}\|^2_{L^2([0,\infty);L^2)} + \|u_{xxt}\|^2_{L^2([0,\infty);L^2)}$$

$$\leq C_4 \|(f, u_0, u_1)\|^2_Y, \tag{15}$$

>r some constant C_4. The estimates (13) and (15) show that the linearization of
' about zero has a bounded inverse, and the proof is complete. ∎

3. The damped wave equation

In this section we use Matsumura's [M9] refinement of the energy method
ɔ establish a small-data global existence theorem for the initial-boundary value
roblem

$$u_{tt} + u_t = \phi(u_x)_x + f, \quad x \in [0,1], \ t \geq 0, \tag{16$_1$}$$

$$u(0,t) = u(1,t) = 0, \quad t \geq 0, \tag{16$_2$}$$

$$u(x,0) = u_0(x), \ u_t(x,0) = u_1(x), \quad x \in [0,1]. \tag{16$_3$}$$

'he same basic approach will be applied to problems from viscoelasticity in the
ext section. Concerning the data u_0, u_1, and f, we impose the following as-
umptions of smoothness and compatibility:

S1) $$u_0 \in H^3(0,1), \quad u_1 \in H^2(0,1);$$

S2) $$f, f_x, f_t \in C_b([0,\infty); L^2(0,1)),$$
$$f, f_x, f_t, f_{tt} \in L^2([0,\infty); L^2(0,1));$$

C) $$u_0(0) = u_0(1) = u_1(0) = u_1(1) = 0,$$
$$\phi'(u_0'(0))u_0''(0) + f(0,0) = \phi'(u_0'(1))u_0''(1) + f(1,0) = 0.$$

. global solution should be expected only if the data are suitably small; to mea-
ure the sizes of u_0, u_1, and f we set

$$U_0(u_0, u_1) := \|u_0\|^2_3 + \|u_1\|^2_2 \tag{17$_1$}$$

139

and

$$F(f) := \sup_{t \geq 0} \int_0^1 \{f^2 + f_x^2 + f_t^2\}(x,t)\ dx$$

$$+ \int_0^\infty \int_0^1 \{f^2 + f_x^2 + f_t^2 + f_{tt}^2\}(x,t)\ dx\ dt. \tag{17}$$

The main result of this section is

Theorem IV.3:

Assume that $\phi \in C^3(\mathbb{R})$ with $\phi'(0) > 0$. Then there is a number $\mu > 0$ with the following property: for each u_0, u_1, and f satisfying (S1), (S2), (C), and

$$U_0(u_0, u_1) + F(f) \leq \mu^2, \tag{18}$$

the initial-boundary value problem (16) has a unique solution $u : [0,1] \times [0, \infty) \to \mathbb{R}$ with

$$u, u_x, u_t, u_{xx}, u_{xt}, u_{tt}, u_{xxx}, u_{xxt}, u_{xtt}, u_{ttt}$$
$$\in C_b([0, \infty); L^2(0,1)) \cap L^2([0, \infty); L^2(0,1)). \tag{19}$$

In addition

$$u, u_x, u_t, u_{xx}, u_{xt}, u_{tt} \to 0 \tag{20}$$

uniformly on $[0,1]$ as $t \to \infty$.

Remarks IV.4:

1. It follows from (19) that $u \in C_b^2([0,1] \times [0, \infty))$.
2. Similar existence theorems hold for Neumann or mixed boundary conditions and for the pure initial value problem on all of space.
3. A small-data global existence result for equation $(16)_1$ was first established by Nishida [N4] using the method of Riemann invariants. Subsequently Matsumura [M9] devised a new proof based on energy methods and extended the result to equations in several space dimensions.

Proof of Theorem IV.3:

We choose a sufficiently small number $\delta > 0$ and modify ϕ outside the interval $[-\delta, \delta]$ in such a way that

$$\phi'(\xi) \geq \underline{\phi} > 0, \quad \forall \xi \in \mathbb{R}, \tag{21}_1$$

$$\phi'; \phi'', \phi''' \in C_b(\mathbb{R}). \tag{21}_2$$

There is no harm in making this modification because we shall show a posteriori that

$$|u_x(x,t)| \leq \delta \quad \forall x \in [0,1],\ t \geq 0. \tag{22}$$

140

Moreover, without loss of generality we assume

$$\phi(0) = 0. \tag{23}$$

It follows from a straightforward local existence theorem that (16) has a unique local solution u on a maximal time interval $[0, T_0)$ with

$$u, u_x, u_t, u_{xx}, u_{xt}, u_{tt}, u_{xxx}, u_{xxt}, u_{xtt}, u_{ttt} \in C([0, T_0); L^2(0, 1)). \tag{24}$$

In addition, if

$$\sup_{t \in [0, T_0)} \{\|u(\cdot, t)\|_3^2 + \|u_t(\cdot, t)\|_2^2\} < \infty \tag{25}$$

then $T_0 = \infty$. Our objective is to show that if (18) is satisfied with μ sufficiently small then (25) holds. This will be accomplished via a chain of energy estimates. For $t \in [0, T_0)$, we set

$$
\begin{aligned}
\mathcal{E}(t) := \max_{s \in [0,t]} \int_0^1 &\{u^2 + u_x^2 + u_t^2 + u_{xx}^2 + u_{xt}^2 \\
&+ u_{tt}^2 + u_{xxx}^2 + u_{xxt}^2 + u_{xtt}^2 + u_{ttt}^2\}(x, s)\, dx \\
&+ \int_0^t \int_0^1 \{u^2 + u_x^2 + u_t^2 + u_{xx}^2 + u_{xt}^2 + u_{tt}^2 \\
&+ u_{xxx}^2 + u_{xxt}^2 + u_{xtt}^2 + u_{ttt}^2\}(x, s)\, dx\, ds
\end{aligned}
\tag{26}_1
$$

and

$$\nu(t) := \max_{\substack{x \in [0,1] \\ s \in [0,t]}} \{u_x^2 + u_{xx}^2 + u_{xt}^2\}^{1/2}(x, s). \tag{26}_2$$

The purpose of the computations which follow is to obtain the estimate

$$
\begin{aligned}
\mathcal{E}(t) \leq &\bar{\Gamma}\{U_0(u_0, u_1) + F(f)\} \\
&+ \bar{\Gamma}\{\nu(t) + \nu(t)^2\}\mathcal{E}(t) \quad \forall t \in [0, T_0),
\end{aligned}
\tag{27}
$$

where $\bar{\Gamma}$ is a constant that can be chosen independently of u_0, u_1, f, and T_0. Because the derivation of (27) is rather involved, we shall first show that the theorem follows from (27) and then give the derivation of this estimate. Assuming now that (27) holds we choose $\bar{\mathcal{E}}$, $\mu > 0$ such that

$$\bar{\mathcal{E}} \leq \frac{1}{2}\delta^2, \quad \bar{\Gamma}\{2(\bar{\mathcal{E}})^{1/2} + 4\bar{\mathcal{E}}\} \leq \frac{1}{4}, \quad \bar{\Gamma}\mu^2 \leq \frac{1}{4}\bar{\mathcal{E}}. \tag{28}$$

Suppose that (18) holds with the above choice of μ. It follows from the Sobolev embedding theorem that

$$\nu(t) \leq 2\sqrt{\mathcal{E}(t)} \quad \forall t \in [0, T_0). \tag{29}$$

141

We therefore conclude from (27), (28), and (29) that for any $t \in [0, T_0)$ with $\mathcal{E}(t) \leq \bar{\mathcal{E}}$, we actually have $\mathcal{E}(t) \leq \frac{1}{2}\bar{\mathcal{E}}$. Consequently, by continuity, we have

$$\mathcal{E}(t) \leq \frac{1}{2}\bar{\mathcal{E}} \quad \forall t \in [0, T_0) \tag{30}$$

provided that $\mathcal{E}(0) \leq \frac{1}{2}\bar{\mathcal{E}}$. We can always choose a smaller $\mu > 0$ (if necessary) such that (18) implies $\mathcal{E}(0) \leq \frac{1}{2}\bar{\mathcal{E}}$. (Observe that (28) still holds if the size of μ is reduced.) Thus, if (18) is satisfied with our revised choice of μ then (30) holds; this implies that (25) holds and consequently $T_0 = \infty$. It follows from the fact that $\mathcal{E}(\cdot)$ is bounded on $[0, \infty)$ and standard embedding inequalities that u and its derivatives through second order converge to zero uniformly on $[0,1]$ as $t \to \infty$. Moreover, we note that

$$|u_x(x,t)| \leq 2\sqrt{\mathcal{E}(t)} \leq \delta \quad \forall x \in [0,1],\ t \geq 0, \tag{31}$$

which justifies our modification of ϕ outside the interval $[-\delta, \delta]$.

We now establish the inequality (27). To simplify the notation we write U_0 and F in place of $U_0(u_0, u_1)$ and $F(f)$, and we use Γ to denote a (possibly large) generic positive constant that can be chosen independently of u_0, u_1, f, and T_0. Three basic types of estimates will be used in the derivation of (27):
(i)   those obtained directly from energy integrals;
(ii) inequalities obtained by using equation $(16)_1$ to express various derivatives of u in terms of quantities for which we already have estimates; and
(iii) interpolation inequalities.

Moreover, we shall make use of a simple form of the Poincaré inequality which asserts that if $w \in H^1(0,1)$ and either $w(0) = 0$ or w has zero mean (i.e. $\int_0^1 w(x)\ dx = 0$) then

$$\int_0^1 w(x)^2\ dx \leq \int_0^1 w'(x)^2\ dx. \tag{32}$$

Differentiation of $(16)_2$ with respect to t yields

$$u_t(0,t) = u_t(1,t) = u_{tt}(0,t) = u_{tt}(1,t) = 0 \quad \forall t \in [0, T_0). \tag{33}$$

Observe that by virtue of $(16)_2$ and (33) the functions u_x and u_{xt} have zero mean spatially.

To obtain the basic energy integral, we multiply $(16)_1$ by u_t and integrate over $[0,1] \times [0,t]$ using integration by parts, (33), and $(16)_3$ to obtain

$$\int_0^1 \{\frac{1}{2}u_t^2 + \Phi(u_x)\}(x,t)\ dx + \int_0^t \int_0^1 u_t^2(x,s)\ dx\ ds$$

$$= \int_0^1 \{\frac{1}{2}u_1(x)^2 + \Phi(u_0'(x))\}\ dx + \int_0^t \int_0^1 f u_t(x,s)\ dx\ ds \quad \forall t \in [0, T_0), \tag{34}$$

where

$$\Phi(\xi) := \int_0^\xi \phi(\eta) \, d\eta \quad \forall \xi \in \mathbb{R}. \tag{35}$$

We note that

$$\Phi(\xi) \geq \frac{1}{2} \phi \xi^2 \quad \forall \xi \in \mathbb{R}, \tag{36}$$

and

$$f u_t \leq \frac{1}{2} f^2 + \frac{1}{2} u_t^2. \tag{37}$$

The inequality

$$\int_0^1 \{u_t^2 + u_x^2\}(x,t) \, dx + \int_0^t \int_0^1 u_t^2(x,s) \, dx \, ds$$

$$\leq \Gamma \int_0^1 \{u_1(x)^2 + \Phi(u_0'(x))\} \, dx$$

$$+ \Gamma \int_0^t \int_0^1 f(x,s)^2 \, dx \, ds \quad \forall t \in [0, T_0) \tag{38}$$

thus follows from (34). It is interesting to observe that (37) yields a time-independent bound for $\|u(\cdot,t)\|_1^2 + \|u_t(\cdot,t)\|_0^2$ even if the data are large. However, such a bound is not sufficient to guarantee global existence and we must obtain estimates for higher order derivatives of u. In contrast with (34), the energy integrals of higher order do not yield time-independent bounds unless the data are small.

Differentiation of $(16)_1$ with respect to t gives

$$u_{ttt} + u_{tt} = \phi'(u_x) u_{xxt} + \phi''(u_x) u_{xt} u_{xx} + f_t. \tag{39}$$

We multiply (39) by u_{tt} and integrate over $[0,t]$ using integrations by parts and (33) to obtain

$$\frac{1}{2} \int_0^1 \{u_{tt}^2 + \phi'(u_x) u_{xt}^2\}(x,t) \, dx + \int_0^t \int_0^1 u_{tt}^2(x,s) \, dx \, ds$$

$$= \frac{1}{2} \int_0^1 \{u_{tt}^2 + \phi'(u_x) u_{xt}^2\}(x,0) \, dx$$

$$+ \int_0^t \int_0^1 \{\frac{1}{2} \phi''(u_x) u_{xt}^3 + f_t u_{tt}\}(x,s) \, dx \, ds \quad \forall t \in [0, T_0). \tag{40}$$

Using equation $(16)_1$ to express $u_{tt}(\cdot,0)$ in terms of u_0, u_1, and $f(\cdot,0)$, we find that

$$u_{tt}^2(\cdot,0) \leq 3[u_1^2 + (\phi'(u_0') u_0'')^2 + f^2(\cdot,0)] \tag{41}$$

and consequently

$$\int_0^1 \{u_{tt}^2 + \phi'(u_x) u_{xt}^2\}(x,0) \, dx$$

$$\leq \int_0^1 \{3[u_1(x)^2 + \phi'(u_0'(x))^2 u_0''(x)^2 + f(x,0)^2] + \phi'(u_0'(x))u_1'(x)^2\} \, dx$$

$$\leq \Gamma\{U_0 + F\} \tag{42}$$

by virtue of (17) and $(21)_2$. We next observe that

$$\int_0^t \int_0^1 \phi''(u_x)u_{xt}^3(x,s) \, dx \, ds \leq \Gamma\nu(t)\mathcal{E}(t) \quad \forall t \in [0, T_0) \tag{43}$$

by virtue of $(21)_2$ and (26), and that

$$\int_0^t \int_0^1 f_t u_{tt}(x,s) \, dx \, ds \leq \frac{1}{2} \int_0^t \int_0^1 \{u_{tt}^2 + f_t^2\}(x,s) \, dx \, ds \quad \forall t \in [0, T_0). \tag{44}$$

The inequality

$$\int_0^1 \{u_{tt}^2(x,t) + u_{xt}^2(x,t)\} \, dx + \int_0^t \int_0^1 u_{tt}^2(x,s) \, dx \, ds$$
$$\leq \Gamma\{U_0 + F\} + \Gamma\nu(t)\mathcal{E}(t) \quad \forall t \in [0, T_0) \tag{45}$$

follows easily from $(21)_1$, (40), (42), (43), and (44).

Combining (38) and (45) we find that

$$\int_0^1 \{u_x^2 + u_t^2 + u_{xt}^2 + u_{tt}^2\}(x,t) \, dx + \int_0^t \int_0^1 \{u_t^2 + u_{tt}^2\}(x,s) \, dx \, ds$$
$$\leq \Gamma\{U_0 + F\} + \Gamma\nu(t)\mathcal{E}(t) \quad \forall t \in [0, T_0). \tag{46}$$

(We note that $\Phi(\xi) \leq \Gamma\xi^2$ for all $\xi \in \mathbb{R}$ by virtue of $(21)_2$.) Next, we observe that

$$u_{xx}^2 \leq \frac{3}{\phi'(u_x)^2}[u_{tt}^2 + u_t^2 + f^2] \tag{47}$$

and consequently

$$\int_0^1 u_{xx}^2(x,t) \, dx \leq 3(\underline{\phi})^{-2} \int_0^1 \{u_{tt}^2 + u_t^2 + f^2\}(x,t) \, dx$$
$$\leq \Gamma\{U_0 + F\} + \Gamma\nu(t)\mathcal{E}(t) \quad \forall t \in [0, T_0). \tag{48}$$

Similarly, we find that

$$\int_0^t \int_0^1 u_{xx}^2(x,s) \, dx \, ds \leq \Gamma\{U_0 + F\} + \Gamma\nu(t)\mathcal{E}(t) \quad \forall t \in [0, T_0). \tag{49}$$

Since u vanishes at the endpoints and u_x has zero mean spatially, the Poincaré inequality gives

$$\int_0^1 u^2(x,s) \, dx \leq \int_0^1 u_x^2(x,s) \, dx \leq \int_0^1 u_{xx}^2(x,s) \, dx \quad \forall s \in [0, T_0). \tag{50}$$

144

Combining (46), (48), (49), and using (50) we now have the estimate

$$\int_0^1 \{u^2 + u_x^2 + u_t^2 + u_{xx}^2 + u_{xt}^2 + u_{tt}^2\}(x,t)\ dx$$

$$+ \int_0^t \int_0^1 \{u^2 + u_x^2 + u_t^2 + u_{xx}^2 + u_{tt}^2\}(x,s)\ dx\ ds$$

$$\leq \Gamma\{U_0 + F\} + \Gamma\nu(t)\mathcal{E}(t) \quad \forall t \in [0, T_0). \tag{51}$$

We could obtain a bound for $\int_0^t \int_0^1 u_{xt}^2$ from (51) and an interpolation inequality. However, we shall not do so because a bound for this integral will follow from a subsequent estimate and the Poincaré inequality.

The next energy identity can be obtained formally by differentiating (39) with respect to t, multiplying the result by u_{ttt} and integrating over $[0,1] \times [0,t]$. Unfortunately, this procedure is not legitimate because fourth derivatives of u do not necessarily exist as functions. To overcome this difficulty, we can work with a discrete analogue of the desired identity and take limits. For $h > 0$, we apply the difference operator Δ_h defined by

$$(\Delta_h v)(x,t) := v(x, t+h) - v(x,t) \tag{52}$$

to (39), multiply the result by $\Delta_h u_{tt}$ and integrate over $[0,1] \times [0,t]$. After appropriate integrations by parts we divide by h^2 and let $h \downarrow 0$. The result of this calculation is

$$\frac{1}{2}\int_0^1 \{u_{ttt}^2 + \phi'(u_x)u_{xtt}^2\}(x,t)\ dx + \int_0^t \int_0^1 u_{ttt}^2(x,s)\ dx\ ds$$

$$= \frac{1}{2}\int_0^1 \{u_{ttt}^2 + \phi'(u_x)u_{xtt}^2\}(x,0)\ dx$$

$$+ \int_0^t \int_0^1 \{\frac{1}{2}\phi''(u_x)u_{xt}u_{xtt}^2 + 2\phi''(u_x)u_{xt}u_{xxt}u_{ttt}$$

$$+ \phi'''(u_x)u_{xt}^2 u_{xx}u_{ttt} + f_{tt}u_{ttt}\}(x,s)\ dx\ ds. \tag{53}$$

Following the procedure used to derive (45) from (40), we conclude from (53) that

$$\int_0^1 \{u_{ttt}^2 + u_{xtt}^2\}(x,t)\ dx + \int_0^t \int_0^1 u_{ttt}^2(x,s)\ dx\ ds$$

$$\leq \Gamma\{U_0 + \Gamma\} + \Gamma\{\nu(t) + \nu(t)^2\}\mathcal{E}(t) \quad \forall t \in [0, T_0). \tag{54}$$

This first integral on the right-hand side of (53) is slightly more complicated to estimate than the analogous term in (40), but the same basic approach works. Indeed, it follows from (39) that

$$u_{ttt}^2(x,0) \leq 4\{u_{tt}^2 + \phi'(u_x)^2 u_{xxt}^2 + \phi''(u_x)^2 u_{xt}^2 u_{xx}^2 + f_t^2\}(x,0). \tag{55}$$

145

Using (41) and (55) we find that

$$\int_0^1 u_{ttt}^2(x,0)\ dx \le \Gamma\{U_0 + F\} + \Gamma\nu(t)^2\,\mathcal{E}\,(t) \quad \forall t \in [0, T_0). \tag{56}$$

Of course, we could use ΓU_0^2 in place of $\Gamma\nu(t)^2\,\mathcal{E}\,(t)$ on the right-hand side of (56) but it simplifies matters slightly to use $\Gamma\nu(t)^2\,\mathcal{E}\,(t)$.

Using (39) and (21)$_1$ we find that

$$u_{xxt}^2 \le \Gamma\{u_{ttt}^2 + u_{tt}^2 + u_{xt}^2 u_{xx}^2 + f_t^2\}. \tag{57}$$

It follows from (57), (51) and (54) that

$$\int_0^1 u_{xxt}^2(x,t)\ dx + \int_0^t \int_0^1 u_{xxt}^2(x,s)\ dx\ ds$$
$$\le \Gamma\{U_0 + F\} + \Gamma\{\nu(t) + \nu(t)^2\}\mathcal{E}\,(t) \quad \forall t \in [0, T_0). \tag{58}$$

We obtain a bound for $\int_0^t \int_0^1 u_{xtt}^2$ by interpolation. The identity

$$\int_0^1 \int_0^1 u_{xtt}^2(x,s)\ dx\ ds = \int_0^t \int_0^1 u_{xxt}u_{ttt}(x,s)\ dx\ ds$$
$$+ \int_0^1 u_{xxt}u_{tt}(x,0)\ dx - \int_0^1 u_{xxt}u_{tt}(x,t)\ dx \quad \forall t \in [0, T_0) \tag{59}$$

can be derived formally via integrations by parts. It is easy to give a rigorous derivation of (59) using difference operators. It follows from (59) that

$$\int_0^t \int_0^1 u_{xtt}^2(x,s)\ dx\ ds \le \frac{1}{2}\int_0^1 \{u_{tt}^2 + u_{xxt}^2\}(x,0)\ dx + \frac{1}{2}\int_0^1 \{u_{tt}^2 + u_{xxt}^2\}(x,t)\ dx$$
$$+ \frac{1}{2}\int_0^t \int_0^1 \{u_{ttt}^2 + u_{xxt}^2\}(x,s)\ dx\ ds \quad \forall t \in [0, T_0). \tag{60}$$

Moreover, since u_{xt} has zero mean, we observe that

$$\int_0^t \int_0^1 u_{xt}^2(x,s)\ dx\ ds \le \int_0^t \int_0^1 u_{xxt}^2(x,s)\ dx\ ds \quad \forall t \in [0, T_0) \tag{61}$$

by the Poincaré inequality. Combining (51), (54), (58), (60), and (61), and recalling (41), we now have the estimate

$$\int_0^1 \{u^2 + u_x^2 + u_t^2 + u_{xx}^2 + u_{xt}^2 + u_{tt}^2 + u_{xxt}^2 + u_{xtt}^2 + u_{ttt}^2\}(x,t)\ dx$$
$$+ \int_0^t \int_0^1 \{u^2 + u_x^2 + u_t^2 + u_{xx}^2 + u_{xt}^2 + u_{tt}^2 + u_{xxt}^2 + u_{xtt}^2 + u_{ttt}^2\}(x,s)\ dx\ ds$$
$$\le \Gamma\{U_0 + F\} + \Gamma\{\nu(t) + \nu(t)^2\}\mathcal{E}\,(t) \quad \forall t \in [0, T_0). \tag{62}$$

146

It remains only to obtain a similar bound for $\int_0^1 u_{xxx}^2 + \int_0^t \int_0^1 u_{xxx}^2$.
Differentiation of $(16)_1$ with respect to x yields

$$u_{xtt} + u_{xt} = \phi'(u_x)u_{xxx} + \phi''(u_x)u_{xx}^2 + f_x,\qquad (63)$$

from which we conclude that

$$u_{xxx}^2 \leq \Gamma\{u_{xtt}^2 + u_{xt}^2 + u_{xx}^4 + f_x^2\}.\qquad (64)$$

It follows from (64) and (62) that

$$\int_0^1 u_{xxx}^2(x,t)\,dx + \int_0^t \int_0^1 u_{xxx}^2(x,s)\,dx\,ds$$
$$\leq \Gamma\{U_0 + F\} + \Gamma\{\nu(t) + \nu(t)^2\}\mathcal{E}(t) \quad \forall t \in [0, T_0). \qquad (65)$$

Combining (62) and (65) we easily obtain the desired inequality (27).

As we saw at the beginning of the proof, it follows from the inequality (27) that if $U_0 + F$ is sufficiently small then $\mathcal{E}(\cdot)$ is bounded on $[0, T_0)$ and that (16) has a unique globally defined solution u which satisfies (19). We complete the proof by showing that

$$u_{xx}(\cdot, t) \to 0 \qquad (66)$$

uniformly on $[0,1]$ as $t \to \infty$. Convergence to zero of u, u_x, u_t, u_{xt} and u_{tt} as $t \to \infty$ follows from the same argument. We first observe that boundedness of $\mathcal{E}(\cdot)$ on $[0, \infty)$ implies

$$u_{xx}, u_{xxt} \in L^2([0, \infty); L^2(0,1)),\qquad (67)_1$$

$$u_{xxx} \in L^\infty([0, \infty); L^2(0,1)).\qquad (67)_2$$

It follows from $(67)_1$ that

$$u_{xx}(\cdot, t) \to 0 \text{ in } L^2(0,1) \text{ as } t \to \infty.\qquad (68)$$

Next we observe that

$$u_{xx}^2(x,t) \leq \int_0^1 u_{xx}^2(\xi,t)\,d\xi + 2\int_0^1 |u_{xx}u_{xxx}(\xi,t)|\,d\xi \quad \forall x \in [0,1],\qquad (69)$$

as can be verified by a simple calculus argument. The first term on the right-hand side of (69) tends to zero as $t \to \infty$ by virtue of (68). Using $(67)_2$, (68), and the inequality

$$\int_0^1 |u_{xx}(\xi,t)u_{xxx}(\xi,t)|\,d\xi \leq \|u_{xx}(\cdot,t)\|_0 \cdot \|u_{xxx}(\cdot,t)\|_0\qquad (70)$$

we find that the second term on the right-hand side of (69) also tends to zero as $t \to \infty$, and consequently $u_{xx}(\cdot, t) \to 0$ uniformly on $[0,1]$ as $t \to \infty$.

4. Viscoelastic materials of integral type

We now present some global existence results for initial value problems of the form

$$u_{tt}(x,t) = \phi(u_x(x,t))_x + \int_0^t a'(t-\tau)\psi(u_x(x,\tau))_x \, d\tau + f(x,t),$$

$$x \in B \subset \mathbb{R}, \quad t \geq 0, \tag{71}_1$$

$$u(x,0) = u_0(x), \quad u_t(x,0) = u_1(x), \qquad x \in B, \tag{71}_2$$

together with appropriate boundary conditions if B is not all of $\mathbb{R}$. Throughout this section we normalize a so that

$$a(\infty) = 0, \tag{72}$$

and we assume that

$$\phi(0) = \psi(0) = 0. \tag{73}$$

It is convenient to define the equilibrium stress function χ by

$$\chi(\xi) = \phi(\xi) - a(0)\psi(\xi) \qquad \forall \xi \in \mathbb{R}. \tag{74}$$

We shall concentrate primarily on situations when the material described by (71) is a solid, in which case it is natural to assume

$$\phi'(0) > 0, \quad \psi'(0) > 0, \quad \chi'(0) > 0. \tag{75}$$

(For a fluid we would have $\chi \equiv 0$.) Precise conditions on the kernel will be given later.

For most of this section we assume that a is smooth on $[0, \infty)$. The procedure which we employ to analyze (71) can be adapted to handle equations describing much more general classes of materials with smooth memory functions. This procedure can also be applied to certain problems involving singular memory functions. Some comments on these matters will be given at the end of the section. We shall also make some remarks on multidimensional problems and problems involving fluids.

Exponential kernel

It is instructive to consider (71) in the special case when the kernel a is an exponential of the form

$$a(t) = \alpha e^{-\lambda t} \tag{76}$$

with $\alpha, \lambda > 0$. When a is given by (76), equation $(71)_1$ can be converted to the third order partial differential equation

$$u_{ttt} + \lambda u_{tt} = \phi(u_x)_{xt} + \lambda\chi(u_x)_x + f_t + \lambda f, \qquad x \in B, \quad t \geq 0, \qquad (77)$$

where the equilibrium stress function χ is now given by $\chi(\xi) = \phi(\xi) - \alpha\psi(\xi)$. (To convert $(71)_1$ to (77), one simply differentiates $(71)_1$ with respect to time and observes that the integral term in the differentiated equation has the same form as the integral term in the original equation.) Observe that the term λu_{tt} on the left-hand side of (77) has a damping effect.

Small-data global existence theorems for (77) can be established by a procedure similar to (but more complicated than) the one used in Section IV.3. Greenberg [G13] has studied equation (77) with $B = [0, 1]$, $f \equiv 0$, and homogeneous Dirichlet boundary conditions. He derived a priori estimates which show that solutions which are smooth and small must decay to zero exponentially as $t \to \infty$. The pure initial value problem for (77) with $B = \mathbb{R}$ is discussed in [H11]. We note that when $B = \mathbb{R}$, exponential decay of solutions no longer holds.

For general kernels, the results concerning (71) make use of a number of ideas from the theory of Volterra integral equations. We therefore recall a few basic concepts.

Properties of Volterra kernels

Let $b \in L^1_{loc}[0, \infty)$ be given and consider the linear scalar Volterra equation

$$w(t) + \int_0^t b(t - \tau)w(\tau) \, d\tau = g(t), \qquad t \geq 0. \qquad (78)$$

For each $g \in L^1_{loc}[0, \infty)$, equation (78) has a unique solution $w \in L^1_{loc}[0, \infty)$. Moreover this solution is given by

$$w(t) = g(t) + \int_0^t r(t - \tau)g(\tau) \, d\tau, \qquad t \geq 0, \qquad (79)_1$$

where r is the resolvent kernel associated with b, i.e. the unique solution of the resolvent equation

$$r(t) + \int_0^t b(t - \tau)r(\tau) \, d\tau = -b(t), \qquad t \geq 0. \qquad (79)_2$$

Observe that if $r \in L^1(0, \infty)$ and $g \in L^p(0, \infty)$ with $1 \leq p \leq \infty$, then $w \in L^p(0, \infty)$ and

$$\|w\|_{L^p} \leq (1 + \|r\|_{L^1}) \cdot \|g\|_{L^p}. \qquad (80)$$

A classical theorem of Paley and Wiener asserts that if $b \in L^1(0, \infty)$ then $r \in L^1(0, \infty)$ if and only if $1 + \hat{b}(z)$ does not vanish for any z with $\mathrm{Re}\, z \geq 0$.

The notion of a strongly positive definite kernel plays a central role in the global existence theory for equation $(71)_1$. A function $b \in L^1_{loc}[0, \infty)$ is said to be positive definite (or of positive type) if

$$\int_0^T w(t) \int_0^t b(t - \tau)w(\tau) \, d\tau \, dt \geq 0 \tag{81}$$

for every $w \in C[0, \infty)$ and every $T > 0$; b is said to be strongly positive definite if there is an $\varepsilon > 0$ such that the function $t \mapsto b(t) - \varepsilon e^{-t}$ is positive definite. As the terminology suggests, strongly positive definite implies positive definite. These definitions are not easy to check directly. Using transform techniques one can show that if $b \in L^1(0, \infty)$ then b is strongly positive definite if and only if there is a constant $\varepsilon > 0$ such that

$$\text{Re } \hat{b}(i\omega) \geq \frac{\varepsilon}{\omega^2 + 1} \qquad \forall \omega \in \mathbb{R}, \tag{82}$$

where $\hat{b}$ is the Laplace transform of b.

From the viewpoint of applications to viscoelasticity, it is useful to know that certain types of sign conditions guarantee strong positive definiteness. More precisely, if $b \in C^2[0, \infty)$ and

$$(-1)^j b^{(j)}(t) \geq 0 \quad \forall t \geq 0, \quad j = 0, 1, 2; \quad b' \not\equiv 0, \tag{83}$$

then b is strongly positive definite. On the other hand, even if b is assumed to be very smooth, strong positive definiteness of b does not imply that (83) holds. (It does not even imply that $b \geq 0$ on $[0, \infty)$.) Indeed, it is easy to verify that the function b given by

$$b(t) = e^{-t} \cos t \tag{84}$$

satisfies (82) with $\varepsilon > 0$ and hence is strongly positive definite.

As the above example shows, strong positive definiteness does not imply any global sign conditions. However, if a strongly positive definite function is sufficiently regular, then statements can be made regarding its pointwise behavior near zero. In particular, if

$$b, b', b'' \in L^1(0, \infty), \, b \text{ is strongly positive definite} \tag{85}$$

then

$$b(0) > 0, \qquad b'(0) < 0. \tag{86}$$

That (85) implies $b(0) > 0$ follows from (82) and the inversion formula

$$b(0) = \frac{1}{\pi} \int_{-\infty}^{\infty} \text{Re } \hat{b}(i\omega) \, d\omega. \tag{87}$$

To see that (85) implies $b'(0) < 0$, observe that

$$\lim_{\omega \to \infty} \omega^2 \text{Re } \hat{b}(i\omega) = -b'(0), \tag{88}$$

as can be verified using two integrations by parts and the Riemann-Lebesgue lemma. This limit must be strictly positive by (82). See, for example, [N5], [S13], and [S14] for more information on strongly positive definite kernels.

The special case $\psi \equiv \phi$

For the special case when $\psi \equiv \phi$, global existence theorems for (71) were established by MacCamy [M3], Dafermos and Nohel [D3], and Staffans [S14]. The following idea of MacCamy shows that when $\psi \equiv \phi$, equation $(71)_1$ is closely related to the frictionally damped wave equation. For simplicity, we assume that $f \equiv 0$, i.e. we consider the problem

$$u_{tt}(x,t) = \phi(u_x(x,t))_x + \int_0^t a'(t-\tau)\phi(u_x(x,\tau))_x \, d\tau, \quad x \in B, \, t \geq 0, \quad (89)_1$$

$$u(x,0) = u_0(x), \, u_t(x,0) = u_1(x), \qquad x \in B \qquad (89)_2$$

with appropriate boundary conditions if $B \neq \mathbb{R}$. Equation $(89)_1$ can be regarded as a Volterra equation of the form (78) with $w = \phi(u_x)_x$ and $g = u_{tt}$. It follows from the representation formula (79) that

$$\phi(u_x(x,t))_x = u_{tt}(x,t) + \int_0^t q(t-\tau)u_{tt}(x,\tau) \, d\tau, \qquad (90)$$

where q is the unique solution of

$$q(t) + \int_0^t a'(t-\tau)q(\tau) \, d\tau = -a'(t). \qquad (91)$$

Integrating the memory term in (90) by parts and using $(89)_2$ we obtain

$$u_{tt}(x,t) + q(0)u_t(x,t) = \phi(u_x(x,t))_x$$
$$+ q(t)u_1(x) - \int_0^t q'(t-\tau)u_t(x,\tau) \, d\tau, \qquad x \in B, \, t \geq 0. \qquad (92)$$

Observe that $q(0) = -a'(0)$. Thus the term $q(0)u_t$ in (92) has a damping effect if $a'(0) < 0$. Observe further that the memory term in (92) is linear and involves only a first order derivative of u.

Equation (92) is convenient for many purposes. One can obtain a priori estimates from (92) (and a related equation) which are sufficient to prove a small-data global existence theorem for (89). (See [M3] and [D3].) We shall not follow this approach here because it relies crucially on the very special form of equation $(89)_1$.

Existence theorems (smooth kernel)

We now give an existence theorem, due to Dafermos and Nohel [D4], for the problem

$$u_{tt}(x,t) = \phi(u_x(x,t))_x + \int_0^t a'(t-\tau)\psi(u_x(x,\tau))_x \, d\tau + f(x,t), \qquad x \in [0,1],\ t \geq 0 \tag{93}_1$$

$$u(0,t) = u(1,t) = 0, \qquad t \geq 0, \tag{93}_2$$

$$u(x,0) = u_0(x), \quad u_t(x,0) = u_1(x), \qquad x \in [0,1]. \tag{93}_3$$

Concerning the kernel, we assume

(a1) $\qquad\qquad a, a', a'' \in L^1(0,\infty),\ a$ is strongly positive definite,

and we impose the following assumptions of smoothness on the coefficients and the data:

(s1) $$\phi, \psi \in C^3(\mathbb{R}),$$

(s2) $$u_0 \in H^3(0,1), \quad u_1 \in H^2(0,1),$$

(s3) $$\begin{aligned} &f, f_x, f_t \in C_b([0,\infty);\ L^2(0,1)), \\ &f, f_x, f_t, f_{tt} \in L^2([0,\infty);\ L^2(0,1)). \end{aligned}$$

The appropriate compatibility conditions read

(c) $$\begin{aligned} &u_0(0) = u_0(1) = u_1(0) = u_1(1) = 0, \\ &\phi'(u_0'(0))u_0''(0) + f(0,0) = \phi'(u_0'(1))u_0''(1) + f(1,0) = 0. \end{aligned}$$

We also assume that

(e) $$\phi'(0) > 0,\ \psi'(0) > 0,\ \phi'(0) - a(0)\psi'(0) > 0,$$

and to measure the sizes of u_0, u_1, and f, we set

$$U_0(u_0, u_1) := \|u_0\|_3^2 + \|u_1\|_2^2, \tag{94}_1$$

$$F(f) := \sup_{t \geq 0} \int_0^1 \{f^2 + f_x^2 + f_t^2\}(x,t)\, dx$$

$$+ \int_0^\infty \int_0^1 \{f^2 + f_x^2 + f_t^2 + f_{tt}^2\}(x,t)\, dx\, dt. \tag{94}_2$$

Theorem IV.5:

Assume that (a1), (s1), and (e) hold. Then there is a constant $\mu > 0$ with e following property: for each u_0, u_1, and f satisfying (s2), (s3), (c), and

$$U_0(u_0, u_1) + F(f) \le \mu^2 \tag{95}$$

e initial-boundary value problem (93) has a unique solution $u : [0,1] \times [0, \infty) \to$ with

$$u, u_x, u_t, u_{xx}, u_{xt}, u_{tt}, u_{xxx}, u_{xxt}, u_{xtt}, u_{ttt}$$

$$\in C_b([0, \infty); L^2(0,1)) \cap L^2([0, \infty); L^2(0,1)). \tag{96}$$

oreover, as $t \to \infty$, we have

$$u, u_x, u_t, u_{xx}, u_{xt}, u_{tt} \to 0 \tag{97}$$

niformly on $[0,1]$ as $t \to \infty$.

Remarks IV.6

1. It follows from (96) that $u \in C_b^2([0,1] \times [0, \infty))$.
2. A similar theorem holds if the Dirichlet boundary conditions $(93)_2$ are replaced by the Neumann conditions

$$u_x(0,t) = u_x(1,t) = 0, \quad t \ge 0. \tag{98}$$

In this case the compatibility relations (c) should be replaced by

(c') $\qquad\qquad u_0'(0) = u_0'(1) = u_1'(0) = u_1'(1) = 0.$

Moreover, the boundary conditions (98) permit nontrivial rigid motions, and the conclusions (96) and (97) must be modified accordingly; throughout (96) and (97) the function u should be replaced by u minus its spatial mean $\langle u \rangle$ which evolves according to the equation

$$\frac{d^2}{dt^2} \langle u \rangle(t) = \langle f \rangle(t). \tag{99}$$

We note that if the initial data and forcing term have zero mean spatially then $\langle u \rangle \equiv 0$ and (96), (97) hold for the solution u.

3. A similar existence result also holds for the mixed boundary conditions

$$u(0,t) = u_x(1,t) = 0, \quad t \ge 0, \tag{100}$$

(or $u_x(0,t) = u(1,t) = 0$). In this case, no nontrivial rigid motions are possible, and the only change needed in Theorem IV.5 is that the compatibility conditions (c) should be replaced with an appropriate mixed version of (c) and (c').

The proof of Theorem IV.5 given in [D4] makes essential use of the Poincaré inequality and consequently is not applicable to the pure initial value problem on all of space. This problem was solved in [H12] by combining the estimates of [D4] with a variant of MacCamy's procedure [M3]. We now state the main result of [H12]. The problem under consideration is

$$u_{tt}(x,t) = \phi(u_x(x,t))_x + \int_0^t a'(t-\tau)\psi(u_x(x,\tau))_x \; d\tau + f(x,t), \quad x \in \mathbb{R}, \; t \geq 0,$$
(101)
$$u(x,0) = u_0(x), \quad u_t(x,0) = u_1(x), \qquad x \in \mathbb{R}.$$
(101)

The assumptions on the kernel a will be slightly stronger than in Theorem IV.5. In addition to (a1) we require

$$(a2) \qquad \int_0^\infty t|a(t)| \; dt < \infty, \quad \mathrm{Re} \; \widehat{a'}(\lambda) \neq 0 \quad \forall \lambda \in \Pi,$$

where ˆ denotes the Laplace transform and $\Pi := \{z \in \mathbb{C} : \mathrm{Re} \; z \geq 0\}$.

Conditions (s1) and (e) are still appropriate for (101). Concerning the data we now assume

$$(s2^*) \qquad u_0 \in L^2_{loc}(\mathbb{R}), \; u_0', u_1 \in H^2(\mathbb{R}),$$

$$(s3^*) \qquad \begin{aligned} f, f_x, f_t &\in C_b([0,\infty); L^2(\mathbb{R})), \\ f &\in L^1([0,\infty); L^2(\mathbb{R})), \\ f_x, f_t, f_{tt} &\in L^2([0,\infty); L^2(\mathbb{R})). \end{aligned}$$

Of course, an analogue of the compatibility relations (c) is not needed for problem (101). To measure the sizes of u_0, u_1, and f we put

$$U_0^*(u_0, u_1) := \|u_0'\|_2^2 + \|u_1\|_2^2,$$
(102)

$$\begin{aligned} F^*(f) := \sup_{t \geq 0} \int_{-\infty}^\infty \{f^2 + f_x^2 + f_t^2\}(x,t) \; dx \\ + \int_0^\infty \int_{-\infty}^\infty \{f_x^2 + f_t^2 + f_{tt}^2\}(x,t) \; dx \; dt \\ \left(\int_0^\infty \left(\int_{-\infty}^\infty f(x,t)^2 \; dx \right)^{1/2} dt \right)^2. \end{aligned}$$
(102)

The additional assumption (a2) on the kernel is not terribly restrictive. We note that if $a \in C^3[0,\infty)$, $(-1)^j a^{(j)}(t) \geq 0$ for all $t \geq 0$, $j = 0, 1, 2, 3$, and

$$0 < \int_0^\infty ta(t) \; dt < \infty$$
(103)

154

then (a1) and (a2) both are satisfied.

Theorem IV.7:

Assume that (a1), (a2), (s1), and (e) hold. Then there is a constant $\mu > 0$ with the following property: for each u_0, u_1, and f satisfying (s2*), (s3*), and

$$U_0^*(u_0, u_1) + F^*(f) \leq \mu^2 \tag{104}$$

the initial value problem (101) has a unique solution $u : \mathbb{R} \times [0, \infty) \to \mathbb{R}$ with

$$u_x, u_t, u_{xx}, u_{xt}, u_{tt}, u_{xxx}, u_{xxt}, u_{xtt}, u_{ttt} \in C_b([0, \infty); L^2(\mathbb{R})), \tag{105}_1$$

$$u_{xx}, u_{xt}, u_{tt}, u_{xxx}, u_{xxt}, u_{xtt}, u_{ttt} \in L^2([0, \infty); L^2(\mathbb{R})). \tag{105}_2$$

Moreover, as $t \to \infty$, we have

$$u_x, u_t, u_{xx}, u_{xt}, u_{tt} \to 0 \quad \text{uniformly on } \mathbb{R}, \tag{106}_1$$

$$u_{xx}, u_{xt}, u_{tt} \to 0 \quad \text{in } L^2(\mathbb{R}). \tag{106}_2$$

Remarks IV.8

1. If it is also assumed that $u_0 \in L^2(\mathbb{R})$ then the solution also satisfies $u \in C([0, \infty); L^2(\mathbb{R}))$. However, one does not obtain a time-independent bound for $\|u(\cdot, t)\|_0$, even if $\|u_0\|_0$ is small.
2. In [H12], a slightly different assumption on f is made, but the proof in [H12] can easily be adapted to fit the present situation.

Existence proofs

We shall prove Theorem IV.5 in detail and then outline the additional steps needed to establish Theorem IV.7. The following lemma, which is based on ideas of Staffans [S14], expresses the dissipative effect of the memory term. Let X be a Hilbert space with inner product $\langle \cdot, \cdot \rangle$ and associated norm $\| \cdot \|$. For $T > 0$ and $w \in C([0, T]; X)$ we set

$$Q(w, t, a) := \int_0^t \left\langle w(s), \int_0^s a(s - \tau) w(\tau) \, d\tau \right\rangle ds \qquad \forall t \in [0, T]. \tag{107}$$

Lemma IV.9:

Assume that (a1) holds. Then there is a constant $\kappa > 0$ such that for every $T > 0$ and every $w \in C([0, T]; X)$ we have

$$\int_0^t \|w(s)\|^2 \, ds \leq \kappa \|w(0)\|^2 + \kappa Q(w, t, a)$$

$$+ \kappa \liminf_{h \downarrow 0} \frac{1}{h^2} Q(\Delta_h w, t, a) \quad \forall t \in [0, T), \tag{108}$$

155

where Δ_h is the forward difference operator of stepsize h, i.e.

$$(\Delta_h w)(t) := w(t+h) - w(t). \tag{109}$$

For a proof of this result see Lemma 2.5 of [H12]. (Lemma 2.5 of [H12] is formulated for the special case $X = L^2(\mathbb{R})$, but the same proof applies in general.)

We need one more preliminary result. The kernel k obtained by solving the integral equation

$$\phi'(0)k(t) + \int_0^t a'(t-\tau)\psi'(0)k(\tau)\,d\tau = -\psi'(0)a'(t), \quad t \geq 0, \tag{110}$$

can be used to express u_{xx} (or u_{xxx}) in terms of u_{tt} (or u_{xtt}) and small correction terms through an equation quite similar to (79). Using (a1), (e), properties of strongly positive definite kernels, and the maximum principle for harmonic functions one can establish

Lemma IV.10:

Assume that (a1) and (e) hold. Then, the solution k of (110) belongs to $L^1(0,\infty)$.

See Lemma 3.2 of [D4] for the details.

A. Proof of Theorem IV.5:

Recall that the equilibrium stress function χ is given by

$$\chi(\xi) := \phi(\xi) - a(0)\psi(\xi) \quad \forall \xi \in \mathbb{R}. \tag{111}$$

We choose a sufficiently small positive number δ and modify ϕ and ψ (and also χ accordingly by (111)) smoothly outside the interval $[-\delta, \delta]$ in such a way that

$$\phi', \phi'', \phi''', \psi', \psi'', \psi''', \chi', \chi'', \chi''' \in C_b(\mathbb{R}) \tag{112}$$

and

$$\phi'(\xi) \geq \underline{\phi} > 0, \ \psi'(\xi) \geq \underline{\psi} > 0, \ \chi'(\xi) \geq \underline{\chi} > 0 \quad \forall \xi \in \mathbb{R} \tag{113}$$

for some positive constants $\underline{\phi}, \underline{\psi}$, and $\underline{\chi}$. (This can always be accomplished by virtue of (e).) There is no harm in making this modification because we shall show a posteriori that

$$|u_x(x,t)| \leq \delta \quad \forall x \in [0,1], \ t \geq 0. \tag{114}$$

It follows from Theorem III.11 that the initial-boundary value problem (93) has a unique solution u on a maximal time interval $[0, T_0)$ with

$$u, u_x, u_t, u_{xx}, u_{xt}, u_{tt}, u_{xxx}, u_{xxt}, u_{xtt}, u_{ttt} \in C([0, T_0); L^2(0,1)). \tag{115}$$

Moreover, if

$$\sup_{t \in [0,T_0)} \{\|u(\cdot,t)\|_3^2 + \|u_t(\cdot,t)\|_2^2\} < \infty \tag{116}$$

then $T_0 = \infty$. Our goal is to show that if (95) is satisfied with μ sufficiently small then (116) holds.

For $t \in [0,T_0)$ we set

$$\mathcal{E}(t) := \max_{s \in [0,t]} \int_0^1 \{u^2 + u_x^2 + u_t^2 + u_{xx}^2 + u_{xt}^2 + u_{tt}^2 + u_{xxx}^2 + u_{xxt}^2 + u_{xtt}^2 + u_{ttt}^2\}(x,s)\, dx$$

$$+ \int_0^t \int_0^1 \{u^2 + u_x^2 + u_t^2 + u_{xx}^2 + u_{xt}^2 + u_{tt}^2 + u_{xxx}^2 + u_{xxt}^2 + u_{xtt}^2 + u_{ttt}^2\}(x,s)\, dx\, ds, \tag{117}_1$$

and

$$\nu(t) := \max_{x \in [0,1],\ s \in [0,t]} \{u_x^2 + u_{xx}^2 + u_{xt}^2\}^{1/2}(x,s). \tag{117}_2$$

The purpose of the computations which follow is to obtain an estimate of the norm

$$\mathcal{E}(t) \leq \Gamma\{U_0(u_0,u_1) + F(f)\} + \Gamma\{\nu(t) + \nu(t)^3\}\mathcal{E}(t) \quad \forall t \in [0,T_0), \tag{118}$$

where Γ is a positive constant that can be chosen independently of u_0, u_1, f, and T_0. Once the inequality (118) has been established, the proof can be completed by an argument similar to one used in the proof of Theorem IV.3. The reason why ν^3 rather than ν^2 (as in the previous section) appears in (118) is that we use multipliers with variable coefficients to derive our energy identities.

An integration by parts in $(93)_1$ produces

$$u_{tt}(x,t) = \chi(u_x(x,t))_x + \int_0^t a(t-\tau)\psi(u_x)_{xt}(x,\tau)\, d\tau + a(t)\psi(u_0'(x))_x + f(x,t). \tag{119}$$

We multiply (119) by $-\psi(u_x)_{xt}$ and integrate over $[0,1] \times [0,t]$. After several integrations by parts we obtain

$$\frac{1}{2}\int_0^1 \{\psi'(u_x)u_{xt}^2 + \chi'(u_x)\psi'(u_x)u_{xx}^2\}(x,t)\, dx + Q(\psi(u_x)_{xt}, t, a)$$

$$= \frac{1}{2}\int_0^1 \{\psi'(u_0'(x))u_1'(x)^2 + \chi'(u_0')\psi'(u_0')u_0''(x)^2\}\, dx$$

$$+ \frac{1}{2}\int_0^t \int_0^1 \{\psi''(u_x)u_{xt}^3 - \chi'(u_x)\psi''(u_x)u_{xt}u_{xx}^2$$

$$+ \chi''(u_x)\psi'(u_x)u_{xt}u_{xx}^2\}(x,s)\, dx\, ds$$

$$- \int_0^t \int_0^1 a(s)\psi(u_0'(x))_x\psi(u_x)_{xt}(x,s)\, dx\, ds$$

$$- \int_0^t \int_0^1 f\psi(u_x)_{xt}(x,s)\, dx\, ds \quad \forall t \in [0,T_0), \tag{120}$$

157

where Q is defined by (107) with $X := L^2(0,1)$.

For $h > 0$, let Δ_h denote the forward difference operator of stepsize h in the time variable, i.e.

$$(\Delta_h w)(x,t) := w(x,t+h) - w(x,t). \tag{121}$$

We apply Δ_h to $(93)_1$ thereby obtaining

$$\begin{aligned}
\Delta_h u_{tt} =& \Delta_h[\chi(u_x)_x] + \int_0^t a(t-\tau)\Delta_h[\psi(u_x)_{xt}(\cdot,\tau)]\, d\tau \\
&+ [a(t+h) - a(t)]\psi(u_0'(x))_x \\
&+ \int_0^h a(t+h-\tau)\psi(u_x)_{xt}(\cdot,\tau)\, d\tau + \Delta_h f.
\end{aligned} \tag{122}$$

We multiply (122) by $-\Delta_h\psi(u_x)_{xt}$ and integrate over $[0,1] \times [0,t]$. After appropriate integrations by parts, we divide the resulting equation by h^2 and let $h \downarrow 0$. The outcome of this rather long (but straightforward) computation is

$$\frac{1}{2}\int_0^1 \{\psi'(u_x)u_{xtt}^2 + \chi'(u_x)\psi'(u_x)u_{xxt}^2\}(x,t)\, dx + \lim_{h \downarrow 0}\frac{1}{h^2}Q(\Delta_h\psi(u_x)_{xt},t,a)$$

$$= \frac{1}{2}\int_0^1\{\psi'(u_x)u_{xtt}^2 + \chi'(u_x)\psi'(u_x)u_{xxt}^2\}(x,0)\, dx$$

$$+ \int_0^1 \chi''(u_0')\psi'(u_0')u_0''u_1'u_1''(x)\, dx - \int_0^1 \chi''(u_x)\psi'(u_x)u_{xx}u_{xt}u_{xxt}(x,t)\, dx$$

$$+ \int_0^t\int_0^1\Big\{\frac{1}{2}\psi''(u_x)u_{xt}u_{xtt}^2 + 2\psi''(u_x)u_{xt}u_{xxt}u_{ttt} + \psi'''(u_x)u_{xx}u_{xt}^2u_{ttt}$$

$$+ \frac{3}{2}\chi''(u_x)\psi'(u_x)u_{xt}u_{xxt}^2 + \chi''(u_x)\psi'(u_x)u_{xx}u_{xxt}u_{xtt} + \chi'''(u_x)\psi'(u_x)u_{xx}u_{xt}^2u_{xxt}$$

$$-\chi''(u_x)\psi''(u_x)u_{xx}u_{xt}^2u_{xxt} - \chi''(u_x)\psi'''(u_x)u_{xx}^2u_{xt}^3 - \frac{3}{2}\chi'(u_x)\psi''(u_x)u_{xt}u_{xxt}^2$$

$$-\chi'(u_x)\psi''(u_x)u_{xx}u_{xxt}u_{xtt} - \chi'(u_x)\psi'''(u_x)u_{xx}u_{xt}^2u_{xxt}$$

$$-\chi''(u_x)\psi''(u_x)u_{xx}^2u_{xt}u_{xtt}\Big\}(x,s)\, dx\, ds$$

$$-a'(t)\int_0^1 \psi(u_0'(x))_x\psi(u_x)_{xt}(x,t)\, dx + a'(0)\int_0^1 \psi(u_0'(x))_x\psi(u_x)_{xt}(x,0)\, dx$$

$$+ \int_0^t\int_0^1 a''(s)\psi(u_0'(x))_x\psi(u_x)_{xt}(x,s)\, dx\, ds$$

$$-a(t)\int_0^1 (\psi'(u_0'(x))u_1'(x))_x\psi(u_x)_{xt}(x,t)\, dx + a(0)\int_0^1 [(\psi'(u_0'(x))u_1'(x))_x]^2\, dx$$

$$+ \int_0^t\int_0^1 a'(s)(\psi'(u_0'(x))u_1'(x))_x\psi(u_x)_{xt}(x,s)\, dx\, ds$$

$$-\int_0^1 f_t \psi(u_x)_{xt}(x,t) \, dx + \int_0^1 f_t(x,0)(\psi'(u_0'(x))u_1'(x))_x \, dx$$

$$+ \int_0^t \int_0^1 f_{tt} \psi(u_x)_{xt}(x,s) \, dx \, ds \quad \forall t \in [0, T_0). \tag{123}$$

It is not evident a priori that $\displaystyle\lim_{h \downarrow 0} \frac{1}{h^2} Q(\Delta_h \psi(u_x)_{xt}, t, a)$ exists. However, the limits of each of the other terms involved in the derivation of (123) exist for all $\in [0, T_0)$. Consequently, the limit in question exists for all $t \in [0, T_0)$. We add (120) to (123) and use Lemma IV.9 with $w = \psi(u_x)_{xt}$ to obtain a lower bound or the left-hand side. After some routine estimations on the right-hand side we obtain the inequality

$$\int_0^1 \{u_{xx}^2 + u_{xt}^2 + u_{xxt}^2 + u_{xtt}^2\}(x,t) + \int_0^t \int_0^1 u_{xxt}^2(x,s) \, dx \, ds$$

$$\leq \Gamma\{U_0 + F\} + \Gamma\{\nu(t) + \nu(t)^3\}\mathcal{E}(t)$$

$$+ \Gamma\{\sqrt{U_0} + \sqrt{F}\}\sqrt{\mathcal{E}(t)} \quad \forall t \in [0, T_0). \tag{124}$$

To give an indication of the steps involved in deriving (124) from (120) and (123), we show the detailed estimation of several typical terms. The reader is cautioned that there are many possible ways to carry out such estimations.

Many of the terms from (120) and (123) can be handled in a very simple manner, e.g.

$$\left| \int_0^t \int_0^1 \psi''(u_x) u_{xt} u_{xtt} u_{ttt}(x,s) \, dx \, ds \right|$$

$$\leq \sup_{x \in [0,1], \, s \in [0,t]} |\psi''(u_x) u_{xt}(x,s)| \cdot \int_0^t \int_0^1 |u_{xtt} u_{ttt}(x,s)| \, dx \, ds$$

$$\leq \Gamma\nu(t) \int_0^t \int_0^1 |u_{xtt} u_{ttt}(x,s)| \, dx \, ds$$

$$\leq \Gamma\nu(t) \int_0^t \int_0^1 \{u_{xtt}^2 + u_{ttt}^2\}(x,s) \, dx \, ds$$

$$\leq \Gamma\nu(t)\mathcal{E}(t) \quad \forall t \in [0, T_0). \tag{125}$$

A similar computation yields

$$\left| \int_0^t \int_0^1 f_{tt} \psi(u_x)_{xt}(x,s) \, dx \, ds \right|$$

$$\leq \int_0^t \int_0^1 |f_{tt} \psi'(u_x) u_{xxt}(x,s)| \, dx \, ds$$

$$+ \int_0^t \int_0^1 |f_{tt} \psi''(u_x) u_{xx} u_{xt}(x,s)| \, dx \, ds$$

$$\leq \Gamma \cdot \left(\int_0^t \int_0^1 f_{tt}^2(x,s) \; dx \; ds \right)^{1/2} \cdot \left(\int_0^t \int_0^1 u_{xxt}^2(x,s) \; dx \; ds \right)^{1/2}$$

$$+ \Gamma \int_0^t \int_0^1 f_{tt}^2(x,s) \; dx \; ds + \Gamma \int_0^t \int_0^1 u_{xx}^2 u_{xt}^2(x,s) \; dx \; ds$$

$$\leq \Gamma \sqrt{F} \cdot \sqrt{\mathcal{E}(t)} + \Gamma F + \Gamma \nu(t)^2 \mathcal{E}(t) \quad \forall t \in [0,T_0). \tag{126}$$

To estimate terms such as the first integral on the right-hand side of (123) we observe that the initial values of derivatives of u can be expressed in terms of u_0, u_1 and $f(\cdot, 0)$, using equation $(93)_1$ if necessary. For example,

$$u_{xtt}(x,0) = \phi'(u_x)u_{xxx}(x,0) + \phi''(u_x)u_{xx}^2(x,0) + f_x(x,0). \tag{127}$$

Therefore we have

$$\int_0^1 u_{xtt}^2(x,0) \; dx \leq \Gamma \int_0^1 u_0'''(x)^2 \; dx + \Gamma \int_0^1 u_{xx}^4(x,0) \; dx$$

$$+ \Gamma \int_0^1 f_x^2(x,0) \; dx \leq \Gamma U_0 + \Gamma \nu(t)^2 \mathcal{E}(t) + \Gamma F \quad \forall t \in [0,T_0), \tag{128}$$

and consequently

$$\int_0^1 \{ \psi'(u_x)u_{xtt}^2 + \chi'(u_x)\psi'(u_x)u_{xxt}^2 \}(x,0) \; dx$$

$$\leq \Gamma \{ U_0 + F \} + \Gamma \nu(t)^2 \mathcal{E}(t) \quad \forall t \in [0,T_0). \tag{129}$$

The other calculations used to derive (124) from (120) and (123) are in the same spirit as those shown above. We note that (a1) implies

$$a, a' \in C_b[0,\infty) \cap L^1(0,\infty). \tag{130}$$

Taking $L^2(0,1)$ norms in $(93)_1$ and squaring the result we find

$$\int_0^1 u_{tt}^2(x,t) \; dx \leq 3 \int_0^1 \{ \phi'(u_x)^2 u_{xx}^2 + f^2 \}(x,t) \; dx$$

$$+ 3 \int_0^1 \left(\int_0^t a'(t-\tau)\psi(u_x(x,\tau))_x \; d\tau \right)^2 \; dx \tag{131}$$

from which it follows easily that

$$\int_0^1 u_{tt}^2(x,t) \; dx \leq \Gamma F + \Gamma \max_{s \in [0,t]} \int_0^1 u_{xx}^2(x,s) \; dx \quad \forall t \in [0,T_0). \tag{132}$$

We use a similar argument to obtain estimates for $\int_0^1 u_{ttt}^2$ and $\int_0^t \int_0^1 u_{ttt}^2$. Differentiation of $(93)_1$ with respect to t yields

$$u_{ttt}(x,t) = \phi'(u_x)u_{xxt}(x,t)$$

$$+ \phi''(u_x)u_{xx}u_{xt}(x,t) + a'(t)\psi(u_0'(x))_x + f_t(x,t)$$

$$+ \int_0^t a'(t-\tau)[\psi'(u_x)u_{xxt} + \psi''(u_x)u_{xx}u_{xt}](x,\tau) \; d\tau. \tag{133}$$

It follows from (133) that

$$\int_0^1 u_{ttt}^2(x,t) \, dx \leq \Gamma\{U_0 + F\} + \Gamma\nu(t)^2 \, \mathcal{E}(t) + \Gamma \max_{s \in [0,t]} \int_0^1 u_{xxt}^2(x,s) \, dx \quad \forall t \in [0, T_0).$$

(134)

Making use of the inequality

$$\int_0^t \| \int_0^s a'(s - \tau)w(\tau) \, d\tau \|^2 \, ds \leq \|a'\|_{L^1}^2 \cdot \int_0^t \|w(s)\|^2 \, ds$$

(135)

(where $\|.\|$ is the norm on $L^2(0,1)$) we also conclude from (133) that

$$\int_0^t \int_0^1 u_{ttt}^2(x,s) \, dx \, ds \leq \Gamma\{U_0 + F\} + \Gamma\nu(t)^2 \, \mathcal{E}(t) + \Gamma \int_0^t \int_0^1 u_{xxt}^2(x,s). \quad (136)$$

Combining (124), (132), and (134) we now have the estimate

$$\int_0^1 \{u_{xx}^2 + u_{xt}^2 + u_{tt}^2 + u_{xxt}^2 + u_{xtt}^2 + u_{ttt}^2\}(x,t) \, dx$$

$$+ \int_0^t \int_0^1 \{u_{xxt}^2 + u_{ttt}^2\}(x,s) \, dx \, ds$$

$$\leq \Gamma\{U_0 + F\} + \Gamma\{\nu(t) + \nu(t)^3\}\mathcal{E}(t)$$

$$+ \Gamma\{\sqrt{U_0} + \sqrt{F}\}\sqrt{\mathcal{E}(t)} \quad \forall t \in [0, T_0).$$

(137)

We can obtain a bound for $\int_0^t \int_0^1 u_{xtt}^2$ by interpolation. Employing the identity

$$\int_0^t \int_0^1 u_{xtt}^2(x,s) \, dx \, ds = \int_0^t \int_0^1 u_{xxt} u_{ttt}(x,s) \, dx \, ds$$

$$+ \int_0^1 u_{tt} u_{xxt}(x,0) \, dx - \int_0^1 u_{tt} u_{xxt}(x,t) \, dx$$

(138)

(which was used in the proof of Theorem IV.3) we find that

$$\int_0^t \int_0^1 u_{xtt}^2(x,s) \, dx \, ds \leq \Gamma\{U_0 + F\} + \Gamma \int_0^1 \{u_{tt}^2 + u_{xxt}^2\}(x,t) \, dx$$

$$+ \Gamma \int_0^t \int_0^1 \{u_{xxt}^2 + u_{ttt}^2\}(x,s) \, dx \, ds \quad \forall t \in [0, T_0).$$

(139)

To obtain our next set of estimates, we put

$$G(x,t) := u_{tt}(x,t) - f(x,t) - [\phi'(u_x) - \phi'(0)]u_{xx}(x,t)$$

$$- \int_0^t a'(t - \tau)[\psi'(u_x) - \psi'(0)]u_{xx}(x,\tau) \, d\tau$$

(140)

161

and observe that $(93)_1$ can be rewritten as

$$\phi'(0)u_{xx}(x,t) + \int_0^t a'(t-\tau)\psi'(0)u_{xx}(x,\tau)\,d\tau = G(x,t). \tag{141}$$

Using the resolvent kernel k defined by equation (110), we solve (141) for u_{xx} to get

$$\phi'(0)u_{xx}(x,t) = G(x,t) + \int_0^t k(t-\tau)G(x,\tau)\,d\tau. \tag{142}$$

Differentiation of (142) with respect to x yields

$$\phi'(0)u_{xxx}(x,t) = G_x(x,t) + \int_0^t k(t-\tau)G_x(x,\tau)\,d\tau. \tag{143}$$

Since $k \in L^1(0,\infty)$ (by Lemma IV.10), it follows from (140), (143), and a routine computation that

$$\int_{-\infty}^\infty u_{xxx}^2(x,t)\,dx \le \Gamma F + \Gamma\nu(t)^2\,\mathcal{E}(t) + \max_{s\in[0,t]}\int_{-\infty}^\infty u_{xtt}^2(x,s)\,dx, \quad \forall t \in [0,T_0), \tag{144}_1$$

and

$$\int_0^t\int_{-\infty}^\infty u_{xxx}^2(x,s)\,dx\,ds \le \Gamma F + \Gamma\nu(t)^2\,\mathcal{E}(t)$$
$$+ \int_0^t\int_{-\infty}^\infty u_{xtt}^2(x,s)\,dx\,ds \quad \forall t \in [0,T_0). \tag{144}_2$$

Up to this point we have made no use of the Poincaré inequality. Anticipating the proof of Theorem IV.7, let us summarize the estimates obtained so far. Combining (137), (139), and (144), we conclude that

$$\int_0^1 \{u_{xx}^2 + u_{xt}^2 + u_{tt}^2 + u_{xxx}^2 + u_{xxt}^2 + u_{xtt}^2 + u_{ttt}^2\}(x,t)\,dx$$
$$+ \int_0^t\int_0^1 \{u_{xxx}^2 + u_{xxt}^2 + u_{xtt}^2 + u_{ttt}^2\}(x,s)\,dx\,ds$$
$$\le \Gamma\{U_0 + F\} + \Gamma\{\nu(t) + \nu(t)^3\}\mathcal{E}(t)$$
$$+ \Gamma\{\sqrt{U_0} + \sqrt{F}\}\sqrt{\mathcal{E}(t)} \quad \forall t \in [0,T_0). \tag{145}$$

Using (140) and (142) we find

$$\int_0^t\int_0^1 u_{xx}^2(x,s)\,dx\,ds \le \Gamma F + \Gamma\nu(t)^2\,\mathcal{E}(t) + \Gamma\int_0^t\int_0^1 u_{tt}^2(x,s)\,dx\,ds \quad \forall t \in [0,T_0). \tag{146}$$

162

Since u, u_x, u_t, u_{xt}, and u_{tt} either vanish at the endpoints or have zero mean spatially, it follows from (145), (146), and the Poincaré inequality that

$$\mathcal{E}(t) \leq \Gamma\{U_0 + F\} + \Gamma\{\nu(t) + \nu(t)^3\}\mathcal{E}(t) + \Gamma\{\sqrt{U_0} + \sqrt{F}\}\sqrt{\mathcal{E}(t)} \quad \forall t \in [0, T_0). \tag{147}$$

Employing the simple algebraic inequality

$$|AB| \leq \eta A^2 + \frac{1}{4\eta} B^2 \quad \forall A, B \in \mathbb{R}, \; \eta > 0 \tag{148}$$

with η sufficiently small, we conclude from (147) that

$$\mathcal{E}(t) \leq \bar{\Gamma}\{U_0 + F\} + \bar{\Gamma}\{\nu(t) + \nu(t)^3\}\mathcal{E}(t) \quad \forall t \in [0, T_0), \tag{149}$$

where $\bar{\Gamma}$ is a positive constant that can be chosen independently of u_0, u_1, f, and T_0. The proof can now be completed by the type of argument used in the proof of Theorem IV.3.

∎

B. Sketch of proof of Theorem IV.7

We now provide a rather complete sketch of the proof of Theorem IV.7. As in the proof of Theorem IV.5 we choose a sufficiently small positive number δ and modify ϕ, ψ (and also χ) outside the interval $[-\delta, \delta]$ in such a way that (112) and (113) hold for some positive constants ϕ, ψ, and χ. It follows from the arguments of Chapter III that if the data satisfy $(s2^*)$ and $(s3^*)$ then the initial value problem (101) has a unique solution u on a maximal time interval $[0, T_0)$ with

$$u_x, u_t, u_{xx}, u_{xt}, u_{tt}, u_{xxx}, u_{xxt}, u_{xtt}, u_{ttt} \in C([0, T_0); L^2(\mathbb{R})). \tag{150}$$

Moreover, if

$$\sup_{t \in [0, T_0)} \{\|u_x(\cdot, t)\|_2^2 + \|u_t(\cdot, t)\|_2^2\} < \infty \tag{151}$$

then $T_0 = \infty$. As in the proof of Theorem IV.5, we use Γ to denote a generic constant which can be chosen independently of u_0, u_1, f, and T_0. Our basic approach is the same as in the proof of Theorem IV.5; in fact, many of the same estimates can be used.

In place of $\mathcal{E}(t)$ and $\nu(t)$ we define

$$\mathcal{E}^*(t) := \max_{s \in [0,t]} \int_{-\infty}^{\infty} \{u_x^2 + u_t^2 + u_{xx}^2 + u_{xt}^2 + u_{tt}^2 + u_{xxx}^2 + u_{xxt}^2 + u_{xtt}^2 + u_{ttt}^2\}(x, s)\, dx\, ds$$

$$+ \int_0^t \int_{-\infty}^{\infty} \{u_{xx}^2 + u_{xt}^2 + u_{tt}^2 + u_{xxx}^2 + u_{xxt}^2 + u_{xtt}^2 + u_{ttt}^2\}(x, s)\, dx\, ds \quad \forall t \in [0, T_0) \tag{152}_1$$

163

and

$$\nu^*(t) := \sup_{x \in \mathbb{R}, \ s \in [0,t]} \{u_x^2 + u_{xx}^2 + u_{xt}^2\}^{1/2}(x,s). \tag{152}_2$$

The procedure used to derive (145) in the proof of Theorem IV.5 yields

$$\int_{-\infty}^{\infty} \{u_{xx}^2 + u_{xt}^2 + u_{tt}^2 + u_{xxx}^2 + u_{xxt}^2 + u_{xtt}^2 + u_{ttt}^2\}(x,t) \ dx$$

$$+ \int_0^t \int_{-\infty}^{\infty} \{u_{xxx}^2 + u_{xxt}^2 + u_{xtt}^2 + u_{ttt}^2\}(x,s) \ dx \ ds$$

$$\le \Gamma\{U_0^* + F^*\} + \Gamma\{\nu^*(t) + \nu^*(t)^3\}\mathcal{E}^*(t)$$

$$+ \Gamma\{\sqrt{U_0^*} + \sqrt{F^*}\}\sqrt{\mathcal{E}^*(t)} \quad \forall t \in [0, T_0). \tag{153}$$

We must obtain a similar bound for

$$\int_{-\infty}^{\infty} \{u_x^2 + u_t^2\}(x,t) \ dx + \int_0^t \int_{-\infty}^{\infty} \{u_{xx}^2 + u_{xt}^2 + u_{tt}^2\}(x,s) \ dx \ ds.$$

The Poincaré inequality is not applicable on all of space and we must therefore use a different approach. To proceed further, we transform $(101)_1$ to a more convenient form. This transformation was motivated by MacCamy's procedure [M3]. It follows from (a1) that $a'(0) < 0$. Therefore, without loss of generality we assume

$$a'(0) = -1. \tag{154}$$

Let r denote the resolvent kernel associated with $-a''$, i.e. the unique solution of

$$r(t) - \int_0^t a''(t - \tau)r(\tau) \ d\tau = a''(t), \quad t \ge 0. \tag{155}$$

It follows from the Paley-Wiener theorem that $r \notin L^1(0, \infty)$. However, it is not difficult to show that

$$r(t) = a(0)^{-1} + R(t), \quad t \ge 0, \tag{156}$$

where

$$R \in L^1(0, \infty). \tag{157}$$

(See Lemma 2.3 of [H12].)

Differentiation of $(101)_1$ with respect to t yields

$$u_{ttt}(x,t) = \phi(u_x(x,t))_{xt} - \psi(u_x(x,t))_x + \int_0^t a''(t - \tau)\psi(u_x(x,\tau))_x \ d\tau + f_t(x,t). \tag{158}$$

Using the representation formulas (79) to solve (158) for $\psi(u_x)_x$ and then rearranging the terms we find that

$$u_{ttt}(x,t) = \phi(u_x(x,t))_{xt} - \psi(u_x(x,t))_x + f_t(x,t)$$

$$+ \int_0^t r(t - \tau)[\phi(u_x)_{xt} - u_{ttt} + f_t](x,\tau) \ d\tau. \tag{159}$$

164

Setting

$$\alpha := a(0)^{-1}$$

and using (156) we rewrite (159) as

$$u_{ttt} + \alpha u_{tt} = \phi(u_x)_{xt} + \alpha\chi(u_x)_x + f_t + \alpha f + [R * (\phi(u_x)_x - u_{tt} + f)]_t \quad (161)$$

where the $*$ denotes convolution on $[0,t]$, i.e.

$$(R * w)(x,t) := \int_0^t R(t-\tau)w(x,\tau) \, d\tau. \quad (162)$$

(In the derivation of (161) we used the fact that $[\phi(u_x)_x - u_{tt} + f](x,0) \equiv 0$ which follows from equation $(101)_1$.)

Let us set

$$W(\xi) := \int_0^\xi \chi(s) \, ds \quad (163)$$

and note that

$$W(\xi) \geq \frac{1}{2}\underline{\chi}\xi^2 \quad \forall\xi \in \mathbb{R}$$

by virtue of (113) and the fact that $\chi(0) = 0$. We multiply (161) by u_t and (119) by u_{tt}, add the resulting equations and integrate over $(-\infty,\infty) \times [0,t]$. After several integrations by parts we obtain

$$\alpha\int_{-\infty}^\infty \{\frac{1}{2}u_t^2 + W(u_x)\}(x,t) \, dx + a(0)\int_0^t\int_{-\infty}^\infty \psi'(u_x)u_{xt}^2(x,s) \, dx \, ds$$

$$+ \int_0^t\int_{-\infty}^\infty u_{tt}[R * (\phi(u_x)_x - u_{tt} + f](x,s) \, dx \, ds$$

$$= \int_{-\infty}^\infty \{\frac{1}{2}\alpha u_t^2 + \alpha W(u_x) + u_t u_{tt} + \chi(u_x)u_{xt} - fu_t\}(x,0) \, dx$$

$$+ \int_{-\infty}^\infty \{fu_t - u_t u_{tt} - \chi(u_x)u_{xt}\}(x,t) \, dx$$

$$+ \int_{-\infty}^\infty u_t[R * (\phi(u_x)_x - u_{tt} + f)](x,t) \, dx$$

$$+ \int_0^t\int_{-\infty}^\infty \{\alpha f u_t + [a(s)\psi(u_0'(x))]_x\}u_{tt}(x,s) \, dx \, ds$$

$$+ \int_0^t\int_{-\infty}^\infty u_{tt}[a * \psi(u_x)_{xt}](x,s) \, dx \, ds \quad \forall t \in [0,T_0). \quad (164)$$

The crucial term to analyze is

$$\Phi(t) := \int_0^t\int_{-\infty}^\infty u_{tt}[R * (\phi(u_x)_x - u_{tt} + f](x,s) \, dx \, ds; \quad (165)$$

165

the other terms in (164) are favorable or can be estimated routinely. It follows from (111) and (119) that

$$\phi(u_x)_x - u_{tt} + f = a(0)\psi(u_x)_x - a(t)\psi(u_0'(x))_x - a * \psi(u_x)_{xt}, \tag{166}$$

and substitution of this relation into (165) yields

$$\begin{aligned}
\Phi(t) = {}& a(0) \int_0^t \int_{-\infty}^\infty u_{tt}[R * \psi(u_x)_x](x,s) \; dx \; ds \\
& - \int_0^t \int_{-\infty}^\infty [R * a](s)\psi(u_0'(x))_x u_{tt}(x,s) \; dx \; ds \\
& - \int_0^t \int_{-\infty}^\infty u_{tt}[R * a * \psi(u_x)_{xt}](x,s) \; dx \; ds.
\end{aligned} \tag{167}$$

Since $R, a \in L^1(0,\infty)$ and we already have estimates for the third order derivatives of u, the last two terms in (167) cause no difficulties. However, the first term on the right-hand side of (167) requires special attention.

Let us set

$$M(t) := -\int_t^\infty R(s) \; ds \quad \forall t \geq 0. \tag{168}$$

One can show that

$$M \in L^1(0,\infty), \quad M(0) < 1. \tag{169}$$

(See Lemma 2.4 of [H12].) Employing the kernel M, we find that

$$R * \psi(u_x)_x = -M(0)\psi(u_x)_x + M(t)\psi(u_0'(x))_x + M * \psi(u_x)_{xt}. \tag{170}$$

We observe further that

$$\begin{aligned}
\int_0^t \int_{-\infty}^\infty \psi(u_x)_x u_{tt}(x,s) \; dx \; ds = {}& \int_0^t \int_{-\infty}^\infty \psi'(u_x)u_{xt}^2(x,s) \; dx \; ds \\
& + \int_{-\infty}^\infty \psi(u_x)u_{xt}(x,0) \; dx - \int_{-\infty}^\infty \psi(u_x)u_{xt}(x,t) \; dx
\end{aligned} \tag{171}$$

as can be verified using two integrations by parts. Combining (167), (168), and (170), we arrive at the following expression for Φ:

$$\begin{aligned}
\Phi(t) = {}& -a(0)M(0) \int_0^t \int_{-\infty}^\infty \psi'(u_x)u_{xt}^2(x,s) \; dx \; ds \\
& + a(0)M(0) \int_{-\infty}^\infty \psi(u_x)u_{xt}(x,0) \; dx - a(0)M(0) \int_{-\infty}^\infty \psi(u_x)u_{xt}(x,t) \; dx \\
& + \int_0^t \int_{-\infty}^\infty [M - (R * a)](s)\psi(u_0'(x))_x u_{tt}(x,s) \; dx \; ds \\
& + \int_0^t \int_{-\infty}^\infty \{[M - (R * a)] * \psi(u_x)_{xt}\}u_{tt}(x,s) \; dx \; ds.
\end{aligned} \tag{172}$$

Since $M(0) < 1$, the first integral in the above expression can be absorbed by the second integral on the left-hand side of (164). Moreover, the remaining terms in (172) can be estimated easily. (Note that $[M - (R * a)] \in L^1(0, \infty)$ since $M, R, a \in L^1(0, \infty)$.) After substitution of (172) into (164), and a rather long computation, we obtain

$$\int_{-\infty}^{\infty} \{u_x^2 + u_t^2\}(x, t) \, dx + \int_0^t \int_{-\infty}^{\infty} u_{xt}^2(x, s) \, dx \, ds$$
$$\leq \Gamma\{U_0^* + F^*\} + \Gamma \nu^*(t) \mathcal{E}^*(t)$$
$$+ \Gamma \Big(\int_0^t \int_{-\infty}^{\infty} u_{tt}^2(x, s) \, dx \, ds \Big)^{1/2} \cdot \Big(\int_0^t \int_{-\infty}^{\infty} u_{xxt}^2(x, s) \, dx \, ds \Big)^{1/2}$$
$$+ \Gamma \max_{s \in [0, t]} \int_{-\infty}^{\infty} \{u_{xx}^2 + u_{xt}^2 + u_{tt}^2\}(x, s) \, dx \, ds \quad \forall t \in [0, T_0). \quad (173)$$

In the derivation of (173) we have made use of the simple algebraic inequality (148) to handle several terms. For example, observe that

$$\Big| \int_{-\infty}^{\infty} \chi(u_x) u_{xt}(x, t) \, dx \Big| \leq \bar{\chi} \int_{-\infty}^{\infty} |u_x u_{xt}(x, t)| \, dx$$
$$\leq \eta \bar{\chi} \int_{-\infty}^{\infty} u_x^2(x, t) \, dx + \frac{\bar{\chi}}{4\eta} \int_{-\infty}^{\infty} u_{xt}^2(x, t) \, dx \quad \forall t \in [0, T_0) \quad (174)$$

for every $\eta > 0$, where

$$\bar{\chi} := \sup_{\xi \in \mathbb{R}} \chi'(\xi). \quad (175)$$

On account of (163), $\eta \bar{\chi} \int_{-\infty}^{\infty} u_x^2$ can be absorbed by the first integral on the left-hand side of (164) if η is sufficiently small. The size of the coefficient $\frac{\bar{\chi}}{4\eta}$ is unimportant because we already have an estimate for $\int_{-\infty}^{\infty} u_{xt}^2$. Moreover, we have made essential use of the assumption $f \in L^1([0, \infty); L^2(\mathbb{R}))$ to estimate $\int_0^t \int_{-\infty}^{\infty} f u_t$ since it does not seem possible to obtain a time-independent bound for $\int_0^t \int_{-\infty}^{\infty} u_t^2$.

We now multiply (119) by u_{tt} and integrate over $(-\infty, \infty) \times [0, t]$. After a routine computation we obtain

$$\int_0^t \int_{-\infty}^{\infty} u_{tt}^2(x, s) \, dx \, ds \leq \Gamma\{U_0^* + F^*\} + \Gamma \nu^*(t) \mathcal{E}^*(t)$$
$$+ \Gamma \Big(\int_0^t \int_{-\infty}^{\infty} u_{tt}^2(x, s) \, dx \, ds \Big)^{1/2} \cdot \Big(\int_0^t \int_{-\infty}^{\infty} u_{xxt}^2(x, s) \, dx \, ds \Big)^{1/2}$$
$$+ \Gamma \max_{s \in [0, t]} \int_{-\infty}^{\infty} u_{tt}^2(x, s) \, dx + \Gamma \int_0^t \int_{-\infty}^{\infty} u_{xt}^2(x, s) \, dx \, ds. \quad (176)$$

Combining (173) and (176) and using (148) (with η sufficiently small) and recalling the estimate (153) we obtain the inequality

$$\int_{-\infty}^{\infty} \{u_x^2 + u_t^2\}(x, t) \, dx + \int_0^t \int_{-\infty}^{\infty} \{u_{xt}^2 + u_{tt}^2\}(x, s) \, dx \, ds$$
$$\leq \Gamma\{U_0^* + F^*\} + \Gamma\{\nu^*(t) + \nu^*(t)^3\} \mathcal{E}^*(t)$$
$$+ \Gamma\{\sqrt{U_0^*} + \sqrt{F^*}\} \sqrt{\mathcal{E}^*(t)} \quad \forall t \in [0, T_0). \quad (177)$$

167

To obtain an estimate for $\int_0^t \int_{-\infty}^\infty u_{xx}^2$ we proceed as in the derivation of (144). Using the resolvent kernel k defined by (110) to solve for u_{xx} in terms of u_{tt} we find that

$$\int_0^t \int_{-\infty}^\infty u_{xx}^2(x,s) \, dx \, ds \le \Gamma F^* + \Gamma \nu^*(t)^2 \, \mathcal{E}^*(t)$$

$$+ \Gamma \int_0^t \int_{-\infty}^\infty u_{tt}^2(x,s) \, dx \, ds \quad \forall t \in [0, T_0). \tag{178}$$

Combining (177) and (178) and adding the result to (153) we obtain the inequality

$$\mathcal{E}^*(t) \le \Gamma\{U_0^* + F^*\} + \Gamma\{\nu^*(t) + \nu^*(t)^3\}\mathcal{E}^*(t)$$

$$+ \Gamma\{\sqrt{U_0^*} + \sqrt{F^*}\}\sqrt{\mathcal{E}^*(t)} \quad \forall t \in [0, T_0). \tag{179}$$

Using the inequality (148) with η small we conclude from (179) that

$$\mathcal{E}^*(t) \le \Gamma\{U_0^* + F^*\} + \Gamma\{\nu^*(t) + \nu^*(t)^3\}\mathcal{E}^*(t) \quad \forall t \in [0, T_0). \tag{180}$$

The proof can now be completed by the type of argument that was used in the proof of Theorem IV.3.

A problem with a singular kernel

The global estimates used in the proof of Theorem IV.5 can be adapted to handle certain problems with singular kernels. We now consider the history value problem of Section III.5, viz.

$$u_{tt}(x,t) = \phi(u_x(x,t))_x + \int_{-\infty}^t a'(t-\tau)\psi(u_x(x,\tau))_x \, d\tau + f(x,t), \quad x \in [0,1], \, t \ge 0,$$
$$\tag{181}_1$$
$$u(0,t) = u(1,t) = 0. \tag{181}_2$$
$$u(x,t) = \tilde{u}(x,t), \, x \in [0,1], \, t \le 0, \quad u(x,0^+) = \tilde{u}(x,0^-), \quad u_t(x,0^+) = \tilde{u}_t(x,0^-).$$
$$\tag{181}_3$$

As in Section III.5, we assume that the given history $\tilde{u}$ satisfies equation $(181)_1$ and the boundary conditions $(181)_2$ for $t \le 0$, and that f is smooth on $[0,1] \times \mathbb{R}$. Our assumptions read

$$a, a' \in L^1(0, \infty), \quad a, -a', a'' \ge 0,$$

($\tilde{a}$)   and a'' has a nontrivial absolutely continuous component.

($\tilde{s}$1) $$\phi, \psi \in C^3(\mathbb{R}),$$

($\tilde{s}$2) $$f, f_x, f_t \in L^\infty(\mathbb{R}; L^2(0,1)), \quad f, f_x, f_t, f_{tt} \in L^2(\mathbb{R}; L^2(0,1)),$$

168

)
$$\phi'(0) > 0, \ \psi'(0) > 0, \ \phi'(0) - a(0)\psi'(0) > 0.$$

)   The given history satisfies $(181)_1$ and $(181)_2$ for $t \leq 0$,

ad to measure the size of the data we set

$$\tilde{F}(f) := \text{ess} - \sup_{t \in \mathbb{R}} \int_0^1 \{f^2 + f_x^2 + f_t^2\}(x,t) \ dx$$

$$+ \int_{-\infty}^{\infty} \int_0^1 \{f^2 + f_x^2 + f_t^2 + f_{tt}^2\}(x,t) \ dx \ dt. \tag{182}$$

Theorem IV.11:

Assume that $(\tilde{a})$, $(\tilde{s}1)$, $(\tilde{s}2)$, (e), and $(\tilde{c})$ hold. Then there is a constant $\mu > 0$ such that if

$$\tilde{F}(f) \leq \mu, \tag{183}$$

he history value problem (181) has a unique solution $u : [0,1] \times \mathbb{R} \to \mathbb{R}$ with

$$u, u_x, u_t, u_{xx}, u_{xt}, u_{tt}, u_{xxx}, u_{xxt}, u_{xtt}, u_{ttt} \in L^{\infty}(\mathbb{R}; L^2(0,1)). \tag{184}$$

foreover, as $t \to \infty$, we have, uniformly on $[0,1]$,

$$u, u_x, u_t, u_{xx}, u_{xt}, u_{tt} \to 0. \tag{185}$$

Remarks IV.12:

1. A similar global existence result holds for Neumann or mixed boundary conditions. In the case of a Neumann problem the decay statement must be modified due to the possibility of nontrivial rigid motions; more precisely u should be replaced by u minus its spatial mean.
2. For unbounded spatial intervals, the problem of global existence (with a singular kernel) is open.

   The proof of Theorem IV.11 (see [H13]) is based on Theorem III.19 (local xistence) and a simple variant of the proof of Theorem IV.5.

Remarks on extensions

   The approach of this section can be adapted to handle equations governing iotions of very general classes of viscoelastic solids. The analysis was carried out i [H8] (for one-dimensional problems) assuming a general functional dependence f the stress on the strain (see also [H9]). In the case of a general functional ependence, the assumptions become rather complicated.

To indicate the procedure for more general constitutive relations, we consid
the equation

$$u_{tt}(x,t) = \int_{-\infty}^{t} h(t - \tau, u_x(x,t), u_x(x,\tau))_x \ d\tau + f(x,t), \quad x \in B, \ t \geq 0, \quad (186$$

and for simplicity we assume that $u \equiv 0$ for $t < 0$. We seek a solution u subje
to the initial conditions

$$u(x,0+) = u_0(x), \ u_t(x,0+) = u_1(x), \quad x \in B, \tag{186}$$

and appropriate boundary conditions if $B \neq \mathbb{R}$. We rewrite $(186)_1$ as its li:
earization about zero plus correction terms, i.e.

$$u_{tt} = \alpha u_{xx} + \int_0^t a'(t - \tau) u_{xx}(x,\tau) \ d\tau + f$$

$$+ \left(\int_{-\infty}^t h_{,2}(t - \tau, u_x(x,t), u_x(x,\tau)) - h_{,2}(t - \tau, 0, 0) \ d\tau \right) u_{xx} \tag{18}$$

$$+ \int_0^t \left(h_{,3}(t - \tau, u_x(x,t), u_x(x,\tau)) - h_{,3}(t - \tau, 0, 0) \right) u_{xx}(x,\tau) \ d\tau,$$

where

$$\alpha := \int_0^\infty h_{,2}(s, 0, 0) \ ds, \quad a(t) := -\int_t^\infty h_{,3}(s, 0, 0) \ ds. \tag{18}$$

If h is smooth on $[0, \infty) \times \mathbb{R} \times \mathbb{R}$ and satisfies appropriate integrability condition
with respect to its first argument, then the last two terms in (187) can be treate
as a perturbation. If, in addition, $\alpha > 0$, the kernel a is smooth on $[0, \infty)$ ar
satisfies the same assumptions as before (i.e. (a1) if B is bounded and (a1), (a2)
$B = \mathbb{R}$), and $\alpha - a(0) > 0$, one can establish global existence of classical solutio:
to (186) provided that u_0, u_1 and f are suitably smooth and small.

In the special case when ϕ is linear (i.e. $\phi(\xi) = \alpha\xi$ with $\alpha > 0$) and ψ'
bounded, one can obtain globally defined classical solutions to (93), or (101) eve
for large initial data. (See Heard [H6] and Hrusa [H10]). Decay of solutions wit
large initial data is discussed in [H10]. Looking back at Example II.4 in Sectic
II.4, we see that the linearity of ϕ is in fact necessary to obtain global smoot
solutions for large data.

The energy method is not tied to one space dimension; however, in mo:
than one space dimension estimates for derivatives of order higher than three a:
needed. The basic approach of this section can be applied to solids in more tha
one space dimension, but the analysis is very complicated and the details ha\
not been carried out.

The energy method can also be applied to problems involving fluids; hov
ever, in this case one cannot expect that u decays to zero. This manifests itse
mathematically in the nonintegrability of the resolvent kernel k of Lemma IV.1
Additional estimates are therefore needed to carry out the analysis. These ca
be obtained for various special problems, but a general existence result for flui
has not been proved. In the next section, we prove a global existence result for
special fluid model in three space dimensions.

5. Three-dimensional motions of a K-BKZ fluid

The energy method can be applied in any number of space dimensions, although the complexity of the estimates increases considerably with the dimension. In this section, we shall derive global estimates for three-dimensional spatially periodic motions of an incompressible K-BKZ fluid. The initial data are assumed to be small. We follow the approach used by Kim [K12] for the initial value problem in all of space in the case of a more special constitutive model.

In order to highlight the main ideas, we shall give only formal derivations of the required energy integrals. The reason that the formal derivation is not rigorous is that it requires smoothness beyond that guaranteed by the local existence theorem. An approximation procedure (e.g. difference quotients, mollifiers, artificial viscosity) can be used for a rigorous justification. See e.g. Kim [K12] who uses mollifiers and artificial viscosity.

Recall that the constitutive relation for an incompressible K-BKZ fluid with separated kernel reads (cf. equation (52) in Chapter I)

$$\mathbf{T}_E (y,t) = 2 \int_{-\infty}^{t} m(t-s) \left\{ \hat{W}_{,1} (I_1, I_2) \mathbf{C}_r^{-1}(s,y,t) - \hat{W}_{,2} (I_1, I_2) \mathbf{C}_r(s,y,t) \right\} \, ds, \tag{189}_1$$

where

$$I_1 := \operatorname{tr} \mathbf{C}_r^{-1}(s,y,t), \quad I_2 := \operatorname{tr} \mathbf{C}_r(s,y,t). \tag{189}_2$$

Kim [K12] studies the special case when $\hat{W} = I_2$. For the purpose of deriving global energy estimates, it is convenient to rewrite (189) in a different form. We introduce the following notations: By $Y(t-\tau,y,t)$ we denote the position at time $t - \tau$ of the particle which occupies the position y at time t, and by $u(\tau,y,t)$ we denote the relative displacement $Y(t - \tau,y,t) - y$. The gradient of $Y(t - \tau,y,t)$ with respect to y is the relative deformation gradient $\mathbf{F}_r(t - \tau,y,t)$ introduced in equation (31) of Chapter I, i.e.

$$(\mathbf{F}_r)^i_j (t - \tau,y,t) = \frac{\partial Y^i(t - \tau,y,t)}{\partial y^j}. \tag{190}$$

By $\mathbf{\Phi}(s,y,t)$ we denote the matrix inverse to $\mathbf{F}_r(s,y,t)$. Observe that $I_1 = \mathbf{\Phi}\mathbf{\Phi}^T$ and $I_2 = \operatorname{tr} \mathbf{\Phi}^{-1}\mathbf{\Phi}^{-T}$, and hence $\hat{W}$ can be expressed as a function of $\mathbf{\Phi}$: $\hat{W}(I_1, I_2) = W(\mathbf{\Phi})$. The constitutive law $(189)_1$ can now be written in the form

$$\mathbf{T}_E^{mn} (y,t) = \int_{-\infty}^{t} m(t-s) \Phi_k^m (s,y,t) \frac{\partial W}{\partial \Phi_k^n}(s,y,t) \, ds. \tag{191}$$

Without loss of generality we assume $W(\mathbf{1}) = 0$.

If we assume that the fluid has unit density and the body force is zero, the equations of motion in Eulerian coordinates read

$$\dot{v} + (v \cdot \nabla_y)v = \operatorname{div}_y \mathbf{T}_E - \nabla_y p, \tag{192}_1$$

171

$$\operatorname{div}_y v = 0. \tag{192}$$

To simplify notation, we shall henceforth omit the subscript y on spatial deriva tives and the subscript E on the extra stress. A dot or the symbol $\frac{\partial}{\partial t}$ denote the Eulerian time derivative (i.e. for fixed y), while $\frac{D}{Dt} = \frac{\partial}{\partial t} + (v \cdot \nabla)$ denotes th material time derivative. We use χ to denote the motion (as in Chapter I) an we write $x = \chi^{-1}(y,t)$ for the material point. We can then write u in the form

$$u(\tau, y, t) = \chi(x, t - \tau) - \chi(x, t), \tag{193}$$

and by taking the material time derivative, we obtain the equation

$$\frac{D}{Dt} u(\tau, y, t) = \frac{D}{Dt} \chi(x, t - \tau) - \frac{D}{Dt} \chi(x, t) = -\frac{\partial}{\partial \tau} u(\tau, y, t) - v(y, t). \tag{194}$$

For $\tau = 0$, the relative displacement vanishes identically. Together with (194) this implies

$$u(0, y, t) = v(y, t) + \frac{\partial}{\partial \tau} u(\tau, y, t)|_{\tau=0} = 0. \tag{194}$$

We seek a spatially periodic solution (with period L_i in the ith coordinate) t equations (192), (191), (194) subject to given periodic initial data

$$v(y, 0) = v_0(y), \ u(\tau, y, 0) = u_0(\tau, y), \ y \in \mathbb{R}^3, \ \tau \in [0, \infty), \tag{195}$$

and we write $Q := [0, L_1] \times [0, L_2] \times [0, L_3]$.

To obtain a first energy integral, we observe that

$$\frac{D}{Dt} \int_{-\infty}^t m(t-s) W(\mathbf{\Phi}(s, y, t)) \, ds = \int_{-\infty}^t m'(t-s) W(\mathbf{\Phi}(s, y, t)) \, ds$$

$$+ \int_{-\infty}^t m(t-s) \frac{\partial W}{\partial \Phi_k^m}(s, y, t) \frac{D\Phi_k^m}{Dt}(s, y, t) \, ds. \tag{196}$$

(Recall that $W(\mathbf{1}) = 0$.) Moreover, noting that $\mathbf{\Phi}(s, y, t) = \mathbf{F}(x, t)\mathbf{F}^{-1}(x, s)$, an using (3)$_1$ of Chapter I, we find

$$\frac{D}{Dt} \Phi_k^m(s, y, t) = \Phi_k^n(s, y, t) \frac{\partial v^m}{\partial y^n}(y, t), \tag{197}$$

and we can therefore rewrite the last term on the right hand side of (196) as

$$\int_{-\infty}^t m(t-s) \frac{\partial W}{\partial \Phi_k^m}(s, y, t) \Phi_k^n(s, y, t) \frac{\partial v^m}{\partial y^n}(y, t) \, ds = \mathbf{T}(y, t) : \nabla v(y, t). \tag{198}$$

We now multiply the equation of motion by v and integrate over Q. After a integration by parts this yields the energy identity

$$\frac{1}{2} \frac{d}{dt} \int_Q |v(y, t)|^2 \, dy = - \int_Q \mathbf{T} : \nabla v \, dy$$

$$= -\frac{d}{dt} \int_Q \int_{-\infty}^t m(t-s)W\big(\Phi(s,y,t)\big)\, ds\, dy + \int_Q \int_{-\infty}^t m'(t-s)W\big(\Phi(s,y,t)\big)\, ds\, dy.$$

$$(199)$$

An identity of this form was obtained by Infante and Walker [I1] in the special case $\hat{W} = I_2$.

To see what type of bounds can be obtained from the energy identity (199), let $\mathbf{H}$ denote the symmetric part of $\Phi - \mathbf{1}$. Using the fact that $\det \Phi = 1$, it can be shown that, in a neighborhood of $\mathbf{1}$, we have

$$I_1 - 3 = 2\mathrm{tr}\,\mathbf{HH}^T + O(|\Phi - \mathbf{1}|^3), \qquad (200)_1$$

$$I_2 - 3 = 2\mathrm{tr}\,\mathbf{HH}^T + O(|\Phi - \mathbf{1}|^3). \qquad (200)_2$$

If Φ lies in a small neighborhood of the identity, we can therefore approximate $W(\Phi)$ by

$$2\Big[\hat{W}_{,1}\,(3,3) + \hat{W}_{,2}\,(3,3)\Big]\mathrm{tr}\,\mathbf{HH}^T. \qquad (201)$$

Moreover, we note that

$$\mathbf{H}(s,y,t) = -\frac{1}{2}\big(\nabla u(t-s,y,t) + (\nabla u(t-s,y,t))^T\big) + O(|\nabla u|^2), \qquad (202)$$

and hence

$$W(\Phi(s,y,t)) = \Big[\hat{W}_{,1}\,(3,3) + \hat{W}_{,2}\,(3,3)\Big]$$

$$\times\, \mathrm{tr}\,\Big[\nabla u(t-s,y,t)(\nabla u(t-s,y,t))^T + (\nabla u(t-s,y,t))^2\Big] + O(|\nabla u|^3). \qquad (203)$$

To proceed further we observe that $\mathrm{tr}\,\big(\nabla u(\nabla u)^T\big) = |\nabla u|^2$, and also that (as can be verified via integration by parts)

$$\int_Q \mathrm{tr}\,(\nabla u)^2\, dy = \int_Q |\mathrm{div}\ u|^2\, dy. \qquad (204)$$

Therefore, to within terms of order $|\nabla u|^3$, we have

$$\int_Q \int_{-\infty}^t m(t-s)W\big(\Phi(s,y,t)\big)\, ds\, dy$$

$$\sim \int_Q \int_0^\infty m(\tau)\big\{|\nabla u(\tau,y,t)|^2 + |\mathrm{div}\ u(\tau,y,t)|^2\big\}\, d\tau\, dy. \qquad (205)$$

Thus, if we assume that ∇u remains uniformly small, and moreover, that $m > 0$, $m' < 0$ and $\hat{W}_{,1}\,(3,3) + \hat{W}_{,2}\,(3,3) > 0$, then the energy identity (199) yields bounds on $\int_Q |v(y,t)|^2\, dy$ and $\int_Q \int_0^\infty m(\tau)|\nabla u(\tau,y,t)|^2\, d\tau\, dy$. Unfortunately such bounds are too weak to establish global existence of a solution or even to show that ∇u does in fact remain uniformly small. As before, the idea is to differentiate the equation of motion and come up with analogues of the energy estimate (199) for the differentiated equations.

To state the existence theorem, we introduce function spaces Y and Z similar to those used by Kim [K12]. Let S denote the space of pairs (v, u) of vector-valued functions that satisfy

$$(v, u) \in [C_p^\infty (Q)]^3 \times [C_{0,p}^\infty ([0, \infty) \times Q)]^3, \qquad (206)_1$$

$$\operatorname{div} v = 0, \qquad (206)_2$$

$$u(0, y) = 0 = v(y) + \frac{\partial}{\partial \tau} u(\tau, y)|_{\tau = 0}. \qquad (207)$$

Here C_p^∞ denotes the space of periodic C^∞-functions, and $C_{0,p}^\infty$ denotes the space of C^∞-functions which are periodic in the spatial direction and have compact support in $[0, \infty)$ in the temporal direction. The spaces Y and Z are the completions of S under the norms

$$\|(v, u)\|_Y^2 := \|v\|_{H^3}^2 + \int_0^\infty m(\tau) \|u(\tau, \cdot)\|_{H^4}^2 \, d\tau + \int_0^\infty m(\tau) \|\frac{\partial u}{\partial \tau}(\tau, \cdot)\|_{H^3}^2 \, d\tau, \quad (208)$$

and

$$\|(v, u)\|_Z^2 := \|v\|_{H^{3/2}}^2 + \int_0^\infty m(\tau) \|u(\tau, \cdot)\|_{H^2}^2 \, d\tau. \qquad (209)$$

The assumptions for the global existence theorem are:
(S1) $\hat{W}$ is of class[1] C^∞.
(S2) $m \in C^2[0, \infty)$.
(M) $m(0) > 0$, and there are constants $d_1, d_2 > 0$ such that $d_1 m(\tau) \leq -m'(\tau) \leq d_2 m(\tau)$ for every $\tau \in [0, \infty)$.
(E) $\hat{W}_{,1}(3,3) + \hat{W}_{,2}(3,3) > 0$.
The main result of this section is

Theorem IV.13:

Assume that (S1),(S2),(M) and (E) hold. Then there is a constant $\mu > 0$ with the following property: For each pair $(v_0, u_0) \in Y$ with $\|(v_0, u_0)\|_Y \leq \mu$, $\sup_{\tau, y} |\nabla u_0(\tau, y)| \leq \mu$, $\sup_\tau \|\Delta u_0(\tau, \cdot)\|_{L^6} \leq \mu$, and $\int_Q v_0(y) \, dy = 0$, the initial value problem (191), (192), (194), (195) has a unique solution $(v, u) \in C([0, \infty); Y) \cap C^1([0, \infty); Z)$. Moreover, $\|(v(t), u(t))\|_Y \to 0$ as $t \to \infty$.

We record without proof the following local existence theorem:

Theorem IV.14:

Assume that (S1),(S2),(M) and (E) hold. Then for every sufficiently small[2] $\delta > 0$ there exists a $T(\delta) > 0$ such that, for each pair $(v_0, u_0) \in Y$ with

---

[1] This assumption is made only for the sake of simplicity. The optimal smoothness assumption on $\hat{W}$ depends on the procedure used for the local existence proof.

[2] Of course, one does not really need small data for a local existence theorem. However, the ellipticity condition looks more complicated than (E) if the relative deformation gradient is not near the identity. See Section V.8.

174

$(v_0, u_0)\|_Y \leq \delta$, $\sup_{r,y} |\nabla u_0(\tau, y)| \leq \delta$, $\sup_r \|\Delta u_0(\tau, \cdot)\|_{L^6} \leq \delta$, and $\int_Q v_0(y)\, dy = 0$, the initial value problem (191), (192), (194), (195) has a unique solution $v, u) \in C([0, T(\delta)]; Y) \cap C^1([0, T(\delta)]; Z)$.

The proof of Theorem V.14 can be carried out by a modification of the procedure used by Kim [K12]. A local existence result under different regularity assumptions for the data is derived in Section V.8; the method of Section III.4 an also be used to obtain a local existence theorem.

We note that it follows from the equation of motion that the average velocity s conserved, and hence solutions for initial data with $\int_Q v_0(y)\, dy \neq 0$ differ from hose considered here only by the superposition of a rigid translation.

We now derive a priori bounds which show that $\|(v, u)\|_Y$, $\sup_{r,y} |\nabla u(\tau, y)|$ nd $\sup_r \|\Delta u(\tau, \cdot)\|_{L^6}$ remain small. Because of the local existence theorem quoted above, such bounds imply that the solution can be continued for all ime. In deriving our estimates, we shall first assume that $\sup_{r,y} |\nabla u(\tau, y)|$ and up$_r \|\Delta u(\tau, \cdot)\|_{L^6}$ are small. By assumption, this is the case initially, and we an later use the energy estimates to show that these quantities do in fact remain small. The procedure is analogous to that of Sections 3 and 4. We note hat $\sup_{r,y} |\nabla u(\tau, y)| + \sup_r \|\Delta u(\tau, \cdot)\|_{L^6}$ plays the role of the quantity called ν n Sections 3 and 4. As before, we let Γ denote a generic positive constant.

We begin by evaluating the L^2-inner product of $(I - \Delta)\nabla \dot{v}$ with $(I - \Delta)\nabla v$. Clearly, we have

$$\langle (I - \Delta)\nabla \dot{v}, (I - \Delta)\nabla v \rangle_{L^2} = \frac{1}{2}\frac{d}{dt}\|(I - \Delta)\nabla v\|^2_{L^2}. \tag{210}$$

Ve note that $\|(I - \Delta)\nabla v\|_{L^2}$ is equivalent to the H^2-norm of ∇v. Let us use the notation

$$\langle a, b \rangle_3 := \langle (I - \Delta)\nabla a, (I - \Delta)\nabla b \rangle_{L^2}. \tag{211}$$

n (210) we now substitute $\dot{v}$ from the equation of motion (192)$_1$, and we analyze he various terms that arise. Because of the incompressibility condition, we have

$$\langle \nabla p, v \rangle_3 = 0. \tag{212}$$

Moreover, it is easily verified that

$$|\langle (v \cdot \nabla)v, v \rangle_3| \leq \Gamma \|v\|^3_{H^3}. \tag{213}$$

The key term to analyze is

$$-\langle \operatorname{div} \mathbf{T}, v \rangle_3, \tag{214}$$

which after an integration by parts assumes the form

$$\int_Q (I - \Delta)\frac{\partial T^{mn}}{\partial y^l}(I - \Delta)\frac{\partial^2 v^m}{\partial y^n \partial y^l}\, dy. \tag{215}$$

175

Using (191), we find that

$$\frac{\partial T^{m\,n}}{\partial y^l}(y,t) = \int_{-\infty}^{t} m(t-s)\frac{\partial^2 W}{\partial \Phi_k^m \partial \Phi_b^a}(\Phi(s,y,t))\frac{\partial \Phi_b^a}{\partial y^l}(s,y,t)\Phi_k^n(s,y,t)\,ds$$

$$+ \int_{-\infty}^{t} m(t-s)\frac{\partial W}{\partial \Phi_k^m}(\Phi(s,y,t))\frac{\partial \Phi_k^n}{\partial y^l}(s,y,t)\,ds. \tag{216}$$

We observe that by the derivation of equation (89) of Chapter III, we have

$$\frac{\partial}{\partial y^n}\Phi_k^n(s,y,t) = 0. \tag{217}$$

Using (217) and an integration by parts, we obtain

$$\int_Q (I-\Delta)\int_{-\infty}^{t} m(t-s)\frac{\partial W}{\partial \Phi_k^m}(\Phi(s,y,t))\frac{\partial \Phi_k^n}{\partial y^l}(s,y,t)\,ds \times (I-\Delta)\frac{\partial^2 v^m}{\partial y^n \partial y^l}(y,t)\,dy$$

$$= -\int_Q (I-\Delta)\int_{-\infty}^{t} m(t-s)\frac{\partial^2 W}{\partial \Phi_k^m \partial \Phi_b^a}(\Phi(s,y,t))\frac{\partial \Phi_b^a}{\partial y^n}(s,y,t)\frac{\partial \Phi_k^n}{\partial y^l}(s,y,t)\,ds$$

$$\times (I-\Delta)\frac{\partial v^m}{\partial y^l}(y,t)\,dy. \tag{218}$$

A straightforward calculation shows that

$$(I-\Delta)\left[\frac{\partial^2 W}{\partial \Phi_k^m \partial \Phi_b^a}\frac{\partial \Phi_b^a}{\partial y^n}\frac{\partial \Phi_k^n}{\partial y^l}\right]$$

$$= O(|\partial^2 u|^2 + |\partial^2 u|^3 + |\partial^2 u|^4 + |\partial^2 u||\partial^3 u| + |\partial^2 u|^2|\partial^3 u| + |\partial^3 u|^2 + |\partial^2 u||\partial^4 u|), \tag{219}$$

where we use the notation $O(|\partial^2 u|^4)$ to denote a product of four second derivatives of u with a coefficient depending only on Φ. Using the Sobolev imbedding theorem and Hölder's inequality, we can estimate the (spatial) L^2-norms of the various terms in (219) as follows:

$$\|(\partial^2 u)^2 + (\partial^2 u)(\partial^3 u) + (\partial^3 u)^2 + (\partial^2 u)(\partial^4 u)\|_{L^2} \leq \Gamma\|u\|_{H^4}^2, \tag{220$_1$}$$

$$\|(\partial^2 u)^3\|_{L^2} = \|\partial^2 u\|_{L^6}^3 \leq \Gamma\|\partial^2 u\|_{L^6}\|u\|_{H^4}^2, \tag{220$_2$}$$

$$\|(\partial^2 u)^4\|_{L^2} \leq \Gamma\|\partial^2 u\|_{L^6}^2\|\partial^2 u\|_{L^{12}}^2 \leq \Gamma\|\partial^2 u\|_{L^6}^2\|u\|_{H^4}^2, \tag{220$_3$}$$

$$\|(\partial^2 u)^2\partial^3 u\|_{L^2} \leq \Gamma\|\partial^2 u\|_{L^6}^2\|\partial^3 u\|_{L^6} \leq \Gamma\|\partial^2 u\|_{L^6}\|u\|_{H^4}^2. \tag{220$_4$}$$

Therefore the integral in (218) can be estimated by

$$\Gamma\left(1 + \sup_\tau \|\Delta u(\tau,\cdot,t)\|_{L^6}^2\right)\int_0^\infty m(\tau)\|u(\tau,\cdot,t)\|_{H^4}^2\,d\tau\,\|v(\cdot,t)\|_{H^3}. \tag{221}$$

176

Next we set

$$V(s,y,t) = \frac{\partial^2 W}{\partial \Phi_k^m \, \partial \Phi_b^a}(\Phi(s,y,t))(I-\Delta)\frac{\partial \Phi_b^a}{\partial y^l}(s,y,t)(I-\Delta)\frac{\partial \Phi_k^m}{\partial y^l}(s,y,t). \quad (222)$$

We want to estimate the terms appearing in

$$\frac{d}{dt}\int_Q \int_{-\infty}^t m(t-s)V(s,y,t)\,ds\,dy$$

$$= \int_Q \int_{-\infty}^t m'(t-s)V(s,y,t)\,ds\,dy + \int_Q \int_{-\infty}^t m(t-s)\frac{D}{Dt}V(s,y,t)\,ds\,dy. \quad (223)$$

Taking the material time derivative in (222), we find

$$\frac{D}{Dt}V = 2\frac{\partial^2 W}{\partial \Phi_k^m \, \partial \Phi_b^a}(I-\Delta)\frac{\partial \Phi_b^a}{\partial y^l}\frac{D}{Dt}\left((I-\Delta)\frac{\partial \Phi_k^m}{\partial y^l}\right)$$

$$+ \frac{D}{Dt}\left(\frac{\partial^2 W}{\partial \Phi_k^m \, \partial \Phi_b^a}\right)(I-\Delta)\frac{\partial \Phi_b^a}{\partial y^l}(I-\Delta)\frac{\partial \Phi_k^m}{\partial y^l}. \quad (224)$$

Regarding the second term on the right-hand side, we note that

$$\frac{D}{Dt}\frac{\partial^2 W}{\partial \Phi_k^m \, \partial \Phi_b^a} = O(|\partial v|), \quad (225)_1$$

$$(I-\Delta)\frac{\partial \Phi_b^a}{\partial y^l} = O(|\partial^2 u| + |\partial^2 u|^2 + |\partial^2 u|^3 + |\partial^3 u| + |\partial^2 u||\partial^3 u| + |\partial^4 u|). \quad (225)_2$$

From this we conclude that

$$\int_Q \int_{-\infty}^t m(t-s)\frac{D}{Dt}\left(\frac{\partial^2 W}{\partial \Phi_k^m \, \partial \Phi_b^a}(s,y,t)\right)(I-\Delta)\frac{\partial \Phi_b^a}{\partial y^l}(s,y,t)$$

$$\times (I-\Delta)\frac{\partial \Phi_k^m}{\partial y^l}(s,y,t)\,ds\,dy \quad (226)$$

can be estimated by

$$\Gamma\|v(\cdot,t)\|_{H^3}\left(1 + \sup_\tau \|\Delta u(\tau,\cdot,t)\|_{L^6}^4\right)\int_0^\infty m(\tau)\|u(\tau,\cdot,t)\|_{H^4}^2\,d\tau. \quad (227)$$

In a very similar fashion, we can estimate the difference between

$$\int_Q \int_{-\infty}^t m(t-s)\frac{\partial^2 W}{\partial \Phi_k^m \, \partial \Phi_b^a}(\Phi(s,y,t))(I-\Delta)\frac{\partial \Phi_b^a}{\partial y^l}(s,y,t)$$

$$\times \frac{D}{Dt}\left((I-\Delta)\frac{\partial \Phi_k^m}{\partial y^l}(s,y,t)\right)\,ds\,dy \quad (228)$$

177

and

$$\int_Q (I - \Delta) \int_{-\infty}^t m(t - s) \frac{\partial^2 W}{\partial \Phi_k^m \partial \Phi_b^a} (\Phi(s, y, t)) \frac{\partial \Phi_b^a}{\partial y^l} (s, y, t) \Phi_k^n (s, y, t) \, ds$$

$$\times (I - \Delta) \frac{\partial^2 v^m}{\partial y^n \partial y^l} (y, t) \, dy. \tag{229}$$

For this purpose, we note that the terms of highest differential order in these expressions agree, as can be verified using (197). For the remainder terms, we use estimates similar to those above. In summary, the estimates derived thus far yield

$$\frac{1}{2} \frac{d}{dt} \left(\|(I - \Delta) \nabla v\|_{L^2}^2 + \int_{-\infty}^t m(t - s) V(s, y, t) \, ds \right)$$

$$= \frac{1}{2} \int_{-\infty}^t m'(t - s) V(s, y, t) \, ds + R_1, \tag{230}$$

where R_1 is a remainder term satisfying

$$|R_1(t)| \leq \Gamma \left\{ \|v(\cdot, t)\|_{H^3}^3 + \|v\|_{H^3} \int_0^\infty m(\tau) \|u(\tau, \cdot, t)\|_{H^4}^2 \, d\tau \right\}. \tag{231}$$

Finally, we note that as long as $\sup_{\tau, y} |\nabla u(\tau, y, t)|$ and $\sup_\tau \|\Delta u(\tau, \cdot, t)\|_{L^6}$ remain small, the functional

$$\int_Q \int_{-\infty}^t m(t - s) V(s, y, t) \, ds \, dy \tag{232}_1$$

is equivalent to

$$\int_0^\infty m(\tau) \|\Delta u(\tau, \cdot, t)\|_{H^2}^2 \, d\tau. \tag{232}_2$$

To see this, we first note that, ignoring cubic terms, we can approximate $\frac{\partial^2 W}{\partial \Phi_k^m \partial \Phi_b^a} (\Phi)$ by $\frac{\partial^2 W}{\partial \Phi_k^m \partial \Phi_b^a} (1)$ in the expression for V. We have (see [R10], Section 2)

$$\frac{\partial^2 W}{\partial \Phi_k^m \partial \Phi_b^a} (1) = 2 \left(\hat{W}_{,1} (3, 3) + \hat{W}_{,2} (3, 3) \right) \delta_{m a} \delta^{kb}$$

$$+ 4 \left(\hat{W}_{,11} (3, 3) - 2\hat{W}_{,12} (3, 3) + \hat{W}_{,22} (3, 3) \right) \delta_m^k \delta_a^b + 4 \hat{W}_{,2} (3, 3) \delta_a^k \delta_m^b. \tag{233}$$

Since $\det \Phi = 1$, the trace of $\Phi - 1$ can be rewritten as a term of order $|\Phi - 1|^2$. Hence the terms involving $\delta_m^k \delta_a^b$ on the right hand side of (233) lead to cubic terms in the expression for V, and after using an appropriate integration by parts, the same can be shown for the terms involving $\delta_a^k \delta_m^b$. Hence the terms of second degree in V read

$$2 \left(\hat{W}_{,1} (3, 3) + \hat{W}_{,2} (3, 3) \right) |(I - \Delta) \frac{\partial \Phi_b^a}{\partial y^l}|^2, \tag{234}$$

178

which yields the claimed equivalence.

Next, we consider the term

$$-\int_Q \int_0^\infty (I - \Delta)\frac{\partial u}{\partial t}(\tau, y, t) \times (I - \Delta)(-\Delta)\frac{\partial}{\partial \tau}\left\{m(\tau)\frac{\partial u}{\partial \tau}(\tau, y, t)\right\} d\tau\, dy. \quad (235)$$

After integrations by parts with respect to τ and y, this expression becomes

$$\frac{1}{2}\frac{d}{dt}\int_Q \int_0^\infty m(\tau)\left|(I - \Delta)\nabla\frac{\partial u}{\partial \tau}(\tau, y, t)\right|^2 d\tau\, dy. \quad (236)$$

The integral in this expression is equivalent to

$$\int_0^\infty m(\tau)\|\nabla\frac{\partial u}{\partial \tau}(\tau, \cdot, t)\|_{H^2}^2\, d\tau. \quad (237)$$

Using $(194)_1$, we find that on the other hand, the expression (235) is equal to

$$\int_Q \int_0^\infty (I - \Delta)\left(v(y, t) + \frac{\partial u}{\partial \tau}(\tau, y, t) + (v(y, t)\cdot\nabla)u(\tau, y, t)\right)$$

$$\times (I - \Delta)(-\Delta)\frac{\partial}{\partial \tau}\left\{m(\tau)\frac{\partial u}{\partial \tau}(\tau, y, t)\right\} d\tau\, dy. \quad (238)$$

Integrating by parts with respect to τ and y and taking into account $(194)_2$, we find

$$\int_Q \int_0^\infty (I - \Delta)\left(v(y, t) + \frac{\partial u}{\partial \tau}(\tau, y, t)\right) \times (I - \Delta)(-\Delta)\frac{\partial}{\partial \tau}\left\{m(\tau)\frac{\partial u}{\partial \tau}(\tau, y, t)\right\} d\tau\, dy$$

$$= -\int_Q \int_0^\infty m(\tau)(I - \Delta)\nabla\frac{\partial^2 u}{\partial \tau^2}(\tau, y, t) \times (I - \Delta)\nabla\frac{\partial u}{\partial \tau}(\tau, y, t)\, d\tau\, dy$$

$$= \frac{1}{2}m(0)\int_Q |(I - \Delta)\nabla v(y, t)|^2\, dy + \frac{1}{2}\int_Q \int_0^\infty m'(\tau)|(I - \Delta)\nabla\frac{\partial u}{\partial \tau}(\tau, y, t)|^2\, d\tau\, dy. \quad (239)$$

The first term after the last equality in (239) is equivalent to $\|\nabla v(\cdot, t)\|_{H^2}^2$ and the second is equivalent to $-\int_0^\infty m(\tau)\|\nabla\frac{\partial u}{\partial \tau}(\tau, \cdot, t)\|_{H^2}^2\, d\tau$. Finally, we find

$$\int_Q \int_0^\infty (I - \Delta)(v(y, t)\cdot\nabla)u(\tau, y, t)$$

$$\times (I - \Delta)(-\Delta)\frac{\partial}{\partial \tau}\left\{m(\tau)\frac{\partial u}{\partial \tau}(\tau, y, t)\right\} d\tau\, dy$$

$$= -\int_Q \int_0^\infty m(\tau)(I - \Delta)\nabla(v(y, t)\cdot\nabla)\frac{\partial u}{\partial \tau}(\tau, y, t)$$

179

$$\times (I - \Delta)\nabla\frac{\partial u}{\partial \tau}(\tau, y, t) \ d\tau \ dy,$$

and a straightforward calculation shows that this term can be estimated by

$$\Gamma\|v(\cdot, t)\|_{H^3} \int_0^\infty m(\tau)\|\frac{\partial u}{\partial \tau}(\tau, \cdot, t)\|_{H^3}^2 \ d\tau. \tag{240}$$

In summary, we have found that

$$\frac{1}{2}\frac{d}{dt}\int_Q \int_0^\infty m(\tau)\left|(I - \Delta)\nabla\frac{\partial u}{\partial \tau}(\tau, y, t)\right|^2 \ d\tau \ dy$$

$$= \frac{1}{2}\int_Q \int_0^\infty m'(\tau)\left|(I - \Delta)\nabla\frac{\partial u}{\partial \tau}(\tau, y, t)\right|^2 \ d\tau \ dy + R_2, \tag{241}$$

where the remainder term R_2 obeys the estimate

$$|R_2(t)| \leq \Gamma\left(\|v\|_{H^3}^2 + \|v\|_{H^3}\int_0^\infty m(\tau)\|\frac{\partial u}{\partial \tau}(\tau, \cdot, t)\|_{H^3}^2 \ d\tau\right). \tag{242}$$

Next, we consider the expression

$$\frac{d}{dt}\int_Q \int_0^\infty m(\tau)(I - \Delta)\nabla v(y, t) \times (I - \Delta)\nabla u(\tau, y, t) \ d\tau \ dy$$

$$= \int_Q \int_0^\infty m(\tau)(I - \Delta)\nabla\frac{\partial v}{\partial t}(y, t) \times (I - \Delta)\nabla u(\tau, y, t) \ d\tau \ dy$$

$$+ \int_Q \int_0^\infty m(\tau)(I - \Delta)\nabla v(y, t) \times (I - \Delta)\nabla\frac{\partial u}{\partial t}(\tau, y, t) \ d\tau \ dy. \tag{243}$$

Let us estimate the first term on the right-hand side. With P denoting the Hodge projection, we have

$$-\int_Q \int_0^\infty m(\tau)(I - \Delta)\nabla P((v \cdot \nabla)v)(y, t) \times (I - \Delta)\nabla u(\tau, y, t) \ d\tau \ dy$$

$$= \int_Q \int_0^\infty m(\tau)(I - \Delta)(v(y, t) \cdot \nabla)v(y, t) \times (I - \Delta)\Delta P u(\tau, y, t) \ d\tau \ dy, \tag{244}$$

which can be estimated by

$$\Gamma\|v\|_{H^3}^2\left[\int_0^\infty m(\tau)\|u\|_{H^4}^2 \ d\tau\right]^{1/2}. \tag{245}$$

Moreover, we have

$$\int_Q \int_0^\infty m(\tau)(I - \Delta)\nabla P[\text{div } \mathbf{T}](y, t) \times (I - \Delta)\nabla u(\tau, y, t) \ d\tau \ dy$$

180

$$= -\int_Q \int_0^\infty m(\tau)(I - \Delta)\mathrm{div}\,\mathbf{T}(y,t) \times (I - \Delta)\Delta \mathcal{P}u(\tau,y,t)\ d\tau\ dy. \qquad (246)$$

Taking into account the expression for $\mathbf{T}$ and the a priori assumption that $|\nabla u(\tau,y,t)|$ and $\|\Delta u(\tau,\cdot,t)\|_{L^6}$ remain uniformly small, we can estimate the right-hand side of (246) by

$$\Gamma\left\{\int_0^\infty m(\tau)\|u(\tau,\cdot,t)\|_{H^4}\ d\tau\right\}^2$$

$$\leq \Gamma \int_0^\infty m(\tau)\|u(\tau,\cdot,t)\|_{H^4}^2\ d\tau. \qquad (247)$$

The second term on the right-hand side of (243) can be written as

$$\int_Q \int_0^\infty m(\tau)(I - \Delta)\nabla v(y,t)$$

$$\times (I - \Delta)\nabla\left\{-(v(y,t)\cdot\nabla)u(\tau,y,t) - \frac{\partial u}{\partial\tau}(\tau,y,t) - v(y,t)\right\}\ d\tau\ dy. \qquad (248)$$

In (248), the term

$$-\int_Q \int_0^\infty m(\tau)(I - \Delta)\nabla v(y,t) \times (I - \Delta)\nabla(v(y,t)\cdot\nabla)u(\tau,y,t)\ d\tau\ dy \qquad (249)$$

can be estimated by

$$\Gamma\|v(\cdot,t)\|_{H^3}^2\left[\int_0^\infty m(\tau)\|u\|_{H^4}^2\ d\tau\right]^{1/2}, \qquad (250)$$

and the term

$$-\int_Q \int_0^\infty m(\tau)(I - \Delta)\nabla v(y,t) \times (I - \Delta)\nabla\frac{\partial u}{\partial\tau}(\tau,y,t)\ d\tau\ dy$$

$$= \int_Q \int_0^\infty m'(\tau)(I - \Delta)\nabla v(y,t) \times (I - \Delta)\nabla u(\tau,y,t)\ d\tau\ dy$$

can be estimated by

$$\Gamma\|v(\cdot,t)\|_{H^3}\left[\int_0^\infty m(\tau)\|u(\tau,\cdot,t)\|_{H^3}^2\ d\tau\right]^{1/2}. \qquad (251)$$

In summary, we have therefore found

$$\frac{d}{dt}\int_Q \int_0^\infty m(\tau)(I - \Delta)\nabla v(y,t) \times (I - \Delta)\nabla u(\tau,y,t)\ d\tau\ dy$$

181

$$= -\int_0^\infty m(\tau)\, d\tau \times \int_Q |(I-\Delta)\nabla v(y,t)|^2 \, dy + R_3, \tag{252}$$

where R_3 satisfies

$$|R_3(t)| \leq \Gamma\Bigg\{ \|v(\cdot,t)\|_{H^3}^2 \left[\int_0^\infty m(\tau)\|u(\tau,\cdot,t)\|_{H^4}^2 \, d\tau\right]^{1/2}$$

$$+\|v(\cdot,t)\|_{H^3} \left[\int_0^\infty m(\tau)\|u(\tau,\cdot,t)\|_{H^3}^2 \, d\tau\right]^{1/2} + \int_0^\infty m(\tau)\|u(\tau,\cdot,t)\|_{H^4}^2 \, d\tau\Bigg\}. \tag{253}$$

The final energy estimate is obtained by considering the time derivative of

$$\frac{1}{2}\int_Q \int_0^\infty m(\tau)\left\{|u(\tau,y,t)|^2 + \left|\frac{\partial u}{\partial \tau}(\tau,y,t)\right|^2\right\} d\tau \, dy, \tag{254}$$

which is equal to

$$\int_Q \int_0^\infty m(\tau)\frac{\partial u}{\partial t}(\tau,y,t) \times \left[u(\tau,y,t) - \frac{1}{m(\tau)}\frac{\partial}{\partial \tau}\left\{m(\tau)\frac{\partial u}{\partial \tau}(\tau,y,t)\right\}\right] d\tau \, dy. \tag{255}$$

By substituting $\frac{\partial u}{\partial t}$ from equation $(194)_1$, we obtain

$$-\int_Q \int_0^\infty m(\tau)(v(y,t)\cdot\nabla)u(\tau,y,t) \times u(\tau,y,t) \, d\tau \, dy$$

$$+\int_Q \int_0^\infty (v(y,t)\cdot\nabla)u(\tau,y,t) \times \frac{\partial}{\partial \tau}\left[m(\tau)\frac{\partial u}{\partial \tau}(\tau,y,t)\right] d\tau \, dy$$

$$-\int_Q \int_0^\infty m(\tau)v(y,t) \times u(\tau,y,t) \, d\tau \, dy$$

$$+\int_Q \int_0^\infty v(y,t) \times \frac{\partial}{\partial \tau}\left[m(\tau)\frac{\partial u}{\partial \tau}(\tau,y,t)\right] d\tau \, dy$$

$$-\int_Q \int_0^\infty m(\tau)\frac{\partial u}{\partial \tau}(\tau,y,t) \times u(\tau,y,t) \, d\tau \, dy$$

$$+\int_Q \int_0^\infty \frac{\partial u}{\partial \tau}(\tau,y,t) \times \frac{\partial}{\partial \tau}\left[m(\tau)\frac{\partial u}{\partial \tau}(\tau,y,t)\right] d\tau \, dy \tag{256}$$

in place of (255). In (256), the first two terms are zero, the third can be estimated by

$$\Gamma\|v(\cdot,,t)\|_{L^2}\left[\int_0^\infty m(\tau)\|u(\tau,\cdot,t)\|_{L^2}^2 \, d\tau\right]^{1/2}, \tag{257}$$

the fourth equals

$$m(0)\|v(\cdot,t)\|_{L^2}^2 \tag{258}$$

by virtue of $(194)_2$, and the last two terms can be rewritten in the form

$$-\frac{1}{2}\int_Q\int_0^\infty m(\tau)\frac{\partial}{\partial\tau}|u(\tau,y,t)|^2\ d\tau\ dy+\frac{1}{2}\int_Q\int_0^\infty m(\tau)\frac{\partial}{\partial\tau}\left|\frac{\partial u}{\partial\tau}(\tau,y,t)\right|^2\ d\tau\ dy$$

$$+\int_Q\int_0^\infty m'(\tau)\left|\frac{\partial u}{\partial\tau}\right|^2\ d\tau\ dy. \tag{259}$$

We integrate by parts with respect to τ in the first two terms and obtain

$$\frac{1}{2}\int_Q\int_0^\infty m'(\tau)\left\{|u(\tau,y,t)|^2+\left|\frac{\partial u}{\partial\tau}(\tau,y,t)\right|^2\right\}\ d\tau\ dy-\frac{1}{2}m(0)\|v(\cdot,t)\|_{L^2}^2, \tag{260}$$

in place of (259). In summary, we have

$$\frac{1}{2}\frac{d}{dt}\int_Q\int_0^\infty m(\tau)\left\{|u(\tau,y,t)|^2+\left|\frac{\partial u}{\partial\tau}(\tau,y,t)\right|^2\right\}\ d\tau\ dy$$

$$=\frac{1}{2}\int_Q\int_0^\infty m'(\tau)\left\{|u(\tau,y,t)|^2+\left|\frac{\partial u}{\partial\tau}(\tau,y,t)\right|^2\right\}\ d\tau\ dy+R_4, \tag{261}$$

where

$$|R_4(t)|\le\Gamma\left[\|v(\cdot,t)\|_{L^2}^2+\|v(\cdot,t)\|_{L^2}\left\{\int_0^\infty m(\tau)\|u(\tau,\cdot,t)\|_{L^2}^2\ d\tau\right\}^{1/2}\right]. \tag{262}$$

Let us now summarize the energy estimates obtained so far.

Summary of energy estimates

$$\frac{d}{dt}\left[\frac{1}{2}\int_Q|v(y,t)|^2\ dy+\int_Q\int_0^\infty m(t-s)W(\Phi(s,y,t))\ ds\ dy\right]$$

$$=\int_Q\int_0^\infty m'(t-s)W(\Phi(s,y,t))\ ds\ dy. \tag{263$_1$}$$

$$\frac{1}{2}\frac{d}{dt}\left(\|(I-\Delta)\nabla v\|_{L^2}^2+\int_{-\infty}^t m(t-s)V(s,y,t)\ ds\right)$$

$$=\frac{1}{2}\int_{-\infty}^t m'(t-s)V(s,y,t)\ ds+R_1, \tag{263$_2$}$$

$$\frac{1}{2}\frac{d}{dt}\int_Q\int_0^\infty m(\tau)\left|(I-\Delta)\nabla\frac{\partial u}{\partial\tau}(\tau,y,t)\right|^2\ d\tau\ dy$$

$$=\frac{1}{2}\int_Q\int_0^\infty m'(\tau)\left|(I-\Delta)\nabla\frac{\partial u}{\partial\tau}(\tau,y,t)\right|^2\ d\tau\ dy+R_2, \tag{263$_3$}$$

$$\frac{d}{dt}\int_Q\int_0^\infty m(\tau)(I-\Delta)\nabla v(y,t)\times(I-\Delta)\nabla u(\tau,y,t)\ d\tau\ dy$$

183

$$= -\int_0^\infty m(\tau)\, d\tau \times \int_Q |(I - \Delta)\nabla v(y,t)|^2\, dy + R_3, \qquad (263)$$

$$\frac{1}{2}\frac{d}{dt}\int_Q \int_0^\infty m(\tau)\Big\{ |u(\tau,y,t)|^2 + \Big|\frac{\partial u}{\partial \tau}(\tau,y,t)\Big|^2 \Big\}\, d\tau\, dy$$

$$= \frac{1}{2}\int_Q \int_0^\infty m'(\tau)\Big\{ |u(\tau,y,t)|^2 + \Big|\frac{\partial u}{\partial \tau}(\tau,y,t)\Big|^2 \Big\}\, d\tau\, dy + R_4. \qquad (263)$$

An appropriate linear combination of these identities easily yields the claime
exponential decay of $\|(v,u)\|_Y$. We must recall, however, that some of the est
mates were derived under the assumption that $\sup_{\tau,y} |\nabla u(\tau,y,t)|$ an
$\sup_\tau \|\Delta u(\tau,\cdot,t)\|_{L^6}$ remain small. Equivalently, we have to show that $|\Phi -$
remains uniformly small and that spatial derivatives of Φ remain small in th
L^6-norm. Because of equation (197), this is the case if

$$\int_0^t \|\nabla v(\cdot,t')\|_{L^\infty}\, dt' \qquad (26$$

and

$$\int_0^t \|\Delta v(\cdot,t')\|_{L^6}\, dt' \qquad (26$$

remain small. This is guaranteed because $\|(v,u)\|_Y$ is small initially and decay
exponentially as long as $\sup_{\tau,y} |\nabla u(\tau,y,t)| + \sup_\tau \|\Delta u(\tau,\cdot,t)\|_{L^6}$ remains small.

6. A global existence result for large data

In this section we present a global existence result, due to Engler [E2], fc
shearing flows of a K-BKZ fluid with a Newtonian contribution to the viscosit
The differential equation under consideration is (cf. Section I.4)

$$u_{tt} = \eta u_{xxt} + \int_0^\infty \big(g(s, u_x(x,t) - u_x(x,t-s))\big)_x\, ds + f(x,t), \quad 0 \leq x \leq 1,\ t > \mathbf{0}$$
$$(266)$$

where $\eta > 0$ is a constant, g is a smooth material function, and f is the bod
force. Engler considers history value problems for $(266)_1$ under Dirichlet as well a
traction boundary conditions; we will confine ourselves to the Dirichlet conditior

$$u(0,t) = \phi_0(t),\ u(1,t) = \phi_1(t),\ t \geq 0, \qquad (266$$

where ϕ_0, ϕ_1 are given functions. In addition, we prescribe the history $\tilde{u}$ of u for ≤ 0, and an initial datum u_1 for u_t:

$$u(x,t) = \tilde{u}(x,t) \text{ for } 0 \leq x \leq 1, \ t \leq 0,$$

$$u(x,0^+) = \tilde{u}(x,0), \ u_t(x,0+) = u_1(x), \ 0 \leq x \leq 1. \tag{266}_3$$

Observe that u is required to be continuous across $t = 0$, but that $u_1(x)$ need not equal $\tilde{u}_t(x,0^-)$.

We make the following assumptions:

(S1) $g : (0,\infty) \times \mathbb{R} \to \mathbb{R}$ is continuously differentiable,

(S2) $g_{,2}(\cdot,0) \in L^1(0,\infty)$, and for every $R > 0$ there is a function $a_R \in L^1(0,\infty)$ such that $|g_{,2}(t,v) - g_{,2}(t,w)| \leq a_R(t)|v - w|$ for $|v|, |w| \leq R$ (as usual, $g_{,2}$ denotes the derivative with respect to the second argument),

(S3) $\tilde{u}, \tilde{u}_x$ and $\tilde{u}_{xx}$ are bounded and continuous on $[0,1] \times (-\infty,0]$,

(S4) $u_1 \in C^1[0,1]$,

(S5) f is continuous on $[0,1] \times [0,T]$ for every $T > 0$,

(S6) ϕ_0 and ϕ_1 are in $C^2[0,T]$,

(C) $\tilde{u}(j,0) = \phi_j(0), \ u_1(j) = \phi_j'(0), \ j = 0,1$,

(E) $\eta > 0$.

Under these assumptions, it follows from Engler's results [E2, Theorems 1.1 and 1.3] that (266) has a unique local solution u such that u_t and u_{xx} are continuous on $[0,1] \times [0,t_0]$ for some $t_0 > 0$; moreover, u_{tt} and $u_{xxt} \in L^p([0,1] \times [\varepsilon,t_0])$ for every $\varepsilon > 0$ and every p with $1 \leq p < \infty$. The proof is obtained by integrating $(266)_1$ and then using an iteration procedure which solves the inhomogeneous heat equation at each step. Under additional smoothness assumptions u will be a classical solution of (266) on $[0,1] \times [0,t_0]$. Since our main interest is in global results, we will not discuss more refined local existence and regularity theorems; we refer to [E2] for several such results and remarks.

As a consequence of the local argument (see [E2, Corollary 1.4]), it follows that under assumptions (S1)-(S6), (C), and (E), the local solution u can be continued for as long as $|u_x|$ remains pointwise bounded. We will now describe how a priori bounds on $|u_x|$ can be obtained under additional assumptions on g, without restricting the size of the data. The analysis combines an energy estimate with a comparison argument; of course, the comparison argument essentially limits the technique of proof to one-dimensional problems. We will need some assumptions on the global behavior of g. Since only $g_{,2}$ appears in equation $(266)_1$, we may assume without loss of generality that $g(t,0) = 0$. We define G as the primitive of g with respect to the second argument,

$$G(t,\xi) := \int_0^\xi g(t,v) \, dv. \tag{267}$$

We make the following additional hypotheses:

(S7) For every $R > 0$, there is a function $d_R \in L^1(0,\infty)$ such that for all $t > 0$, $|\xi| \leq R$, one has $|G_{,1}(t,\xi)| \cdot \min(1,t) \leq d_R(t)$ (i.e. $G_{,1}$ is integrable at $t = \infty$ and $tG_{,1}$ is integrable at $t = 0$).

185

(G1) There is a constant $C_0 \geq 0$ and a function $C_1 \in L^1(0, \infty)$ such that for all $t > 0$, $\xi \in \mathbb{R}$ we have

$$G(t, \xi) \geq -C_1(t)(\xi^2 + 1), \quad G_{,1}(t, \xi) \leq C_0\left(G(t, \xi) + C_1(t)(\xi^2 + 1)\right), \quad (268)_1$$

$$|g(t, \xi)| \leq C_0\left(G(t, \xi) + C_1(t)(\xi^2 + 1)\right). \quad (268)_2$$

(G2) There is a function $L \in L^1(0, \infty)$ such that

$$g(t, \xi) = g_0(t, \xi) + L(t)\xi, \quad (269)$$

where g_0 is nondecreasing with respect to its second argument.

Remark IV.15:

Assumptions (S7), the first part of (G1), and (G2) are satisfied if, e.g., $g(t, \xi) = a(t)h(\xi)$, where a is nonnegative, nonincreasing, a and a' are integrable on $[0, \infty)$, and h is smooth and nondecreasing on $\mathbb{R}$ (up to an affine function). The second part of (G1) then imposes a mild restriction on the behavior of h at infinity (for example power law behavior with any power of ξ is permitted).

Theorem IV.16:

Let $T > 0$ be given. Assume that (S1)-(S7), (C), (E) and (G1)-(G2) hold and let u be a solution of (266) which is defined on $[0, 1] \times [0, t_0]$ for some $t_0 \in (0, T]$. Then there is a constant C^*, depending on the data and on T, but not on t_0, such that

$$|u_x(x, t)| \leq C^* \ \forall x \in [0, 1], \ t \in [0, t_0]. \quad (270)$$

As a consequence, u can be extended to the time-interval $[0, T]$, and since T is arbitrary, the solution exists globally in time. To prove Theorem IV.16, we first derive an energy estimate. Let w be the linear extension of the boundary data into the interior, i.e.

$$w(x, t) := (\phi_1(t) - \phi_0(t))x + \phi_0(t), \quad 0 \leq x \leq 1, 0 \leq t \leq T. \quad (271)$$

We now multiply equation $(266)_1$ by $u_t - w_t$, and integrate over $[0, 1]$. Let $\langle \cdot, \cdot \rangle$ denote the inner product and $\| \cdot \|$ the norm in $L^2(0, 1)$; integrating by parts, we obtain

$$\langle u_{tt}, u_t - w_t \rangle + \eta \langle u_{xt}, u_{xt} - w_{xt} \rangle$$

$$+ \langle \int_0^\infty g(s, u_x(\cdot, t) - u_x(\cdot, t - s)) \, ds, u_{xt} - w_{xt} \rangle = \langle f, u_t - w_t \rangle. \quad (272)$$

Noting the identity

$$\langle \int_0^\infty g(s, u_x(\cdot, t) - u_x(\cdot, t - s)) \, ds, u_{xt} \rangle$$

186

$$= \frac{d}{dt} \int_0^\infty \int_0^1 G(s, u_x(x,t) - u_x(x, t-s)) \, dx \, ds$$

$$- \int_0^\infty \int_0^1 G_{,1}(s, u_x(x,t) - u_x(x, t-s)) \, dx \, ds,$$

 d rearranging the terms in (272), we find

$$\frac{d}{dt}\left\{ \frac{1}{2}\|u_t - w_t\|^2 + \int_0^\infty \int_0^1 G(s, u_x(x,t) - u_x(x, t-s)) \, dx \, ds\right\} + \eta\|u_{xt}\|^2$$

$$= \eta\langle u_{xt}, w_{xt}\rangle + \int_0^\infty \int_0^1 G_{,1}(s, u_x(x,t) - u_x(x, t-s)) \, dx \, ds$$

$$+\langle \int_0^\infty g(s, u_x(\cdot,t) - u_x(\cdot, t-s)) \, ds, w_{xt}\rangle + \langle f - w_{tt}, u_t - w_t\rangle, \quad 0 \le t \le t_0. \quad (273)$$

 ext, it is convenient to define the quantities

$$E(t) := \frac{1}{2}\|u_t(\cdot,t) - w_t(\cdot,t)\|^2 + \int_0^\infty \int_0^1 G(s, u_x(x,t) - u(x, t-s)) \, dx \, ds, \quad (274)_1$$

$$E_1(t) := E(t) + \int_0^\infty C_1(s)\{\|u_x(\cdot,t) - u_x(\cdot, t-s)\|^2 + 1\} \, ds, \quad (274)_2$$

 here C_1 is the function in assumption (G1); observe that $E_1(t) \ge 0$ (by $(268)_1$).
 y Λ we shall denote a generic positive constant which may depend on the data as
 ell as on T, but not on t_0. (Λ need not have the same meaning even within the
 me formula). The first term on the right-hand side of (273) can be estimated
 follows:

$$|\langle u_{xt}, w_{xt}\rangle| = \left| \int_0^1 u_{xt}(x,t)(\dot{\phi}_1(t) - \dot{\phi}_0(t)) \, dx\right|$$

$$= |(\dot{\phi}_1(t) - \dot{\phi}_0(t))(u_t(1,t) - u_t(0,t))| = |(\dot{\phi}_1(t) - \dot{\phi}_0(t))|^2 \le \Lambda. \quad (275)_1$$

 sing assumption $(268)_1$ and definition $(274)_2$ we find that

$$\int_0^\infty \int_0^1 G_{,1}(s, u_x(x,t) - u_x(x, t-s)) \, dx \, ds$$

$$\le C_0 \left(\int_0^\infty \int_0^1 G(s, u_x(x,t) - u_x(x, t-s)) \, dx \, ds \right.$$

$$+ \int_0^\infty C_1(s)(\|u_x(\cdot,t) - u_x(\cdot, t-s)\|^2 + 1) \, ds \bigg) \le \Lambda E_1(t). \quad (275)_2$$

 oreover, definition (271) and assumption $(268)_2$ imply

$$\left| \langle \int_0^\infty g(s, u_x(\cdot,t) - u_x(\cdot, t-s)) \, ds, w_{xt}\rangle \right|$$

187

$$\leq \Lambda \int_0^\infty \int_0^1 |g(s, u_x(x,t) - u_x(x, t-s))| \, dx \, ds$$

$$\leq \Lambda \left(\int_0^\infty \int_0^1 G(s, u_x(x,t) - u_x(x, t-s)) \, dx \, ds \right.$$

$$\left. + \int_0^\infty C_1(s) \left(\|u_x(\cdot, t) - u_x(\cdot, t-s)\|^2 + 1 \right) ds \right) \leq \Lambda E_1(t). \tag{275}$$

Finally, assumptions (S5), (S6), and $(268)_1$ yield

$$\langle f - w_{tt}, u_t - w_t \rangle \leq \Lambda \|u_t - w_t\| \leq \Lambda \sqrt{E_1(t)}. \tag{275}$$

By combining the estimates (275) and using $(274)_2$, we find that (273) become

$$\frac{d}{dt} E(t) + \eta \|u_{xt}\|^2 \leq \Lambda(E_1(t) + 1), \quad 0 \leq t \leq t_0. \tag{27}$$

By integrating (276) from 0 to t, we find that

$$E_1(t) + \eta \int_0^t \|u_{xt}(\cdot, s)\|^2 \, ds$$

$$\leq \Lambda \left(\int_0^t E_1(s) \, ds + 1 \right) + \int_0^\infty C_1(s) \{ \|u_x(\cdot, t) - u_x(\cdot, t-s)\|^2 + 1 \} \, ds$$

$$\leq \Lambda \left(\int_0^t E_1(s) \, ds + \|u_x(\cdot, t)\|^2 + 1 \right) + 2 \int_0^t C_1(t-s) \|u_x(\cdot, s)\|^2 \, ds, \ 0 \leq t \leq t_0$$

$$\tag{27}$$

Next, we note that

$$\|u_x(\cdot, t)\|^2 = \int_0^t u_x(\cdot, s) u_{xt}(\cdot, s) \, ds + \|u_x(\cdot, 0)\|^2$$

$$\leq \Lambda + \delta \int_0^t \|u_{xt}(\cdot, s)\|^2 \, ds + \frac{1}{4\delta} \int_0^t \|u_x(\cdot, s)\|^2 \, ds$$

$$\leq \Lambda + \delta \int_0^t \|u_{xt}(\cdot, s)\|^2 \, ds + K(\delta) \int_0^t \int_0^s \|u_{xt}(\cdot, \tau)\|^2 \, d\tau \, ds, \ 0 \leq t \leq t_0. \tag{27}$$

We insert (278) in (277), choose δ sufficiently small and use Gronwall's inequali to obtain the a priori estimate

$$0 \leq E_1(t) + \int_0^t \|u_{xt}(\cdot, s)\|^2 \, ds \leq \Lambda, \quad 0 \leq t \leq t_0. \tag{27}$$

For fixed x and t, we now integrate equation $(266)_1$ from y to x and the from 0 to 1 with respect to y. Setting

$$p(x,t) := \int_0^1 \int_y^x u_t(\xi, t) \, d\xi \, dy, \tag{28}$$

188

we easily obtain the identity

$$p_t(x,t) = \eta\left(u_{xt}(x,t) - \dot{\phi}_1(t) + \dot{\phi}_0(t) \right) + \int_0^\infty g(s, u_x(x,t) - u_x(x,t-s))\, ds$$

$$- \int_0^\infty \int_0^1 g(s, u_x(y,t) - u_x(y,t-s))\, dy\, ds$$

$$+ \int_0^1 \int_y^x f(\xi,t)\, d\xi\, dy, \quad 0 \le x \le 1,\ 0 \le t \le t_0. \tag{281}$$

According to $(275)_3$, the term

$$\int_0^\infty \int_0^1 g(s, u_x(y,t) - u_x(y,t-s))\, dy\, ds$$

can be bounded in terms of $E_1(t)$ which, in turn, is a priori bounded (see (279)). We can therefore write (281) in the form

$$(\eta u_x(x,t) - p(x,t))_t + \int_0^\infty g(s, u_x(x,t) - u_x(x,t-s))\, ds = k(x,t),$$

$$0 \le x \le 1,\ 0 \le t \le t_0, \tag{282}_1$$

$$k(x,t) := -\int_0^1 \int_y^x f(\xi,t)d\xi dy + \eta(\dot{\phi}_1(t) - \dot{\phi}_0(t))$$

$$+ \int_0^\infty \int_0^1 g(s, u_x(y,t) - u_x(y,t-s))\, dy\, ds, \quad 0 \le x \le 1,\ 0 \le t \le T; \tag{282}_2$$

we note that k is a priori bounded on $[0,1] \times [0,t_0]$. The energy estimate (279) also yields a pointwise a priori bound for p on $[0,1] \times [0,t_0]$. We now set $q := \eta u_x - p$. To obtain the a priori bound (270) for u_x it suffices to show that q is also bounded independently of t_0. We suppress the x-dependence and rewrite equation $(282)_1$ in the form

$$\dot{q}(t) + \int_0^\infty g\left(s, \frac{1}{\eta}(q(t) - q(t-s) + p(t) - p(t-s))\right)\, ds = k(t), \quad 0 \le t \le t_0. \tag{283}$$

We now use assumption (G2) and set $L_0 := \int_0^\infty L(s)\, ds$. Then (283) can be rewritten as

$$\dot{q}(t) + \int_{-\infty}^t g_0\left(t-s, \frac{1}{\eta}(q(t)-q(s)+p(t)-p(s))\right)\, ds + \frac{L_0}{\eta}q(t) - \frac{1}{\eta}\int_{-\infty}^t L(t-s)q(s)\, ds$$

$$= k(t) - \frac{L_0}{\eta}p(t) + \frac{1}{\eta}\int_{-\infty}^t L(t-s)p(s)\, ds, \quad 0 \le t \le t_0. \tag{284}_1$$

Noting the a priori bound for p and the fact that p and q are given in terms of the data for $t < 0$, we write $(284)_1$ in the form

$$\dot{q}(t) + \int_0^t g_0\left(t-s, \frac{1}{\eta}(q(t)-q(s)+p(t)-p(s))\right)\, ds + \frac{L_0}{\eta}q(t)$$

189

$$+ \int_0^\infty g_0\left(t+s, \frac{1}{\eta}(q(t)+p(t)-k_0(s))\right) ds = k_1(t) + \frac{1}{\eta}\int_0^t L(t-s)q(s)\, ds, \quad 0 \le t \le t_0 \tag{284}_2$$

where

$$k_0(s) := q(-s) + p(-s), \quad s > 0,$$

$$k_1(t) := k(t) - \frac{L_0}{\eta}p(t) + \frac{1}{\eta}\int_0^\infty L(t+s)k_0(s)\, ds, \quad 0 \le t \le T; \tag{284}_3$$

we note that k_0 and k_1 are a priori bounded on $[0, t_0]$. To analyze the second and fourth terms on the left side of $(284)_2$, it is easily seen that, without loss of generality, it suffices to consider the two cases: (i) $q(t) > 0$ and $q(t) > q(s)$ for $0 \le s < t$ and (ii) $q(t) < 0$ and $q(t) < q(s)$ for $0 \le s < t$. In case (i), $(284)_2$ $(284)_3$ and the monotonicity of g_0 imply that q satisfies a differential inequality of the form

$$\dot{q}(t) \le \frac{|L_0|}{\eta}q(t) + \frac{1}{\eta}\int_0^t |L(t-s)q(s)|\, ds + k_2(t), \tag{285}_1$$

where

$$k_2(t) := k_1(t) - \int_0^t g_0\left(t-s, \frac{1}{\eta}(p(t)-p(s))\right) ds$$

$$- \int_0^\infty g_0\left(t+s, \frac{1}{\eta}(p(t)-k_0(s))\right) ds, \quad 0 \le t \le t_0; \tag{285}_2$$

we note that k_2 is a priori bounded on $[0, t_0]$. In case (ii) a similar analysis yields $(285)_1$ with the inequality reversed. A Gronwall-type argument is then used to obtain an a priori pointwise bound for q on $[0, t_0]$, and hence for u_x on $[0, 1] \times [0, t_0]$; this proves (270).

∎

If further assumptions on g are made, it can be shown that solutions not only exist globally in time, but also decay. Since equations (266) describe the motion of a fluid, the decay is not to zero, but to a time-independent displacement field which cannot be determined in advance. Engler [E2] proves such a result and obtains a rate for the decay. The rate is expressed in terms of the decay of the memory function, the lowest eigenvalue of the heat operator and the decay of the forcing term (for simplicity, it is assumed that the boundary data are independent of time). The essential additional hypotheses are
(G3) $0 = G(t, 0) \le G(t, \xi)$ for all $\xi \in \mathbb{R}$, $t > 0$.
(G4) There is a constant $\delta > 0$ such that for all $\xi \in \mathbb{R}$, $t > 0$ we have

$$G_{,1}(t, \xi) + \delta G(t, \xi) \le 0. \tag{286}$$

Observe that assumption (G4) implies $G(t, \xi) \le G(0, \xi)e^{-\delta t}$; in particular, if $g(t, \xi) = a(t)h(\xi)$, then, under mild assumptions on a and h, (G4) implies that a must decay exponentially.

190

Theorem IV.17:

Assume that (S1)-(S7), (C), (E) and (G1)-(G4) hold. Moreover, assume that ϕ_0 and ϕ_1 are time-independent. Let $b : [0, \infty) \to [0, \infty)$ be a continuously differentiable function such that

$$b(0) = 1, \tag{287}_1$$

$$0 \le b'(t) \le \kappa b(t) \ \text{for some} \ \kappa \le \delta, \ \kappa < 2\eta\pi^2, \tag{287}_2$$

$$\int_0^\infty \frac{1}{b(s)} \, ds < \infty. \tag{287}_3$$

Assume that the forcing term f is such that

$$\int_0^\infty \|f(\cdot, s)\| b(s)^{1/2} \, ds < \infty. \tag{288}$$

Then there exists $u_\infty \in H^1(0, 1)$ such that

$$\|u_x(\cdot, t) - u_{\infty, x}(\cdot)\|^2 = O\left(\int_t^\infty \frac{1}{b(s)} \, ds \right) \tag{289}$$

as $t \to \infty$.

Remark IV.18

Observe that if $\|f(\cdot, t)\|$ decays exponentially as $t \to \infty$, then we may take $b(t) = \exp(\lambda t)$ with $\lambda > 0$ sufficiently small, and (289) yields convergence to equilibrium at an exponential rate.

The proof of Theorem IV.17 is based on establishing the following estimates. Defining

$$C(\tilde{u}, u_1) := \|u_1\|^2 + 2 \int_0^\infty \int_0^1 G(s, \tilde{u}_x(x, 0) - \tilde{u}_x(x, -s)) \, dx \, ds, \tag{290}$$

it is shown that

$$\|u_t(\cdot, t)\| b(t)^{1/2} \le C(\tilde{u}, u_1)^{1/2} + \int_0^t \|f(\cdot, s)\| b(s)^{1/2} \, ds, \ 0 \le t < \infty, \tag{291}$$

$$\int_0^t b(s) \|u_{xt}(\cdot, s)\|^2 \, ds \le \left(\eta - \frac{\kappa}{2\pi^2} \right)^{-1} \left\{ C(\tilde{u}, u_1) \right.$$

$$\left. + \left(\int_0^t \|f(\cdot, s)\| b(s)^{1/2} \, ds \right)^2 \right\}, \ 0 \le t < \infty. \tag{292}$$

The theorem follows by noting that

$$\|u_x(\cdot, t) - u_x(\cdot, s)\|^2 = \left\| \int_t^s u_{xt}(\cdot, \tau) \, d\tau \right\|^2, \tag{293}$$

and then using the Cauchy-Schwarz inequality, $(287)_2$, $(287)_3$, and (292).

To establish (291), we multiply $(266)_1$ by $b(t)u_t(x,t)$ and integrate over the interval $[0,1]$ obtaining

$$\frac{1}{2}\frac{d}{dt}\left\{b(t)\|u_t(\cdot,t)\|^2\right\} - \frac{1}{2}b'(t)\|u_t(\cdot,t)\|^2 + \eta b(t)\|u_{xt}(\cdot,t)\|^2$$

$$+ \int_0^\infty b(t)\langle g(s,u_x(\cdot,t)-u_x(\cdot,t-s)), u_{xt}(\cdot,t)\rangle\,ds = b(t)\langle f(\cdot,t), u_t(\cdot,t)\rangle,\ 0 \le t < \infty.$$
(294)

A simple calculation using (267) yields

$$\frac{d}{dt}\left\{\int_0^\infty b(t)\int_0^1 G(s,u_x(x,t) - u_x(x,t-s))\,dx\,ds\right\}$$

$$= \int_0^\infty b'(t)\int_0^1 G(s,u_x(x,t) - u_x(x,t-s))\,dx\,ds$$

$$+ \int_0^\infty b(t)\langle g(s,u_x(\cdot,t) - u_x(\cdot,t-s)), u_{xt}(\cdot,t)\rangle\,ds$$

$$+ \int_0^\infty b(t)\int_0^1 G_{,1}(s,u_x(x,t) - u_x(x,t-s))\,dx\,ds,\ 0 \le t < \infty.$$
(295)

Because of (G3),(G4) and our assumptions on b, the sum of the first and third terms on the right side of (295) is negative. Moreover, $(287)_2$ and the Poincaré inequality (with the optimal constant $1/\pi^2$) imply that

$$-\frac{1}{2}b'(t)\|u_t(\cdot,t)\|^2 + \eta b(t)\|u_{xt}(\cdot,t)\|^2$$

$$\ge b(t)\left(\eta\|u_{xt}(\cdot,t)\|^2 - \frac{1}{2}\kappa\|u_t(\cdot,t)\|^2\right)$$

$$\ge b(t)\left(\eta - \frac{\kappa}{2\pi^2}\right)\|u_{xt}(\cdot,t)\|^2 \ge 0,\ 0 \le t < \infty.$$
(296)

Next, substituting (295) and (296), integrating the resulting inequality from 0 to t, and using $(287)_1$, (290) one obtains

$$\frac{1}{2}b(t)\|u_t(\cdot,t)\|^2 + \left(\eta - \frac{\kappa}{2\pi^2}\right)\int_0^t b(s)\|u_{xt}(\cdot,s)\|^2\,ds$$

$$+ \int_0^\infty b(t)\int_0^1 G(s,u_x(x,t) - u_x(x,t-s))\,dx\,ds$$

$$\le \frac{1}{2}C(\tilde u,u_1) + \int_0^t \|f(\cdot,s)\|b(s)^{1/2}\|u_t(\cdot,s)\|b(s)^{1/2}\,ds,\ 0 \le t < \infty.$$
(297)

To complete the proof of (291), we make use of the following lemma (see [B12], p. 157).

192

Lemma IV.19:

Let $m \in L^1(0,T)$ be such that $m \geq 0$ and let a be a nonnegative constant. Let ϕ be a continuous function $[0,T] \to \mathbb{R}$ such that

$$\frac{1}{2}\phi^2(t) \leq \frac{1}{2}a^2 + \int_0^t m(s)\phi(s) \ ds \ \forall t \in [0,T]. \tag{298}$$

Then

$$|\phi(t)| \leq a + \int_0^t m(s) \ ds, \ 0 \leq t \leq T. \tag{299}$$

The second and third terms on the left-hand side of (297) are nonnegative by virtue of $(287)_2$ and (G3). Dropping them in (297) and applying Lemma IV.19 with $\phi(t) := b(t)^{1/2}\|u_t(\cdot,t)\|$ yields (291).

Finally, dropping the first and third terms on the left side of (297) and applying (291) to the right side of the resulting inequality easily yields (292), and the proof is complete.

∎

V Applications of methods of semigroup theory

The term "semigroup theory" as used in this chapter is not quite what the name suggests, rather it describes a subject that might better be titled "Evolution problems in Banach spaces". The name "semigroup theory" is motivated by the fact that the solution of such problems involves semigroups of operators. In this chapter, we discuss the application of semigroup theory to problems in viscoelasticity. The first two sections present a review of basic concepts and results concerning evolution problems in Banach spaces. In Section 3, we discuss various ways of applying semigroup theory to equations involving memory. The remaining five sections discuss applications to specific problems in viscoelasticity.

1. Review of linear evolution problems

In this section, we shall state without proofs some results about linear autonomous evolution problems in Banach spaces which are used later in this chapter. For details and further references to the literature we refer to the recent book of Pazy [P1].

We are concerned with linear ordinary differential equations in a Banach space X. More precisely, we consider initial value problems of the form

$$\dot{u}(t) = A(t)u(t) + f(t), \ u(0) = u_0. \tag{1}$$

Here $A(t)$ is a closed, generally unbounded, linear operator in X with dense domain $D(A(t))$, f is a function with values in X, and u_0 is a given element of X. The objective of linear semigroup theory is to find conditions on A, f and u_0 which guarantee the existence and uniqueness of solutions to (1) for $t > 0$.

In the special case when $f = 0$ and A is independent of t, (1) reads

$$\dot{u} = Au, \ u(0) = u_0. \tag{2}$$

Of course we expect that the solution of (2) will be given by $e^{At}u_0$, but we must decide how to interpret e^{At} if A is unbounded. Elementary calculus would suggest a definition by the exponential series, or by the formula

$$e^{At} = \lim_{n \to \infty} (I + \frac{At}{n})^n, \tag{3}$$

where I denotes the identity operator. It is easy to show that both of these definitions "work" if A is bounded. For unbounded operators, however, powers of A are defined on smaller and smaller domains, and, since (3) as well as the exponential series involve powers of arbitrary order, they are not useful as a definition. However, "reasonable" operators of mathematical physics have resolvents, and this suggests replacing (3) by

$$e^{At} = \lim_{n \to \infty} (I - \frac{At}{n})^{-n}. \tag{4}$$

Observe that equation (4) corresponds to the implicit Euler scheme for solving $\dot{u} = Au$.

In order to characterize a class of operators for which this procedure is meaningful, we first require:

Definition V.1:

A one parameter family $T(t)$, $0 \le t < \infty$ of bounded linear operators in X is called a **strongly continuous semigroup** *or* C_0-**semigroup** *if the following conditions hold:*
(i) $T(0) = I$,
(ii) $T(t + s) = T(t)T(s)$ for every $t, s \ge 0$,
(iii) $\lim_{t \to 0+} T(t)x = x$ for every $x \in X$.
The **infinitesimal generator** A *of the semigroup* $T(t)$ *is the linear operator defined by*

$$Ax := \lim_{t \to 0+} \frac{T(t)x - x}{t}, \tag{5}$$

and its domain $D(A)$ is the set of all $x \in X$ for which this limit exists.

It is not obvious that $D(A)$ contains any nonzero elements. However, it follows from (i)-(iii) that $D(A)$ is actually dense in X.

One of the principal points of interest is to find conditions which guarantee that a given operator A is the infinitesimal generator of a C_0-semigroup. Necessary and sufficient conditions are given by the Hille-Yosida theorem. Before stating this result, we record some basic properties of semigroups.

If $T(t)$ is a C_0-semigroup, then there exist $M > 0$ and $\omega \in \mathbb{R}$ such that

$$\|T(t)\| \le Me^{\omega t} \quad \forall t \ge 0; \tag{6}$$

moreover, for each $x \in X$, the mapping $t \to T(t)x$ is continuous on $[0, \infty)$. This latter continuity property is often used in place of (iii) in the definition of a C_0-semigroup. It is obvious that the infinitesimal generator of a semigroup is unique.

Conversely, a given operator cannot be the infinitesimal generator of two distinct C_0-semigroups. Since there is a one-to-one correspondence between generators and semigroups, we use the notation e^{At} for the semigroup generated by A. If A is the infinitesimal generator of a C_0-semigroup, then $D(A)$ is dense and A is closed.

The fundamental result on generation of semigroups as well as some additional results are contained in the following theorem.

Theorem V.2:

Let $M, \omega \in \mathbb{R}$ be given.

(i) *A linear operator A is the infinitesimal generator of a C_0-semigroup satisfying (6) if and only if A is closed, $D(A)$ is dense in X, and the resolvent $R(\lambda, A) := (\lambda I - A)^{-1}$ exists for every real $\lambda > \omega$ and satisfies the estimate*

$$\|R(\lambda, A)^n\| \le \frac{M}{(\lambda - \omega)^n} \quad \forall n \ge 1. \tag{7}$$

Assume now that A is the infinitesimal generator of a C_0-semigroup. Then

(ii) *for each $x \in X$ we have*

$$e^{At}x = \lim_{n \to \infty} \left(I - \frac{At}{n}\right)^{-n} x, \ 0 \le t < \infty;$$

(iii) *if $u_0 \in D(A)$, then $e^{At}u_0 \in D(A) \ \forall t \ge 0$, the mapping $t \to e^{At}u_0$ is continuously differentiable on $[0, \infty)$, and*

$$\frac{d}{dt}e^{At}u_0 = e^{At}Au_0 = Ae^{At}u_0.$$

Moreover, $e^{At}u_0$ is the unique solution of (2).

For given $\omega \in \mathbb{R}$ and $M > 0$, we denote by $G(X, M, \omega)$ the set of all closed densely defined linear operators which satisfy the resolvent estimate (7). It is obvious from (6) that M cannot be less than 1. If (7) holds with $M = 1$ for $n = 1$, then it automatically holds with $M = 1$ for all larger values of n. In this case, we call the semigroup generated by A a **quasi-contraction semigroup**. It is called a **contraction semigroup** if in addition $\omega \le 0$, i.e., a contraction semigroup satisfies $\|T(t)\| \le 1$. We remark that if $A \in G(X, M, \omega)$, then the resolvent set of A contains the half-plane $\text{Re } \lambda > \omega$ and the estimate (7) automatically extends to this half-plane, i.e.

$$\|R(\lambda, A)^n\| \le \frac{M}{(\text{Re } \lambda - \omega)^n} \tag{8}$$

for $\text{Re } \lambda > \omega$ and $n \ge 1$.

A simple criterion for generators of contraction semigroups is given by the **Lumer-Phillips theorem**, which we state only for the case of a Hilbert space H with inner product $\langle \cdot, \cdot \rangle$. To motivate the theorem, observe that if e^{At} is a contraction semigroup, then $t \to \|e^{At}u_0\|$ is a decreasing function. For $u_0 \in D(A)$,

196

we have $\frac{d}{dt}\|e^{At}u_0\|^2 = 2 \operatorname{Re} \langle e^{At}Au_0, e^{At}u_0\rangle \leq 0$. Putting $t = 0$, we conclude $\operatorname{Re} \langle Au_0, u_0\rangle \leq 0$ for all $u_0 \in D(A)$. Operators with this property are called **dissipative**.

Theorem V.3:

Let A be a densely defined linear operator in a Hilbert space H. Assume that A is dissipative and that the operator $A - \lambda_0 I$ is surjective for some $\lambda_0 > 0$. Then A is the infinitesimal generator of a contraction semigroup.

If A is a bounded operator, we can replace $\operatorname{Re} \lambda - \omega$ in (8) by $|\lambda - \omega|$. If A is unbounded, this need not be the case. As an example, consider $X = L^2(\mathbb{R})$ and $A = \frac{d}{dx}$ with $D(A) = H^1(\mathbb{R})$. It is easy to see that A generates a contraction semigroup, but that (8) (with $M = 1$ and $\omega = 0$) is in fact an equality. If, on the other hand, we take $A = \frac{d^2}{dx^2}$, $D(A) = H^2(\mathbb{R})$ (i.e. the heat equation), then we can replace $\operatorname{Re} \lambda$ by $|\lambda|$ for λ in the right half plane. If this stronger type of resolvent estimate holds, we obtain what is called an **analytic semigroup**. Such semigroups have nicer properties regarding the regularity of solutions to (2) as well as regarding the stability under perturbation of the operator A. As in the examples above, analytic semigroups are often associated with parabolic PDEs, while C_0-semigroups which are not analytic are often associated with hyperbolic PDEs. It has therefore become usual to refer to evolution problems involving analytic semigroups as "parabolic", while evolution problems involving nonanalytic C_0-semigroups are called "hyperbolic". We emphasize that this definition is not equivalent to the definition in terms of characteristics. For example, the Schrödinger equation and the Korteweg-de Vries equation are parabolic in the sense of characteristics, but hyperbolic in the sense of semigroup theory.

We now give the definition of analytic semigroups, and a theorem characterizing their infinitesimal generators.

Definition V.4:

A family of bounded linear operators $T(z)$ in X, defined for z in a sector of the form $\Delta = \{0\} \cup \{z \in \mathbb{C} | \phi_1 < \arg z < \phi_2\}$, with $\phi_1 < 0 < \phi_2$ is called an **analytic semigroup** *if the following conditions hold:*

(i) $T(z)$ is an analytic function of z for z in the interior of Δ.
(ii) $T(0)x = \lim_{z \to 0, z \in \Delta} T(z)x = x$ for every $x \in X$.
(iii) $T(z_1 + z_2) = T(z_1)T(z_2)$ for $z_1, z_2 \in \Delta$.

The following theorem characterizes the infinitesimal generators of analytic semigroups.

Theorem V.5:

Let A be the infinitesimal generator of a C_0-semigroup. The semigroup e^{At} can be extended to an analytic semigroup (in some sector Δ) if and only if the following strengthened version of (8) holds: There are constants $C > 0$ and $\omega \in \mathbb{R}$ such that the resolvent set of A contains the half-plane $\operatorname{Re} \lambda > \omega$, and we have

$$\|R(\lambda, A)\| \leq \frac{C}{|\lambda - \omega|} \text{ for } \operatorname{Re} \lambda > \omega. \tag{9}$$

197

It is easy to show that if A is a self-adjoint operator in a Hilbert space and $\langle Ax, x \rangle \leq \omega \langle x, x \rangle$, then (9) holds with $C = 1$ and hence A generates an analytic semigroup. We remark that it is not obvious, but not hard to prove either (Theorem 1.7.7 in [P1]), that (9) really implies (7). Moreover, elementary perturbation theory [K3] implies that (9) actually holds (with a larger constant C) in a sector $\{\lambda \in \mathbb{C} \mid |\arg(\lambda - \omega)| < \frac{\pi}{2} + \delta\}$ with δ strictly positive. One can then express the semigroup as a contour integral involving the resolvent:

$$e^{At} = \frac{1}{2\pi i} \int_\Gamma e^{\lambda t} R(\lambda, A) \, d\lambda, \quad t > 0, \tag{10}$$

where $0 < \theta < \delta$ and Γ is any contour from $-ie^{-i\theta}\infty$ to $ie^{i\theta}\infty$ which lies entirely to the right of the spectrum of A. Since the exponential function decays rapidly at the ends of the contour and the resolvent stays bounded, the integral in (10) converges. In a purely formal sense, (10) is obtained by solving (2) using the Laplace transform and then deforming the contour for the inverse Laplace integral. Alternatively, we can think of (10) as an extension of Dunford's definition of functions of a bounded operator [D12] in terms of integrals over contours encircling the spectrum.

Such contour integrals can also be used to define **fractional powers of** operators. Let A be the generator of an analytic semigroup, and let ω in (9) be negative. Then we define

$$(-A)^{-\alpha} = \frac{1}{2\pi i} \int_\Gamma (-\lambda)^{-\alpha} R(\lambda, A) \, d\lambda, \tag{11}$$

where Γ is as in (10) and, in addition, lies to the left of the origin. The integral converges for any positive α. It can be shown that $(-A)^{-\alpha}$ has the properties that the notation suggests (i.e. it reduces to a power of $(-A)^{-1}$ when α is an integer, it converges strongly to the identity when $\alpha \to 0$ and $(-A)^{-\alpha}(-A)^{-\beta} = (-A)^{-(\alpha+\beta)}$). $(-A)^\alpha$ is defined as the inverse of $(-A)^{-\alpha}$; it is a closed, densely defined operator and the domain decreases with increasing α ($D((-A)^\alpha) \subset D((-A)^\beta)$ for $\alpha > \beta$). In applications, it is relevant to characterize operators which are bounded relative to $(-A)^\alpha$. The closed graph theorem implies that a closed operator B is bounded relative to $(-A)^\alpha$ if $D(B)$ includes $D((-A)^\alpha)$. A precise characterization of $D((-A)^\alpha)$ is not easy in most applications. Fortunately, it is not required; instead, the applications are based on the following theorem.

Theorem V.6:

Assume that A generates an analytic semigroup in X such that (9) holds with $\omega < 0$. Let Y be a Banach space such that $D(A) \subset Y \subset X$ with continuous embeddings (here $D(A)$ is equipped with the graph norm). If, for some $\gamma \in (0,1)$ and every $\rho \geq \rho_0 > 0$, we have

$$\|x\|_Y \leq C(\rho^\gamma \|x\| + \rho^{\gamma-1}\|Ax\|) \tag{12}$$

for every $x \in D(A)$, then $Y \supset D((-A)^\alpha)$ for every $\alpha > \gamma$.

As an example, let Ω be a bounded domain in $\mathbb{R}^n$ with smooth boundary, $X = L^p(\Omega)$ and $A = \Delta$ with domain $D(A) = \{u \in W^{2,p}(\Omega) \mid u|_{\partial\Omega} = 0\}$. It is well known that this operator generates an analytic semigroup (see e.g. Theorem 7.3.5 in [P1]). Assume that $p > n$, and let $0 < \nu < 1 - n/p$, and $\gamma = (1+\nu)/2 + n/(2p)$. Then it is known that (see equation (3.20) in [S8])

$$\|u\|_{C^{1,\nu}(\Omega)} \leq C'\|u\|_{L^p}^{1-\gamma}\|u\|_{W^{2,p}}^{\gamma} \leq C''\|u\|_{L^p}^{1-\gamma}\|Au\|_{L^p}^{\gamma}$$

$$\leq C(\rho^{\gamma}\|u\|_{L^p} + \rho^{\gamma-1}\|Au\|_{L^p}) \tag{13}$$

for every $\rho > 0$. Hence the domain of $(-A)^{\alpha}$ is contained in $C^{1,\nu}(\Omega)$ for $\alpha > \gamma$. The inhomogeneous problem corresponding to (2),

$$\dot{u} = Au + f(t), \; u(0) = u_0, \tag{14}$$

can be "solved" in the obvious way by the variation of constants formula

$$u(t) = e^{At}u_0 + \int_0^t e^{A(t-s)}f(s)\,ds, \quad t \geq 0. \tag{15}$$

This formal solution is called a **mild solution** of equation (14). Without some restrictions on u_0 and f it need not be a solution in the strict sense because the terms $\dot{u}$ and Au may not be meaningful as functions valued in X. If u is the primitive of a function in $L^1([0,T];X)$, $Au \in L^1([0,T];X)$ and the equation is satisfied a.e. on $[0,T]$, we call u a **strong solution**. The following results give conditions sufficient for a mild solution to be a strong solution. Such conditions are weaker in the case of analytic semigroups than they are for C_0-semigroups. We first state the relevant results for the case of a C_0-semigroup.

Theorem V.7:

Let A be the infinitesimal generator of a C_0-semigroup and let u be given by (15). If $u_0 \in D(A)$ and either $f \in C^1([0,T];X)$ or $f \in L^1([0,T];D(A)) \cap C([0,T];X)$ (again $D(A)$ is equipped with the graph norm), then $\dot{u}$ and Au are in $C([0,T];X)$. If $u_0 \in D(A)$, f is the primitive of a function $\dot{f} \in L^1([0,T];X)$, then u is the primitive of a function $\dot{u} \in L^1([0,T];X)$, Au is in $L^1([0,T];X)$ and (14) is satisfied almost everywhere.

Basically what is assumed here is that the initial condition is in $D(A)$ and that the inhomogeneous term either takes values in $D(A)$ or is differentiable with respect to time. In the case of analytic semigroups a strong solution for $t > 0$ can be obtained without any assumption on u_0. Instead of requiring f to be differentiable only a Hölder condition is needed, and instead of requiring f to take values in the domain of A we need only require that it takes values in the domain of some fractional power. More precisely:

Theorem V.8:

Let A be the infinitesimal generator of an analytic semigroup. If $f \in C^{\theta}([0,T];X)$ for some $\theta \in (0,1)$, then u as given by (15) is a strong solution of (14) for $t \in (0,T]$ and Au, $\dot{u} \in C^{\theta}([\delta,T];X)$ for every $\delta > 0$. If, in addition, $u_0 \in D(A)$, then $\dot{u}$ and Au are continuous at $t = 0$. If f is continuous on $[0,T]$, takes values in $D((-A+\lambda_0)^{\alpha})$ for some $\alpha \in (0,1]$ (where λ_0 is chosen large enough so that the spectrum of $A-\lambda_0$ is in the left half plane), and $\|(-A+\lambda_0)^{\alpha}f\|$ is bounded on $[0,T]$, then $\dot{u}$ and Au are continuous for $t \in (0,T]$.

2. Time-dependent coefficients and nonlinear problems

In this section, we discuss problems involving time-dependent linear operators,

$$\dot{u}(t) = A(t)u(t) + f(t), \quad u(0) = u_0, \tag{16}$$

and "quasilinear" problems of the form

$$\dot{u}(t) = A(t,u(t))u(t) + f(t,u(t)), \quad u(0) = u_0. \tag{17}$$

For the latter class of problems, the idea is that the first term on the right-hand side contains the leading-order terms in the equation which occur in a linear fashion, while their coefficients and the remainder term f are of "lower order". In a concrete situation, the right hand side of (17) might have the form

$$a(t,u,u_x)u_{xx} + f(t,u,u_x). \tag{18}$$

The study of (17) is closely connected to that of (16), because under appropriate assumptions an iterative scheme

$$\dot{u}^{n+1} = A(t,u^n)u^{n+1} + f(t,u^n), \quad u^{n+1}(0) = u_0 \tag{19}$$

can be used to solve (17). At each step of the iteration we then have a linear problem of the form (16).

Since the theories for the "parabolic" and "hyperbolic" cases are quite different, we shall discuss them separately. Again we present theorems without proofs; for details, we refer to the books of Pazy [P1] and Friedman [F3], and also the papers by Sobolevskii [S8], Kato [K4-K9] and Hughes, Kato and Marsden [H17]. We begin with the parabolic case for (16) with $f = 0$. We require that $A(t)$ is the generator of an analytic semigroup for every $t \geq 0$ and that a Hölder condition with respect to time is satisfied. More precisely, we assume

(P1) The domain $D = D(A(t))$ of $A(t)$ is dense in X and independent of t for $0 \leq t \leq T$.

P2) There are constants $M > 0$ and $\omega \in \mathbb{R}$ such that the resolvent $R(\lambda, A(t))$ exists for Re $\lambda \geq \omega$ and satisfies

$$\|R(\lambda, A(t))\| \leq \frac{M}{|\lambda - \omega| + 1} \tag{20}$$

uniformly for $0 \leq t \leq T$.

P3) There are constants L and $0 < \epsilon \leq 1$ such that

$$\|(A(t) - A(s))(A(0) - \omega I)^{-1}\| \leq L|t - s|^{\epsilon}, \tag{21}$$

for $t, s \in [0, T]$.

These assumptions guarantee the existence of a family of evolution operators which is an analogue of the fundamental matrix for a system of ODEs (see [P1],[S8],[F3]).

Theorem V.9:

Let (P1)-(P3) be satisfied. Then there is a family of bounded operators $U(t, s)$ in X, which is defined and strongly continuous for $0 \leq s \leq t \leq T$. For every positive δ, this operator function is uniformly differentiable in t for $t \geq s + \delta$, and

$$\frac{\partial U(t, s)}{\partial t} = A(t)U(t, s). \tag{21}$$

The identity
$$U(t, s) = U(t, r)U(r, s) \tag{22}$$

holds for $0 \leq s \leq r \leq t \leq T$ and we have $U(t, t) = I$. The unique strong solution of (16) (with $f = 0$) for $t \in (0, T]$ is given by

$$u(t) = U(t, 0)u_0; \tag{23}$$

$u(t)$ is continuously differentiable for $t > 0$, and if $u_0 \in D$, $u(t)$ is also continuously differentiable at $t = 0$.

The "mild" solution of the inhomogeneous problem is defined by

$$u(t) := U(t, 0)u_0 + \int_0^t U(t, s)f(s) \, ds. \tag{24}$$

As in the case of time-independent coefficients, a Hölder condition on f suffices to guarantee that a mild solution is a strong solution.

Theorem V.10:

Let (P1)-(P3) hold and let the function f satisfy the Hölder condition

$$\|f(t) - f(s)\| \leq C|t - s|^{\rho}$$

201

for $t, s \in [0, T]$ and some $\rho \in (0, 1]$. Then the formula (24) gives a unique strong solution of (16), which is continuous for $t \in [0, T]$, and continuously differentiable for $t \in (0, T]$. If $u_0 \in D$, then u is also continuously differentiable at $t = 0$.

The assumptions of Theorems V.9 and V.10 yield estimates which will be useful for the discussion of quasilinear problems (see [F3]).

Lemma V.11:

Under the assumptions (P1)-(P3), the following estimates hold uniformly for $0 \leq s \leq t \leq t + \Delta t \leq T$ *and* $0 \leq \varsigma \leq T$:

$$\left\| (\omega - A(t))^\alpha U(t, s)(\omega - A(s))^{-\beta} \right\| \leq C(\alpha, \beta)(t - s)^{\beta - \alpha}, \quad 0 \leq \beta \leq \alpha < 1 + \epsilon, \quad (25)$$

$$\left\| (\omega - A(\varsigma))^\alpha \left[U(t + \Delta t, s) - U(t, s) \right](\omega - A(s))^{-\beta} \right\| \leq C(\alpha, \beta, \gamma)(\Delta t)^{\gamma - \alpha}(t - s)^{\beta - \gamma},$$

$$0 \leq \alpha \leq 1, \ 0 \leq \beta \leq \gamma < 1 + \epsilon, \ 0 < \gamma - \alpha \leq 1, \quad (26)$$

$$\left\| (\omega - A(\varsigma))^\alpha \left[A(t + \Delta t)U(t + \Delta t, s) - A(t)U(t, s) \right](\omega - A(s))^{-\beta} \right\|$$

$$\leq C(\alpha, \beta, \gamma, \delta)(\Delta t)^{\gamma - \alpha}(t - s)^{\beta - \gamma - 1 - \delta}, \ 0 \leq \alpha < \gamma < \epsilon, \ 0 \leq \beta \leq 1, \ \delta > 0. \quad (27)$$

Moreover, under the assumptions of Theorem V.10, the following estimates hold uniformly for $0 \leq s \leq t \leq t + \Delta t \leq T$ *and* $0 \leq \varsigma \leq T$:

$$\left\| (\omega - A(\varsigma))^\alpha \left[\int_s^{t + \Delta t} U(t + \Delta t, r)f(r) \ dr - \int_s^t U(t, r)f(r) \ dr \right] \right\|$$

$$\leq C(\alpha)(\Delta t)^{1 - \alpha} \left[|\ln \Delta t| + 1 \right] \max_{s \leq r \leq t + \Delta t} \| f(r) \|, \quad 0 \leq \alpha < 1, \quad (28)$$

$$\left\| (\omega - A(\varsigma))^\beta \left[\frac{\partial}{\partial t} \int_s^{t + \Delta t} U(t + \Delta t, r)f(r) \ dr - \frac{\partial}{\partial t} \int_s^t U(t, r)f(r) \ dr \right] \right\|$$

$$\leq C(\alpha, \beta, \delta)(t - s)^{-\alpha - \delta}(\Delta t)^{\alpha - \beta} \left\{ \| f(t) \| + \max_{s \leq r \leq t} \left\| \frac{f(t) - f(r)}{(t - r)^\rho} \right\| \right.$$

$$+ \max_{s \leq r \leq t + \Delta t} \left\| \frac{f(t + \Delta t) - f(r)}{(t + \Delta t - r)^\rho} \right\| \Bigg\}, \quad 0 \leq \beta \leq \alpha < \min \{ \epsilon, \rho \}, \ \delta > 0. \quad (29)$$

An existence result for quasilinear equations is obtained by showing that the iteration (19) defines a contraction in an appropriate subset of a Banach space. In Section 5, we shall present a modified version of this argument for quasilinear integrodifferential equations which model certain viscoelastic materials in one space dimension. For equations without delay terms, we quote the following result from [S8].

Theorem V.12:

Assume that the domain D of the linear operator $A_0 = A(0, u_0)$ is dense in and that for every λ with $\operatorname{Re} \lambda \geq 0$ thee operator $A_0 - \lambda I$ has a bounded inverse satisfying

$$\|(A_0 - \lambda I)^{-1}\| \leq \frac{C}{|\lambda| + 1}. \tag{30}$$

or some $\alpha \in [0, 1)$, $R > 0$, and for every $u, v \in X$ with $\|u\|, \|v\| \leq R$ and every $t \in [0, T]$, let the linear operator $A(t, (-A_0)^{-\alpha} u)$ be defined on D and satisfy the Hölder condition

$$\left\| \left[A(t, (-A_0)^{-\alpha} u) - A(s, (-A_0)^{-\alpha} v) \right] A_0^{-1} \right\| \leq C \left[|t - s|^\epsilon + \|u - v\| \right] \tag{31}$$

or some $\epsilon \in (0, 1]$. With α, s, t, u, v as above, assume that $f(t, (-A_0)^{-\alpha} u)$ is efined and satisfies

$$\left\| f(t, (-A_0)^{-\alpha} u) - f(s, (-A_0)^{-\alpha} v) \right\| \leq C \left[|t - s|^\epsilon + \|u - v\| \right]. \tag{32}$$

Moreover, assume that $u_0 \in D((-A_0)^\beta)$ for some $\beta > \alpha$ and $\|(-A_0)^\alpha u_0\| < R$. Then there is a unique solution to equation (17), which exists on some time interval $[0, t_0]$, is continuous for $t \in [0, t_0]$, continuously differentiable for $t > 0$, and can be obtained by the iteration (19).

To continue the solution in Theorem V.12, it suffices to obtain bounds on $\|(-A(t, u(t)))^\beta u(t)\|$, see [S8].

We now turn our attention to the hyperbolic case. Let us recall that, for autonomous problems, the definition of the evolution operator was based on the formula (4)

$$e^{At} = \lim_{n \to \infty} \left(I - \frac{At}{n} \right)^{-n}. \tag{33}$$

or time-dependent operators, it is natural to generalize (33) as

$$U(t, 0) = \lim_{n \to \infty} \prod_{j=1}^{n} \left(I - \frac{t}{n} A(\frac{jt}{n}) \right)^{-1} \tag{34}$$

with the convention that the factors in the product are ordered from right to left). One of the crucial aspects of the theory is therefore the possibility of bounding products such as the right hand side of (34). The following definition characterizes families of operators for which this is possible.

Definition V.13:

We denote by $G(X)$ the set of all infinitesimal generators of C_0-semigroups in the Banach space X. A family $\{A(t)\}$ of elements of $G(X)$, defined for $t \in$

$[0, T]$, *is called* **stable** *if there are constants $\beta \in R$ and $M > 0$ such that for every choice $0 \leq t_1 \leq \ldots \leq t_k \leq T$ and $\lambda_j > \beta$ we have*

$$\| \prod_{j=1}^{k} (A(t_j) - \lambda_j)^{-1} \| \leq M \prod_{j=1}^{k} (\lambda_j - \beta)^{-1}. \tag{35}$$

Unfortunately, it is in general not easy to characterize stable families of C_0-generators. There is a trivial special case, which is when $A(t)$ generates a quasi-contraction semigroup for every t and the constant ω appearing in (7) has a uniform upper bound on $[0, T]$. We then have (35) with $M = 1$ and $\beta = \sup_{t \in [0, T]} \omega(t)$. For applications in continuum mechanics, this is a little too restrictive. It can be shown, however, that it is sufficient that $A(t)$ generate a quasi-contraction semigroup in a norm of X which is allowed to vary with t, provided that the t-dependence of the norm satisfies a Lipschitz estimate. We quote the following result from [K4].

Lemma V.14:

For each $t \in [0, T]$, let $\| \cdot \|_t$ be a norm in X equivalent to the original one and let $\| \cdot \|_t$ be Lipschitz continuous in t in the sense that

$$\|x\|_t / \|x\|_s \leq e^{c|t-s|} \tag{36}$$

for $x \in X$ and $t, s \in [0, T]$. Let X_t be the space X equipped with norm $\| \cdot \|_t$. For each $t \in [0, T]$, assume that $A(t) \in G(X_t, 1, \beta)$. Then the family $\{A(t)\}$ is stable with respect to each of the norms $\| \cdot \|_s$, with the stability constants of Definition V.13 given by β and $M = e^{2cT}$.

In applications to hyperbolic partial differential equations, it is frequently not the case that the domain of $A(t)$ is independent of t, and a different assumption is required. This assumption is phrased in terms of a second Banach space Y, which is dense in X. This space Y must be contained in the domain of $A(t)$ for every t. Moreover, we make an assumption which insures that the restriction of $A(t)$ to Y generates a C_0-semigroup on Y and we impose some smoothness conditions on the t-dependence. The following theorem [K5] gives conditions that guarantee the existence of evolution operators in the hyperbolic case.

Theorem V.15:

Assume that

(H1) The family $\{A(t)\} : [0, T] \to G(X)$ is stable.

(H2) There is a Banach space Y, continuously and densely embedded in X, and a family $\{S(t)\}$ of isomorphisms of Y onto X such that for almost every $t \in [0, T]$ there is a bounded operator $B(t)$ on X such that

$$S(t)A(t)S(t)^{-1} = A(t) + B(t) \tag{37}$$

(this relation includes equality of domains). The function $B : [0,T] \to B(X)$ is strongly measurable (meaning that, for every $x \in X$, $B(t)x$ is strongly measurable, and $\|B(\cdot)\|_X$ is upper-integrable. Furthermore, S is the indefinite strong integral of a strongly measurable function $\dot{S} : [0,T] \to B(Y,X)$, and $\|\dot{S}(\cdot)\|_{Y,X}$ is upper-integrable.

$\mathcal{13})$ $Y \subset D(A(t))$ for every $t \in [0,T]$ and $A : [0,T] \to B(Y,X)$ is norm-continuous.

Then there exists a unique evolution operator $U = U(t,s)$, defined on the set $\Delta : 0 \le s \le t \le T$ with the following properties.

(i) U is strongly continuous from Δ to $B(X)$ and $U(s,s) = I$.

(ii) $U(t,s)U(s,r) = U(t,r)$.

iii) $U(t,s)Y \subset Y$ and U is strongly continuous from Δ to $B(Y)$.

'iv) $\partial U(t,s)/\partial t = A(t)U(t,s)$, $\partial U(t,s)/\partial s = -U(t,s)A(s)$, where the derivatives exist in the strong sense in $B(Y,X)$ and are strongly continuous from Δ to $B(Y,X)$.

We shall state some estimates which are important in dealing with nonlinear problems. These estimates involve various Banach space and operator-valued functions. We use the following straightforward notations. For a function $f : [0,T] \to X$, let

$$\|f\|_{\infty,X} := \operatorname*{ess\,sup}_{t \in [0,T]} \|f(t)\|_X , \quad \|f\|_{1,X} := \int_0^T \|f(t)\|_X \, dt, \tag{38}$$

For operator-valued functions $F : [0,T] \to B(X,Y)$, we define $\|F\|_{\infty,X,Y}$ and $\|F\|_{1,X,Y}$ similarly. However, in the definition of $\|F\|_{1,X,Y}$, we replace the integral by the upper integral. We write $\|F\|_{\infty,X}$ for $\|F\|_{\infty,X,X}$. For later use, we record some estimates for the norm of the evolution operator [K5].

Lemma V.16:

Assume the assumptions of Theorem V.15 hold. With M and β denoting the constants in the stability bound for $\{A(t)\}$, we have

$$\|U\|_{\infty,X} \le Me^{\beta T}. \tag{39}$$

Moreover,

$$\|U\|_{\infty,Y} \le \|S\|_{\infty,Y,X} \|S^{-1}\|_{\infty,X,Y} M \exp[\beta T + M\|B - \dot{S}S^{-1}\|_{1,X}]. \tag{40}$$

For the mild solution given by (24), we have the following lemma concerning regularity and estimates [K5].

Lemma V.17:

Let the assumptions of Theorem V.15 hold and let u be given by (24). If

(a) $u_0 \in X$ *and* $f \in L^1((0,T);X)$, *then* $u \in C([0,T];X)$,

(b) $u_0 \in Y$ *and* $f \in L^1((0,T);Y)$, *then* $u \in C([0,T];Y)$,

(c) $u_0 \in Y$ *and* $f \in L^1((0,T);Y) \cap C([0,T];X)$, *then* $u \in C([0,T];Y) \cap C^1([0,T];X)$, *and* u *is a strong solution.*

Moreover, we have the bounds

$$\|u\|_{\infty,X} \le \|U\|_{\infty,X} (\|u_0\|_X + \|f\|_{1,X}) \tag{41}$$

in case (a),

$$\|u\|_{\infty,Y} \le \|U\|_{\infty,Y} (\|u_0\|_Y + \|f\|_{1,Y}) \tag{42}$$

in cases (b) and (c), and

$$\|\dot{u}\|_{\infty,X} \le \|f\|_{\infty,X} + \|A\|_{\infty,Y,X} \|U\|_{\infty,Y} (\|u_0\|_Y + \|f\|_{1,Y}) \tag{43}$$

in case (c).

We shall also need a perturbation result. For this, we consider a second equation of the same form as (1), which also satisfies the assumptions of Theorem V.15.

$$\dot{u}^*(t) = A^*(t)u^*(t) + f^*(t), \quad u^*(0) = u_0^*. \tag{44}$$

The next lemma provides an estimate for the difference of the solutions of (1) and (44).

Lemma V.18:

Let u denote the solution of (1), and u^* the solution of (44), where both equations satisfy the assumptions of Theorem V.15 with the same spaces X and Y. Let $u_0 \in Y$, $u_0^* \in X$, and $f \in L^1((0,T);Y)$, $f^* \in L^1([0,T];X)$. Then

$$\|u^* - u\|_{\infty,X} \le \|U^*\|_{\infty,X} \left[\|u_0^* - u_0\|_X + \|f^* - f\|_{1,X} + \|(A^* - A)u\|_{1,X} \right]. \tag{45}$$

The preceding results, together with the iteration (19), can be used to obtain a local existence theorem for quasilinear problems. We quote the following result from Hughes, Kato and Marsden [H17].

Theorem V.19:

Let $Y \subset Z \subset Z' \subset X$ be four real Banach space, all of them reflexive and separable, with continuous and dense inclusions. We assume that

(Z) Z' is an interpolation space between Y and X (i.e. linear operators which are bounded on both Y and X are also bounded on Z').

Let $N(X)$ be the set of all norms on X equivalent to the given one. On $N(X)$ we introduce a distance function

$$d(\|\cdot\|_\alpha, \|\cdot\|_\beta) := \ln \max\{\sup_{z \neq 0} \|z\|_\alpha / \|z\|_\beta, \sup_{z \neq 0} \|z\|_\beta / \|z\|_\alpha\}. \tag{46}$$

Let W be an open set in Y. We assume that there is a real number β and positive numbers λ_N, μ_N, ... such that the following hold for all $t, t' \in [0, T]$ and $w, w' \in W$.

(N) $N(t, w) \in N(X)$, and

$$d(N(t, w), \|\cdot\|_X) \leq \lambda_N, \quad d(N(t', w'), N(t, w)) \leq \mu_N [|t' - t| + \|w' - w\|_Z]. \tag{47}$$

(S) *There is an isomorphism $S(t, w) \in B(Y, X)$, with*

$$\|S(t, w)\|_{Y, X} \leq \lambda_S, \quad \|S(t, w)^{-1}\|_{X, Y} \leq \lambda_S',$$

$$\|S(t', w') - S(t, w)\|_{Y, X} \leq \mu_S [|t' - t| + \|w' - w\|_Z]. \tag{48}$$

(A1) $A(t, w) \in G(X_{N(t, w)}, 1, \beta)$.

(A2) $S(t, w)A(t, w)S(t, w)^{-1} = A(t, w) + B(t, w)$, *where $B(t, w)$ is a bounded operator in X and*

$$\|B(t, w)\|_X \leq \lambda_B. \tag{49}$$

(A3) $A(t, w) \in B(Y, Z)$ *with*

$$\|A(t, w)\|_{Y, Z} \leq \lambda_A, \quad \|A(t, w') - A(t, w)\|_{Y, Z'} \leq \mu_A \|w' - w\|_{Z'}. \tag{50}$$

Moreover, the mapping $t \to A(t, w) \in B(Y, X)$ is continuous in norm.

(f) $f(t, w) \in Y$, *with*

$$\|f(t, w)\|_Y \leq \lambda_f, \quad \|f(t, w') - f(t, w)\|_{Z'} \leq \mu_f \|w' - w\|_{Z'}, \tag{51}$$

and the mapping $t \to f(t, w) \in X$ is continuous.

If all of the above assumptions are satisfied, and $u_0 \in W$, then there is a $T' \in (0, T]$ such that (17) has a unique solution u on $[0, T']$ with $u \in C([0, T']; W) \cap C^1([0, T']; X)$. Here T' may depend on all the constants involved in the assumptions and on the distance between u_0 and the boundary of W. The mapping $u_0 \to u(t)$ is Lipschitz continuous in the Z'-norm, uniformly for $t \in [0, T']$. The solution is obtained by the iteration (19).

In [H17], an additional condition (A4) is assumed, but it is pointed out in [K8] that this condition is unnecessary. In any case, it is trivial to verify (A4) in the applications. In our notation, we have reversed the roles of X and Z in [H17] because we wanted X and Y to have the same significance as in the results stated earlier for linear problems. We remark that a stronger continuous dependence result can be obtained under additional assumptions, see [H17],[K7]. Also, it is easy to obtain continuation results, since we have a characterization of the quantities on which the length of the interval of existence depends.

3. Various alternatives for treating memory terms

The results discussed in the last two sections are not formulated in such a way that memory or delay terms are included in the differential equations. If we want to apply them in viscoelasticity, we must find a way to allow for delay terms. There are several possible approaches for this. One possibility is to follow a procedure similar in spirit to what we have done in Chapter III. If the highest order derivatives appear only through their present values, an iteration procedure can be used which involves solving only problems without delays at each step of the iteration. The results of the last two sections can be combined with a contraction argument to prove convergence of the iteration. We shall give examples in Sections 5 and 6. We also refer to the papers of Heard [H5],[H6] and Schumacher [S4].

Another possibility is to treat the equation as an evolution problem on a space of functions that involves not only the values of the unknown variables at a fixed instant of time, but their histories as well. Consider a differential-delay equation of the form

$$\dot{u}(t) = \mathcal{F}(u^t, t). \tag{52}$$

Here u takes values in some space X, and u^t denotes the history of u up to time t. That is, u^t is a function $[0, \infty) \to X$, and $u^t(s)$ denotes the value of u at the time $t - s$. The functional $\mathcal{F}$ takes values in X. The idea is now to regard (52) as an evolution problem not on X, but on a suitable space of functions from $[0, \infty)$ to X. The unknown variable is then not $u(t)$, but u^t.

In the context of ODEs with delays, a rather extensive discussion of the relation between the differential-delay equation and an associated evolution problem on a history space is given in the papers by Hale and Kato [H1] and Kappel and Schappacher [K2]. The topology of the function space chosen for u^t turns out to be of crucial importance.

One of the classical approaches to differential-delay equations uses the space of bounded uniformly continuous functions from $[0, \infty)$ to X. Although we shall not use this approach in the remainder of this chapter, we briefly discuss it for purposes of comparison. Under suitable assumptions the differential-delay equation (52) is equivalent to the following problem

$$\frac{d}{dt} u^t = A(t) u^t, \tag{53}$$

where the operator $A(t)$ is defined by

$$A(t)\phi(s) := -\frac{d}{ds}\phi(s) \tag{54}$$

with domain

$$D(A(t)) := \{\phi \in C^1_{bu}([0, \infty); X) \mid -\phi'(0) = \mathcal{F}(\phi, t)\}. \tag{55}$$

The operator $A(t)$ is the operator of differentiation, and all information about the equation (52) is contained in the choice of the domain for $A(t)$; the domain is in general not a linear subspace. The results discussed in the previous two sections are not suited to this type of problem. Let us demonstrate, however, how linear semigroup theory can be applied to equations (53)-(55) in the linear autonomous case when $\mathcal{F}(u^t, t) = Lu^t$, where L is a bounded linear operator from $C_{bu}([0,\infty); X)$ into X. We want to show that the operator A defined by (54), (55) satisfies the conditions of Theorem V.2, part (i). It is easy to see that A is closed and densely defined in $C_{bu}([0,\infty); X)$. To determine the resolvent, we consider the equation

$$\lambda u(s) + \frac{du(s)}{ds} = f(s), \quad -\frac{du}{ds}\Big|_{s=0} = Lu. \tag{56}$$

Using the variation of constants method, we find

$$u(s) = \int_0^s e^{\lambda(\tau - s)} f(\tau)\, d\tau + \int_{-\infty}^0 e^{\lambda(\tau - s)} f(0)\, d\tau + \frac{e^{-\lambda s}}{\lambda} Lu, \tag{57}$$

and for $\lambda > 0$ this yields the estimate

$$\sup_{s \in [0,\infty)} |u(s)| \le \frac{1}{\lambda} \sup_{s \in [0,\infty)} |f(s)| + \frac{1}{\lambda}\|L\| \sup_{s \in [0,\infty)} |u(s)|. \tag{58}$$

This implies (7) with $\omega = \|L\|$ and $M = 1$, and hence A is the infinitesimal generator of a quasi-contraction semigroup.

As we have seen in earlier chapters, many models in viscoelasticity involve the past history of the undetermined variable in the form of integrals, but they also involve the present value. This suggests using the past history and the present value as separate variables in an abstract evolution problem and using an integral type norm to topologize the history space. If we let $u(t)$ denote the present value, and $y^t(s) = u(t - s)$ the history, then the equation has the form

$$\frac{d}{dt} u(t) = \tilde{\mathcal{F}}(u(t), y^t, t), \tag{59}$$

and the associated evolution problem becomes

$$\frac{d}{dt} u(t) = \tilde{\mathcal{F}}(u(t), y^t, t),$$

$$\frac{d}{dt} y^t(s) = -\frac{d}{ds} y^t(s). \tag{60}$$

The second equation in (60) implies that $y^t(s)$ is a function of the difference $t - s$. In order to have a solution of the original problem, we want $y^t(s) = u(t - s)$. Hence we impose the boundary condition $y^t(0) = u(t)$, which becomes part of the definition of the domain of the operator $\frac{d}{ds}$. This type of formulation was used

209

by Slemrod [S5] to treat the equations of linear viscoelasticity. We shall discuss his results in detail in Section 4.

The formulations discussed so far assign an evolution equation in an abstract space to the original differential-delay equation, and this is done in such a way that the abstract evolution equation does not depend on the choice of the prescribed initial history. If we relax this constraint, and are content with an abstract evolution equation that is equivalent to the differential-delay equation for a specific initial history rather than for every initial history, then we have much more freedom and there are many more possible formulations. Here we discuss two such formulations which were used by Renardy [R6-R8],[R10] for problems in viscoelasticity. Some of these results are discussed in Sections 7 and 8.

Consider an equation of the form (52). We associate the functional $\mathcal{F}$ with an operator taking histories to histories in the following way: For any function ψ from $[0, \infty)$ into the state space X and any $r \geq 0$, we define $T_r \psi$ by $T_r \psi(s) = \psi(s + r)$, and the operator $\hat{\mathcal{F}}$ is defined by

$$\hat{\mathcal{F}}(\psi, t)(s) = \mathcal{F}(T_s \psi, t - s). \tag{61}$$

The equation

$$\frac{d}{dt} u^t(s) = \hat{\mathcal{F}}(u^t, t)(s) \tag{62}$$

extends (52) into the past; in other words, (62) holds if (52) is satisfied not only starting from some time $t = 0$, but throughout the whole past history as well. Hence the initial value problem for (62) is not really equivalent to the initial-history value problem for (52) because the usual interpretation of the initial-history value problem does not require that the history satisfy the equation. Even if the initial history does not satisfy the equation, however, we can add a correction term to the right hand side of (52) which is such that it makes the equation satisfied for $t < 0$, but vanishes for $t > 0$. Therefore, in place of (52), we study the modified equation

$$\dot{u}(t) = \mathcal{F}(u^t, t) + f(t), \tag{63}$$

and we replace (62) by the modified equation

$$\frac{d}{dt} u^t(s) = \hat{\mathcal{F}}(u^t, t)(s) + f(t - s). \tag{64}$$

Of course, the choice of f (and hence equation (64)) depends on the initial history. We also note that assumptions on the continuity and smoothness of f across $t = 0$ represent compatibility conditions between the initial history and the solution for positive time. Finally we note that instead of (63) we can use any equation that is satisfied by the initial history for $t < 0$ and is equivalent to the given differential-delay equation for $t > 0$.

One must be a little careful in claiming that a solution to (62) actually yields a solution to (52). The essential point is that in (62) we have the two independent variables t and s, and a solution of (62) yields a solution for (52) only if it depends on t and s only through the combination $t - s$. In fact, this is exactly what goes wrong if we solve (62) with an initial history that does not satisfy the differential-delay equation. We thus have to make sure that for every $r > 0$, the solution of (62) satisfies

$$u^{t-r}(s) = u^t(s+r), \quad s \in [0, \infty). \tag{65}$$

Here we define $u^{t-r}(s)$ to be $u^0(s+r-t)$, i.e. to be given by the initial history, if $t - r$ is negative. It is then obvious that both $u^{t-r}(s)$ and $u^t(s+r)$ satisfy the same differential equation with the same initial condition at $t = 0$. What is required is therefore a uniqueness theorem for the initial value problem associated with (62), and the techniques we will use to treat (62) in the applications will automatically yield such a uniqueness result.

Another possible approach is to split the history of u up to time t into a known part from $-\infty$ to 0 and a part from 0 to t. The latter can be viewed as a function $\hat{u} : [0,1] \to X$, if we use the rescaling $\hat{u}(\sigma,t) = u(\sigma t)$, $\sigma \in [0,1]$. We can then rewrite (52) in the form

$$\dot{u}(t) = \mathcal{G}(\hat{u}(\cdot,t), t). \tag{66}$$

Here the functional $\mathcal{G}$ depends on the given initial history. With the functional $\mathcal{G}$ we can associate an operator $\hat{\mathcal{G}}$ on a suitable space of functions from $[0,1]$ to X in the following way: For any $\rho \in [0,1]$ let $S_\rho \hat{u}(\sigma,t) = \hat{u}(\sigma\rho,t)$ and let

$$\hat{\mathcal{G}}(\hat{u},\sigma,t) = \mathcal{G}(S_\sigma \hat{u}, \sigma t). \tag{67}$$

We then have

$$\dot{u}(\sigma t) = \hat{\mathcal{G}}(\hat{u},\sigma,t), \tag{68}$$

and finally we note that

$$\frac{d}{dt}\hat{u}(\sigma,t) = \frac{d}{dt}u(\sigma t) = \sigma\dot{u}(\sigma t). \tag{69}$$

Thus we can write the equation in the form

$$\frac{d}{dt}\hat{u}(\sigma,t) = \sigma\hat{\mathcal{G}}(\hat{u},\sigma,t). \tag{70}$$

The unknown in this latter equation is now a function from $[0,1]$ to X. The appropriate initial condition is the constant function

$$\hat{u}(\sigma,0) = u(0), \quad \sigma \in [0,1]. \tag{71}$$

For problems with singular kernels, an abstract operator theoretic approach for solving the initial-value problem has yet to be developed. For nonintegrable memory functions, analytic resolvent operators [G16],[P8] arise in the solution of the linear problem; they have many of the properties of analytic semigroups.

It appears feasible that a theory of quasilinear equations analogous to that of Sobolevskii might be developed.

4. Linear viscoelasticity

In this section we shall discuss the results obtained by Slemrod [S5] for an incompressible linearly viscoelastic fluid. Assuming the constitutive relation (46) of Chapter I with $\beta = \eta = 0$, we obtain the following equations of motion for the velocity field

$$\frac{\partial v}{\partial t}(y,t) = -\nabla p(y,t) + \Delta \int_0^\infty G(s)v(y,t-s) \, ds,$$

$$\operatorname{div} v = 0. \tag{72}$$

Here G is the stress relaxation modulus which is related to the memory function m by

$$G(s) = \int_s^\infty m(\xi) \, d\xi. \tag{73}$$

Throughout this section, we assume that $G \in C^2[0, \infty)$, $G > 0$, $G' < 0$, $G'' \geq 0$ and that $G(s) \to 0$ as $s \to \infty$. Equation (72) is considered in a bounded domain $\Omega \subset \mathbb{R}^3$ with smooth boundary, and Dirichlet conditions are prescribed on the boundary,

$$v = 0 \text{ on } \partial\Omega. \tag{74}$$

We seek solutions of (72), (74) for $t > 0$ subject to a prescribed initial history,

$$v(y,\tau) = v^0(y,\tau) \text{ for } y \in \Omega, \ -\infty < \tau \leq 0,$$

$$v(y,0+) = v^0(y,0-) \text{ for } y \in \Omega. \tag{75}$$

The method used by Slemrod follows the prescription of (60). That is, we introduce a pair (v, w) of unknown variables, where w stands for the history of v: $w(x, s, t) = v(x, t-s)$. We define an inner product by

$$\langle (v,w), (v^*,w^*) \rangle_H := \int_\Omega \nabla \{ \int_0^\infty G(s)w(y,s) \, ds \} : \nabla \{ \int_0^\infty G(s)w^*(y,s) \, ds \} \, dy$$

$$- \int_\Omega \int_0^\infty G'(s)(w(y,s) - v(y)) \cdot (w^*(y,s) - v^*(y)) \, ds \, dy. \tag{76}$$

The Hilbert space H is defined as the completion of the set of all divergence-free vector fields $v \in C_0^\infty(\Omega)$, $w \in C_0^\infty([0,\infty) \times \Omega)$. In a generalized sense elements of H satisfy the boundary conditions $\int_0^\infty G(s)w(y,s)\, ds = 0$ and $v \cdot n = 0$, $w \cdot n = 0$ on $\partial\Omega$. Let P denote the orthogonal projection in $L^2(\Omega)$ onto the space of divergence-free vector fields with vanishing normal component on the boundary (see e.g. [L1]). Then we can write (72)-(75) as an abstract evolution problem in H as follows:

$$\frac{d}{dt}(v,w) = A(v,w), \quad (v(0), w(0)) = (v^0, w^0), \tag{77}$$

where the operator A is defined by

$$A(v,w) := \left(P\Delta \int_0^\infty G(s)w(y,s)\, ds, \; -\frac{dw}{ds}(y,s) \right) \tag{78}$$

with domain

$$D(A) := \{(v,w) \in H; \; A(v,w) \in H \text{ and } w(y,0) = v(y) \text{ for } y \in \Omega\}. \tag{79}$$

We note that $D(A)$ is dense in H. It is not obvious how the boundary condition (74) has been incorporated in this formulation. Recall that the definition of H implies that, in a generalized sense, $\int_0^\infty G(s)w(y,s)\, ds = 0$ on $\partial\Omega$, i.e. for a solution of the equation we will have $\int_0^\infty G(s)v(y, t-s)\, ds = 0$ on the boundary for every t, and from this it follows that $v = 0$ provided that the initial history satisfies this boundary condition.

Slemrod [S5] proves the following result.

Theorem V.20:

The operator A generates a C_0-semigroup of contractions on the space H.

Proof:

The proof is based on the Lumer-Phillips theorem V.3. Let us first verify that A is dissipative. For $(v,w) \in D(A)$, we have

$$\langle A(v,w), (v,w) \rangle_H = \int_\Omega \nabla\left\{ -\int_0^\infty G(s)\frac{dw(y,s)}{ds}\, ds \right\} : \nabla\left\{ \int_0^\infty G(s)w(y,s)\, ds \right\} dy$$

$$- \int_\Omega \int_0^\infty G'(s)\left(-\frac{dw(y,s)}{ds} - P\Delta \int_0^\infty G(\sigma)w(y,\sigma)\, d\sigma \right) \cdot (w(y,s) - v(y))\, ds\, dy. \tag{80}$$

After an integration by parts in the first term and rearrangement of terms this leads to

$$\langle A(v,w), (v,w) \rangle_H$$

$$= \int_\Omega \Delta\left(\int_0^\infty G(\sigma)w(y,\sigma)\, d\sigma \right) \cdot \left(\int_0^\infty G(s)\frac{dw(y,s)}{ds} + G'(s)(w(y,s) - v(y))\, ds \right) dy$$

$$+ \int_\Omega \int_0^\infty G'(s)(w(y,s) - v(y))\frac{dw(y,s)}{ds}\, ds\, dy. \tag{81}$$

213

The first term on the right hand side is zero, after integration by parts with respect to s in the second term, we obtain

$$\langle A(v,w),(v,w)\rangle_H$$

$$= -\frac{1}{2}\int_\Omega \int_0^\infty G''(s)(w(y,s)-v(y))\cdot(w(y,s)-v(y))\ ds\ dy \leq 0. \qquad (82)$$

It remains to show that $I-A$ is surjective. Since the operator A is closed it is sufficient to show that the range of $I-A$ is dense in H. Hence, let us look at the problem $(I-A)(v,w) = (f,g)$, i.e.

$$v(y) - \mathcal{P}\Delta \int_0^\infty G(s)w(y,s)\ ds = f(y),$$

$$w(y,s) + \frac{dw(y,s)}{ds} = g(y,s). \qquad (83)$$

We assume that f and g are smooth functions of compact support such that $(f,g) \in H$, and we seek a solution $(v,w) \in D(A)$. The solution of the second equation is easily written down as

$$w(y,s) = e^{-s}v(y) + e^{-s}\int_0^s e^\tau g(y,\tau)\ d\tau. \qquad (84)$$

By inserting this into the first equation, we obtain

$$\mathcal{P}\Delta\big[Mv(y) + h(y)\big] - v(y) = -f(y), \qquad (85)$$

where we have used the abbreviations

$$M := \int_0^\infty G(s)e^{-s}\ ds, \qquad (86)$$

and

$$h(y) := \int_0^\infty G(s)e^{-s}\int_0^s e^\tau g(y,\tau)\ d\tau\ ds. \qquad (87)$$

The boundary condition

$$\int_0^\infty G(s)w(y,s)\ ds = 0 \text{ on } \partial\Omega \qquad (88)$$

assumes the form

$$Mv(y) + h(y) = 0 \text{ on } \partial\Omega. \qquad (89)$$

We can rewrite (85) as

$$M\mathcal{P}\Delta\big[Mv(y) + h(y)\big] - \big[Mv(y) + h(y)\big] = -Mf(y) - h(y). \qquad (90)$$

214

Thus we obtain a Stokes problem for $Mv + h$ with Dirichlet boundary conditions. The solvability of this problem is well known [L1].

■

It can be shown that a solution of the abstract evolution problem with initial data in $D(A)$ satisfies (72) in a strong enough sense such that all the terms appearing in (72) take values in $L^2(\Omega)$. This follows from the next lemma.

Lemma V.21:

a) Let $M := \int_0^\infty G(s)e^{-s}\, ds$. Then, for $(f,g) \in H$, we have the estimate

$$\left\| \int_0^\infty G(s)e^{-s} \int_0^s e^\tau g(\cdot,\tau)\, d\tau\, ds + Mf \right\|_{L^2} \le \text{const.}\ \|(f,g)\|_H. \tag{91}$$

b) For $(f,g) \in D(A)$ we have

$$\left\| P\Delta \int_0^\infty G(s)g(\cdot,s)\, ds \right\|_{L^2} \le \text{const.}\ (\|(f,g)\|_H + \|A(f,g)\|_H). \tag{92}$$

Proof:
a) An elementary calculation shows that

$$\int_0^\infty G(s)e^{-s} \int_0^s e^\tau g(y,\tau)\, d\tau\, ds + Mf(y)$$

$$= \int_0^\infty G(\tau)g(y,\tau)\, d\tau + \int_0^\infty e^\tau \left[\int_\tau^\infty e^{-s} G'(s)\, ds \right] (g(y,\tau) - f(y))\, d\tau. \tag{93}$$

The L^2-norm of the first term on the right hand side of (93) is clearly bounded by a constant times $\|(f,g)\|_H$. Using the fact that $-G'$ is positive and monotone decreasing, and hence $|G'(s)| \le |G'(\tau)|$, we can estimate the second term on the right of (93) by

$$- \int_0^\infty G'(\tau) |g(y,\tau) - f(y)|\, d\tau. \tag{94}$$

Using Schwarz's inequality, we can bound this latter quantity by

$$\sqrt{G(0)} \sqrt{- \int_0^\infty G'(\tau) |g(y,\tau) - f(y)|^2\, d\tau}. \tag{95}$$

In view of the definition H, this is square integrable in Ω, and the proof is complete.

b) For $(f,g) \in D(A)$, we have

$$\left(P\Delta \int_0^\infty G(s)g(\cdot,s)\, ds, -\frac{dg}{ds} \right) \in H, \tag{96}$$

215

which, by part a, implies

$$\left\| -\int_0^\infty G(s)e^{-s}\left[\int_0^s e^\tau \frac{dg}{d\tau}(\cdot,\tau)\,d\tau\right]\,ds + MP\Delta\int_0^\infty G(s)g(\cdot,s)\,ds\right\|_{L^2}$$

$$\leq \text{const.}\,\|A(f,g)\|_H. \tag{97}$$

We integrate the first term on the left by parts and obtain

$$\left\|\int_0^\infty G(s)e^{-s}\int_0^s e^\tau g(\cdot,\tau)\,d\tau\,ds - \int_0^\infty G(s)g(\cdot,s)\,ds + \int_0^\infty G(s)e^{-s}\,ds\,f\right.$$

$$\left. +MP\Delta\int_0^\infty G(s)g(\cdot,s)\,ds\right\|_{L^2} \leq \text{const.}\,\|A(f,g)\|_H. \tag{98}$$

Applying the triangle equality, one finds

$$\left\|MP\Delta\int_0^\infty G(s)g(\cdot,s)\,ds\right\|_{L^2} \leq \left\|\int_0^\infty G(s)e^{-s}\int_0^s e^\tau g(\cdot,\tau)\,d\tau\,ds + Mf\right\|_{L^2}$$

$$+\left\|\int_0^\infty G(s)g(\cdot,s)\,ds\right\|_{L^2} + \text{const.}\,\|A(f,g)\|_H. \tag{99}$$

The first term on the right can be bounded by a constant times $\|(f,g)\|_H$ according to part a, and the second term can be bounded by a constant time $\|(f,g)\|_H$ by using the definition of the norm in H and the Poincaré inequalit. This completes the proof.

The following result on asymptotic stability of the rest state is more general than the one proved by Slemrod [S5] and we give a simpler proof.

Theorem V.22:

In addition to the previous assumptions, assume that $G \in L^1(0,\infty)$. Then for any initial data $(\tilde{v},\tilde{w}) \in H$, we have

$$\lim_{t\to\infty}\|(v(t),w(t))\|_H = 0, \tag{100}$$

where $(v(t),w(t)) = e^{At}(\tilde{v},\tilde{w})$.

Proof:
Since e^{At} is a contraction, it is sufficient to show that (100) holds for a dense set of initial conditions $(\tilde{v},\tilde{w})$. Such a dense set is obtained if we take $\tilde{v}$ an $\tilde{w}$ such that their spatial dependence is given by a finite linear combination ϵ eigenfunctions of the Stokes operator. In addition, we may take $\tilde{w}$ such that it support with respect to time is bounded. If we are dealing with an eigenfunctio of the Stokes operator, associated with an eigenvalue $-\Lambda$, the differential equatio reduces to

$$\dot{v} = -\Lambda\int_0^\infty G(s)v(t-s)\,ds. \tag{101}$$

216

n order to show that the solution of this equation tends to zero and is integrable it infinity, we take the Laplace transform and use Proposition 2.3 of [J3] (see Section II.3). It is an easy consequence that the norm in H tends to zero.

If a certain decay rate for G at infinity is assumed, one can obtain a corresponding decay rate for the solution. For details we refer to Slemrod [S6].

5. A parabolic problem

As an example of a "parabolic" equation with delay, we consider

$$u_{tt}(x,t) = \int_{-\infty}^{t} h(t - \tau, u_{xt}(x,t), u_x(x,t), u_x(x,\tau))_x \, d\tau. \tag{102}$$

Such an equation arises, for example, in describing one-dimensional motions of Curtiss-Bird fluids [C13] or of K-BKZ fluids with a Newtonian viscosity (i.e. with a contribution proportional to $-\delta'$ in the kernel in equation (65) or (70) of Chapter). We denote by $h_{,i}$ the derivative of h with respect to the ith argument. We assume that h is a smooth function, and to guarantee the parabolic character of the equation we assume that, at least on the range of data being considered, $h_{,2}$ is positive. We shall consider (102) on the interval $[0,1]$ with Dirichlet boundary conditions

$$u(0,t) = u(1,t) = 0. \tag{103}$$

We seek a solution u to (102), (103) for $t > 0$ subject to a given initial history and initial conditions for $u(x,0+)$ and $u_t(x,0+)$ which are assumed compatible with the initial history.

We split the integral in (102) into a part from $-\infty$ to 0 and a part from 0 to . Since we always consider the history of u up to time 0 as given, the first part of the integral can be written as

$$\alpha(x,t,u_{xt}(x,t),u_x(x,t))_x, \tag{104}$$

where α is a known function (which depends on the given initial history), and $\alpha_{,3}$ the derivative of α with respect to the third argument) is positive. Moreover, we introduce the variables $v = u_t$, $p = u_x$, $q = u_{xx}$, and rewrite (102) in the form of a system

$$p_t(x,t) = v_x(x,t),$$

$$q_t(x,t) = v_{xx}(x,t),$$

217

$$v_t(x,t) = \alpha_{,1}\left(x,t,v_x(x,t),p(x,t)\right) + \alpha_{,3}\left(x,t,v_x(x,t),p(x,t)\right)v_{xx}(x,t) \quad (105)$$

$$+\alpha_{,4}\left(x,t,v_x(x,t),p(x,t)\right)q(x,t) + \int_0^t h_{,2}\left(t-\tau,v_x(x,t),p(x,t),p(x,\tau)\right)\,d\tau\,v_{xx}(x,t)$$

$$+\int_0^t h_{,3}\left(t-\tau,v_x(x,t),p(x,t),p(x,\tau)\right)\,d\tau\,q(x,t)$$

$$+\int_0^t h_{,4}\left(t-\tau,v_x(x,t),p(x,t),p(x,\tau)\right)q(x,\tau)\,d\tau.$$

The boundary conditions (103) are now replaced by

$$v(0,t) = v(1,t) = 0. \quad (106)$$

We want to find solutions of (105), (106) on some time interval $[0,t_0]$ with given initial conditions at $t=0$

$$p(x,0) = p_0(x),\ \ q(x,0) = q_0(x),\ \ v(x,0) = v_0(x). \quad (107)$$

In order to get back from (105), (106) to (102), (103), it is of course necessary that $(p_0)_x = q_0$, and $\int_0^1 p_0(x)\,dx = 0$, but these constraints are not needed to solve (105), (106).

We shall make the following hypotheses:

(S1) $p_0 \in W^{1,p}(0,1)$, $q_0 \in L^p(0,1)$, $v_0 \in W^{2,p}(0,1)$. Here p is any given number in $(1,\infty)$.

(S2) On an open subset of $\mathbb{R}^4$ which contains the range of the initial data (i.e. the set $\{(x,0,(v_0)_x(x),p_0(x)) \mid x \in [0,1]\}$), $\alpha_{,1}$, $\alpha_{,3}$ and $\alpha_{,4}$ are continuous; moreover, they are uniformly Hölder continuous with respect to the second argument and uniformly Lipschitz continuous with respect to the last two arguments.

(S3) On an open subset of $\mathbb{R}^4$ which contains the range of the initial data (i.e. the set $\{(0,(v_0)_x(x),p_0(x),p_0(x)) \mid x \in [0,1]\}$, $h_{,2}$, $h_{,3}$ and $h_{,4}$ are continuous; moreover, they are uniformly Hölder continuous with respect to the first argument and uniformly Lipschitz continuous with respect to the last three arguments.

(C) v_0 satisfies the boundary condition (106).

(E) $\alpha_{,3} > 0$.

Theorem V.23:

Let assumptions (S1)-(S3), (C) and (E) hold. Then there exists a $T' > 0$ such that (105)-(107) has a unique solution on the time interval $[0,T']$ with the regularity $(p,q,v) \in C^1([0,T'];W^{1,p}(0,1) \times L^p(0,1) \times L^p(0,1))$, $v \in C([0,T'];W^{2,p}(0,1))$.

For the proof, we regard (105) as an evolution problem on the space $X := W^{1,p}(0,1) \times L^p(0,1) \times L^p(0,1)$, to which we shall apply the abstract methods of Section 2. In the space X, we define the operator $\tilde{A}_0$ as follows. Let

$$D(\tilde{A}_0) := \{y = (p,q,v) \in W^{1,p} \times L^p \times W^{2,p} \mid v(0) = v(1) = 0\}, \quad (108)$$

nd for $y \in D(\tilde{A}_0)$ let

$$\tilde{A}_0 y(x) := (v_x(x), v_{xx}(x), \alpha_{,3}(x, 0, (v_0)_x(x), p_0(x))v_{xx}(x)). \tag{109}$$

ence assumptions (S1),(C) above say that the initial datum is in the domain
$\tilde{A}_0$. Assumption (S2) implies that $\alpha_{,3}(x, 0, (v_0)_x(x), p_0(x))$ is a continuous
nction of x (recall that $W^{1,p}(0,1) \subset C[0,1]$), and since we have also assumed
at $\alpha_{,3}$ is positive, it is well known [P1],[S8] that the operator $v \to \alpha_{,3} v_{xx}$
enerates an analytic semigroup on L^p. Let us now consider the problem

$$(\tilde{A}_0 - \lambda)y = f = (f_1, f_2, f_3), \tag{110}$$

e.

$$v_x - \lambda p = f_1,$$
$$v_{xx} - \lambda q = f_2, \tag{111}$$
$$\alpha_{,3} v_{xx} - \lambda v = f_3.$$

ince the operator $v \to \alpha_{,3} v_{xx}$ generates an analytic semigroup, and all its eigen-
alues are real and negative, there is some constant C such that, for $\operatorname{Re} \lambda \geq 0$,
e have

$$|\lambda|\|v\|_{L^p} + \|v\|_{W^{2,p}} \leq C\|f_3\|_{L^p}. \tag{112}$$

follows immediately from the first two equations of (111) that

$$|\lambda|\|p\|_{W^{1,p}} \leq \|f_1\|_{W^{1,p}} + \|v\|_{W^{2,p}} \leq \|f_1\|_{W^{1,p}} + C\|f_3\|_{L^p}, \tag{113}$$

$$|\lambda|\|q\|_{L^p} \leq \|f_2\|_{L^p} + \|v\|_{W^{2,p}} \leq \|f_2\|_{L^p} + C\|f_3\|_{L^p}.$$

equalities (112) and (113) imply that

$$|\lambda|\|y\|_X \leq (3C+1)\|f\|_X, \tag{114}$$

nd hence $\tilde{A}_0$ generates an analytic semigroup on X. The operator $A_0 := \tilde{A}_0 - I$
lso generates an analytic semigroup. Its spectrum lies in the left half plane, so we
an define fractional powers of $-A_0$ as discussed in Section 1. As in the example
ollowing Theorem V.6, if $(p, q, v) \in D(-A_0)^\alpha$ and $\alpha > 1/2 + 1/(2p)$, then v_x is
Hölder continuous.

With $y := (p, q, v)$ and $\hat{y} := (\hat{p}, \hat{q}, \hat{v})$, we now introduce the following nota-
ons:

$$A_y(t)\hat{y} := \left(\hat{v}_x - \hat{p}, \hat{v}_{xx} - \hat{q}, \right.$$

$$\left. \{\alpha_{,3}(x, t, v_x, p) + \int_0^t h_{,2}(t - \tau, v_x(t), p(t), p(\tau)) \, d\tau\}\hat{v}_{xx} - \hat{v} \right), \tag{115}$$

here $D(A_y(t)) := D(A_0)$, and

$$_y(t) := \left(p, q, \alpha_{,1}(x, t, v_x, p) + \alpha_{,4}(x, t, v_x, p)q + q \int_0^t h_{,3}(t - \tau, v_x(t), p(t), p(\tau)) \, d\tau \right.$$

219

$$+ \int_0^t h_{,4}\left(t - \tau, v_x(t), p(t), p(\tau)\right)q(\tau)\,d\tau + v\right). \tag{116}$$

The system (105) now has the form

$$\dot{y}(t) = A_y(t)y(t) + f_y(t), \tag{117}$$

and we solve (117) iteratively,

$$\dot{y}^{n+1}(t) = A_{y^n}(t)y^{n+1}(t) + f_{y^n}(t). \tag{118}$$

We shall show that the mapping $y^n \to y^{n+1}$ defined by (118) with appropriat initial conditions is a contraction in an appropriate space of functions. The proc follows closely that given on pp. 48-52 of [S8], the only necessary modificatio arises from having to allow integral terms in the equation.

We choose a fixed α such that $1/2 + 1/(2p) < \alpha < 1$. We denote by $Q(t_0, K, \eta$ the set of all functions $z(t)$, continuous from $[0, t_0] \to X$, such that

$$z(0) = (-A_0)^\alpha y_0, \tag{119}$$

where $y_0 = (p_0, q_0, v_0)$, and for any $t, \tau \in [0, t_0]$,

$$\|z(t) - z(\tau)\| \le K|t - \tau|^\eta. \tag{120}$$

Here η is some number in $(0, 1)$. If we choose t_0 small enough, then the value c $z(t)$ lies in a small neighborhood of $(-A_0)^\alpha y_0$ for all $t \in [0, t_0]$. It is easy to se that $Q(t_0, K, \eta)$ is a closed, bounded and convex subset of $C([0, t_0]; X)$.

Let $z \in Q(t_0, K, \eta)$ be given. We set $y = (-A_0)^{-\alpha} z$. It is easy to verify tha $A_y(t)$ and $f_y(t)$ are well-defined and that they satisfy Hölder conditions of th form

$$\|(A_y(t) - A_y(\tau))A_0^{-1}\| \le C|t - \tau|^\mu,$$
$$\|f_y(t) - f_y(\tau)\| \le C|t - \tau|^\mu, \tag{121}$$

where μ is the minimum of η and the Hölder constants for $\alpha_{,1}$, $\alpha_{,3}$, $\alpha_{,4}$, $h_{,2}$, $h_{,3}$ and $h_{,4}$. Theorem V.10 guarantees that for $z \in Q(t_0, K, \eta)$ and $y = (-A_0)^{-\alpha} z$ the equation

$$\dot{\hat{y}}(t) = A_y(t)\hat{y}(t) + f_y(t), \quad \hat{y}(0) = y_0 \tag{122}$$

has a solution. We now want to show that, for appropriate choice of t_0, K and η $\hat{z} = (-A_0)^\alpha \hat{y}$ is also in $Q(t_0, K, \eta)$ and that the mapping $z \to \hat{z}$ is a contraction

We denote by U_y the evolution operator associated with the family $\{A_y(t)$ so that

$$\hat{y}(t) = U_y(t, 0)y_0 + \int_0^t U(t, s)f_y(s)\,ds. \tag{123}$$

The operator function U_y satisfies the inequalities of Lemma V.11 with constant independent of $z = (-A_0)^\alpha y \in Q(t_0, K, \eta)$. In particular, by setting $\gamma = \beta = 1$ we obtain from (26) that, for $0 \le \alpha < 1$ and $0 \le t \le t + \Delta t \le t_0$,

$$\|(-A_0)^\alpha (U_y(t + \Delta t, 0) - U_y(t, 0))A_0^{-1}\| \le C(\Delta t)^{1-\alpha}. \tag{124}$$

220

From (28) we obtain

$$\left\| (-A_0)^\alpha \left[\int_0^{t+\Delta t} U(t+\Delta t,s) f_y(s)\ ds - \int_0^t U(t,s) f_y(s)\ ds \right] \right\|$$

$$\leq C(\Delta t)^{1-\alpha} [|\ln \Delta t| + 1]. \tag{125}$$

By combining (124), (125) with (123) and the equation $\hat{z} = (-A_0)^\alpha \hat{y}$, we find

$$\|\hat{z}(t+\Delta t) - \hat{z}(t)\| \leq C(\Delta t)^{1-\alpha} \|A_0 y_0\| + C(\Delta t)^{1-\alpha} (|\ln \Delta t| + 1). \tag{126}$$

we choose $\eta < 1-\alpha$ and t_0 sufficiently small, then the right hand side becomes less than $K(\Delta t)^\eta$. Since $\hat{z}(0) = (-A_0)^\alpha y_0$, it follows that $\hat{z} \in Q(t_0, K, \eta)$.

Consider now two elements z_1 and z_2 in $Q(t_0, K, \eta)$, and let $y_i = (-A_0)^{-\alpha} z_i$. et $\hat{y}_1$ and $\hat{y}_2$ be the solutions of

$$\hat{y}_1(0) = \hat{y}_2(0) = y_0, \tag{127}$$

nd

$$\dot{\hat{y}}_1(t) = A_{y_1}(t)\hat{y}_1(t) + f_{y_1}(t),$$

$$\dot{\hat{y}}_2(t) = A_{y_2}(t)\hat{y}_2(t) + f_{y_2}(t). \tag{128}$$

By taking the difference of the two equations, we obtain

$$\frac{d}{dt}(\hat{y}_1 - \hat{y}_2) = A_{y_1}(t)(\hat{y}_1 - \hat{y}_2) + (A_{y_1}(t) - A_{y_2}(t))\hat{y}_2 + f_{y_1}(t) - f_{y_2}(t). \tag{129}$$

Let us now recall that

$$\hat{y}_2(t) = U_{y_2}(t,0)y_0 + \int_0^t U_{y_2}(t,s) f_{y_2}(s)\ ds. \tag{130}$$

In (27), we set $\alpha = 0$ and choose $\beta = 1 - \epsilon$, where ϵ is small. Moreover, in (29), we set $\beta = 0$ and choose $\alpha = \mu$ small. Applying these to (130), we find that

$$\|\dot{\hat{y}}_2(t+\Delta t) - \dot{\hat{y}}_2(t)\| \leq C(\Delta t)^\gamma t^{-\epsilon-\gamma-\delta} + C(\Delta t)^\mu t^{-\mu-\delta}, \tag{131}$$

i.e., $\dot{\hat{y}}_2$ satisfies a Hölder condition for positive t, and it can be bounded in norm by an inverse power of t. Because of the differential equation and (121), a similar bound holds for $A_{y_2}(t)\hat{y}_2(t)$. Moreover, the first part of (121) says that a Hölder condition holds for $A_{y_2}(t)A_0^{-1}$ and hence also for $A_0(A_{y_2}(t))^{-1}$. Hence $A_0\hat{y}_2(t)$ satisfies a Hölder condition for positive t and can be bounded by a small inverse power of t. Applying the first part of (121) again, we find that the same is true for $(A_{y_1}(t) - A_{y_2}(t))\hat{y}_2(t)$.

We can therefore conclude from (129) and Theorem V.10 that, for $0 < \tau \leq t \leq t_0$, we have

$$\hat{y}_1(t) - \hat{y}_2(t) = U_{y_1}(t,\tau)(\hat{y}_1(\tau) - \hat{y}_2(\tau))$$

221

$$+ \int_\tau^t U_{y_1}(t,s)[(A_{y_1}(s) - A_{y_2}(s))\hat{y}_2(s) + f_{y_1}(s) - f_{y_2}(s)] \, ds. \tag{132}$$

Because of the bounds we have on the integrand in (132), we can pass to the limit $\tau \to 0$ to obtain

$$\hat{y}_1(t) - \hat{y}_2(t) = \int_0^t U_{y_1}(t,s)[(A_{y_1}(s) - A_{y_2}(s))\hat{y}_2(s) + f_{y_1}(s) - f_{y_2}(s)] \, ds. \tag{133}$$

Because of the Lipschitz conditions in the assumptions (S2) and (S3) above, we have a constant C such that

$$\|(A_{y_1}(s) - A_{y_2}(s))\hat{y}_2(s)\| \leq C \max_{t\in[0,t_0]} \|(-A_0)^\alpha (y_1(t) - y_2(t))\| \, \|A_0 y_2(s)\|$$

$$\leq C s^{-\rho} \max_{t\in[0,t_0]} \|(-A_0)^\alpha (y_1(t) - y_2(t))\|, \tag{134}$$

and

$$\|f_{y_1}(s) - f_{y_2}(s)\| \leq C \max_{t\in[0,t_0]} \|(-A_0)^\alpha (y_1(t) - y_2(t))\|. \tag{134}$$

Here ρ can be chosen arbitrarily small. By combining (134) with (25) (setting $\beta = 0$ in that inequality), we find

$$\|\hat{z}_1(t) - \hat{z}_2(t)\| = \|(-A_0)^\alpha (\hat{y}_1(t) - \hat{y}_2(t))\|$$

$$\leq C\|(-A_0)^\alpha (-A_{y_1}(t))^{-\alpha-\epsilon}\| \, \|(-A_{y_1}(t))^{\alpha+\epsilon}(\hat{y}_1(t) - \hat{y}_2(t))\|$$

$$\leq C \max_{t\in[0,t_0]} \|z_1(t) - z_2(t)\| \int_0^t |t-s|^{-\alpha-\epsilon}(s^{-\rho} + 1) \, ds. \tag{135}$$

This easily yields the desired contraction estimate if we choose t_0 small enough. For this purpose, we use the fact that, for any $\epsilon > 0$, the operator $(-A_0)^\alpha (-A_{y_1}(t))^{-\alpha-\epsilon}$ is bounded. For a proof of this last assertion, we refer to [S8].

6. A hyperbolic problem

As an example of a hyperbolic equation, we consider

$$u_{tt}(x,t) = \int_{-\infty}^t h(t-\tau, u_x(x,t), u_x(x,\tau))_x \, d\tau, \tag{136}$$

222

which arises, for example, in one-dimensional motions of K-BKZ fluids (without Newtonian viscosity). This equation differs from (102) in the previous section by the absence of the argument u_{xt} in h. It is assumed that the derivative of h with respect to the second argument, again denoted by $h_{,2}$, is positive. We want to show that the Cauchy problem consisting of (136), together with a given initial history and initial conditions for $u(x,0+)$ and $u_t(x,0+)$, can be treated by the methods[1] of Section 2 (we do not require that the initial conditions for $t \to 0+$ are compatible with the given history). In order to present the equation in a suitable form, we must first go through some transformations. First we differentiate the equation with respect to time to obtain

$$u_{ttt} = \int_{-\infty}^t h_{,2}\left(t - \tau, u_x(t), u_x(\tau)\right) d\tau \, u_{xxt} + h_{,2}\left(0, u_x, u_x\right)u_{xx}$$

$$+ \int_{-\infty}^t h_{,21}\left(t - \tau, u_x(t), u_x(\tau)\right) d\tau \, u_{xx} + \int_{-\infty}^t h_{,22}\left(t - \tau, u_x(t), u_x(\tau)\right) d\tau \, u_{xx}u_{xt}$$

$$+ h_{,3}\left(0, u_x, u_x\right)u_{xx} + \int_{-\infty}^t h_{,31}\left(t - \tau, u_x(t), u_x(\tau)\right)u_{xx}(\tau) \, d\tau$$

$$+ \int_{-\infty}^t h_{,32}\left(t - \tau, u_x(t), u_x(\tau)\right)u_{xx}(\tau) \, d\tau \, u_{xt}. \tag{137}$$

Here we have suppressed the arguments x and t except where they are needed to avoid ambiguity. We transform (137) to a system by setting $a = u_{tt}$, $b = u_{xt}$, $c = u_{xx}$, and $d = u_x$. We then obtain

$$a_t = \int_{-\infty}^t h_{,2}\left(t - \tau, d(t), d(\tau)\right) d\tau \, b_x + \left(h_{,2}\left(0, d, d\right) + h_{,3}\left(0, d, d\right)\right)c$$

$$+ \int_{-\infty}^t h_{,21}\left(t - \tau, d(t), d(\tau)\right) d\tau \, c + \int_{-\infty}^t h_{,22}\left(t - \tau, d(t), d(\tau)\right) d\tau \, bc$$

$$+ \int_{-\infty}^t h_{,31}\left(t - \tau, d(t), d(\tau)\right)c(\tau) \, d\tau + \int_{-\infty}^t h_{,32}\left(t - \tau, d(t), d(\tau)\right)c(\tau) \, d\tau \, b,$$

$$b_t = a_x, \tag{138}$$

$$c_t = b_x$$

$$d_t = b.$$

---

[1] As presented here, these methods are not suited for the treatment of hyperbolic initial-boundary value problems; for a method to treat such problems, which is in many respects similar to the approach used in Chapter III, see [K8],[K9].

Next we introduce a further transformation which diagonalizes the matrix multiplying the x-derivatives on the right hand side of[2] (138). Let us set

$$\alpha^2 := \int_{-\infty}^{t} h_{,2}\left(t - \tau, d(t), d(\tau)\right) d\tau. \tag{139}$$

We abbreviate the right hand side in the first equation of (138) as $\alpha^2 b_x + \phi$. We now set $e := a + \alpha b$, $f := a - \alpha b$, $g := a - \alpha^2 c$. Since $h_{,2}$ is positive, the transformation $(a, b, c) \rightarrow (e, f, g)$ is invertible. The system (138) now assumes the form

$$e_t = \alpha e_x + (\alpha_t - \alpha\alpha_x)b + \phi,$$
$$f_t = -\alpha f_x - (\alpha\alpha_x + \alpha_t)b + \phi, \tag{140}$$
$$g_t = -2\alpha\alpha_t c + \phi,$$
$$d_t = \frac{e - f}{2\alpha}.$$

We note that

$$\alpha_t = \frac{1}{2\alpha}\Big[h_{,2}\left(0, d(t), d(t)\right) + \int_{-\infty}^{t} h_{,21}\left(t - \tau, d(t), d(\tau)\right) d\tau$$

$$+ \int_{-\infty}^{t} h_{,22}\left(t - \tau, d(t), d(\tau)\right) d\tau \, \frac{e - f}{2\alpha}\Big], \tag{141}$$

and

$$\alpha_x = \frac{1}{2\alpha}\Big[\int_{-\infty}^{t} h_{,22}\left(t - \tau, d(t), d(\tau)\right) d\tau \, \frac{e + f - 2g}{2\alpha^2}$$

$$+ \int_{-\infty}^{t} h_{,23}\left(t - \tau, d(t), d(\tau)\right)\frac{e(\tau) + f(\tau) - 2g(\tau)}{2\alpha^2(\tau)} \, d\tau\Big]. \tag{142}$$

We now make the following smoothness hypotheses:

(S1) h is of class C^4 on $[0, \infty) \times \mathbb{R}^2$. Moreover, for any compact subset K of $\mathbb{R}^2$, there is a function $f_K \in L^1(0, \infty)$ such that $|h(t, d_1, d_2)| \le f_K(t)$ for $t \in [0, \infty)$, $(d_1, d_2) \in K$. An analogous condition holds for the derivatives of h up to the fourth order.

(S2) The prescribed history for u on $(-\infty, 0]$ is such that the corresponding histories of c and d are continuous and bounded functions of time taking values in $H^2(\mathbb{R})$.

(S3) At time $t = 0+$, the initial values for a, b, c and d (or equivalently, for e, f, g and d) lie in $H^2(\mathbb{R})$.

(E) $h_{,2} > 0$.

We shall prove the following result.

---

[2] It is only necessary to make this matrix symmetric, but in this example we can actually diagonalize it.

Theorem V.24:

Assume that (E) and (S1)-(S3) hold. Then for sufficiently small $t_0 > 0$, (140) has a unique solution on $[0, t_0]$ which assumes the given initial data and has the regularity $e, f \in C([0, t_0]; H^2(\mathbb{R})) \cap C^1([0, t_0]; H^1(\mathbb{R}))$, $(g, d) \in C^1([0, t_0]; H^2(\mathbb{R}))$.

It is easily checked that in terms of the original problem, this regularity statement means that $u \in C([0, t_0]; H^4(\mathbb{R})) \cap C^1([0, t_0]; H^3(\mathbb{R})) \cap C^2([0, t_0]; H^2(\mathbb{R})) \cap C^3([0, t_0]; H^1(\mathbb{R}))$.

To prove the theorem, we set $y = (e, f, g, d)$ and denote by $y_0 = (e_0, f_0, g_0, d_0)$ the initial values at time $t = 0$. We define spaces $X := (L^2(\mathbb{R}))^4$ and $Y := (H^2(\mathbb{R}))^4$ and we denote by $Q(t_0, R)$ the set of all functions $y : [0, t_0] \to Y$ such that $\|y(t) - y_0\|_Y \leq R$ for $t \in [0, t_0]$ and y is continuous from $[0, t_0]$ to X. Since X and Y are reflexive, a closed ball in Y is also closed in X, and it follows that the set $Q(t_0, R)$ is complete in the norm of $C([0, t_0]; X)$. For $y \in Q(t_0, R)$, we define

$$A_y \hat{y} := (\alpha(y)\hat{e}_x, -\alpha(y)\hat{f}_x, 0, 0), \tag{143}$$

and we let f_y denote the remaining terms on the right hand side of (140).

The goal of the following is to show that for appropriate choices of t_0 and R the mapping $T : y \to \hat{y}$ defined by the equation

$$\dot{\hat{y}} = A_y \hat{y} + f_y, \tag{144}$$

together with the appropriate initial conditions, is a contraction on $Q(t_0, R)$. The proof follows Section 9 in [K7].

We first show that for every $y \in Q(t_0, R)$ the family of operators $A_y(t)$ satisfies the conditions of Theorem V.15. The stability condition (H1) follows from the integration by parts

$$(A_y(t)\hat{y}(t), \hat{y}(t))_X = \int_{\mathbb{R}} \alpha(y; t)\hat{e}_x \hat{e} - \alpha(y; t)\hat{f}_x \hat{f} \, dx$$

$$= -(A_y(t)\hat{y}(t), \hat{y}(t)) + \int_{\mathbb{R}} \alpha_x(y; t)(\hat{f}^2 - \hat{e}^2) \, dx, \tag{145}$$

and the Lumer-Phillips theorem V.3. Condition (H3) is easy to verify. For (H2), we set $S := \frac{d^2}{dx^2} - 1$. We find that

$$(SA_y(t) - A_y(t)S)\hat{y} = (\alpha_{xx}(y; t)\hat{e}_x + 2\alpha_x(y; t)\hat{e}_{xx}, -\alpha_{xx}(y; t)\hat{f}_x - 2\alpha_x(y; t)\hat{f}_{xx}, 0, 0), \tag{146}$$

and this operator is clearly bounded from Y to X. Hence $B_y(t) = SA_y(t)S^{-1} - A_y(t)$ is bounded from X to X. It follows from (H3) that, for $\hat{y} \in Y$, $S^{-1}B_y(t)\hat{y} = A_y(t)S^{-1}\hat{y} - S^{-1}A_y(t)\hat{y}$ depends continuously on t in the X-norm. Since $S^{-1}B_y(t)$ is bounded from X to X, and Y is dense in X, it follows that $S^{-1}B_y(t)\hat{y}$ in the norm of X depends continuously on t for every $\hat{y} \in X$. Moreover, $S^{-1}B_y(t)$ is also bounded from X to Y. It now follows from Lemma 7.4 of [K7] that

225

$S^{-1}B_y(t) : X \to Y$ is strongly measurable, and hence $B_y(t) : X \to X$ is strongly measurable.

We note that the following bounds are uniform for $y \in Q(t_0, R)$: the exponent β in the stability estimate (35) (moreover, $M = 1$), the bound on $\|B_y(t)\|_X$ and the bound on $\|A_y(t)\|_{Y,X}$. Moreover, it is easy to show that $f_y(t)$ is bounded in Y and depends continuously on t is the X-norm. Again this implies that $f_y(t) \in Y$ is strongly measurable as a function of t. The bound on the Y-norm is uniform for $y \in Q(t_0, R)$.

It now follows from Lemma V.17, part (c) that the solution $\hat{y}$ of (144) exists. To show that it lies in $Q(t_0, R)$, we need an estimate for $\hat{y}(t) - y_0$. For this, we note that

$$\hat{y}(t) = U_y(t,0)y_0 + \int_0^t U_y(t,s)f_y(s) \ ds. \tag{147}$$

The Y-norm of the integral can be estimated by

$$t_0 \max_{0 \leq s \leq t \leq t_0} \|U_y(t,s)\|_Y \max_{0 \leq s \leq t_0} \|f_y(s)\|_Y, \tag{148}$$

i.e. by t_0 times a constant. To estimate $\|U_y(t,0)y_0 - y_0\|_Y$, choose a small $\epsilon > 0$ and $y_1 \in (H^3(\mathbb{R}))^4$ such that $\|y_0 - y_1\| \leq \epsilon$. It follows that

$$\|U_y(t,0)y_0 - y_0\|_Y \leq \|U_y(t,0)(y_1 - y_0)\|_Y + \|y_1 - y_0\|_Y + \|U_y(t,0)y_1 - y_1\|_Y. \tag{149}$$

The first two terms can be estimated by a constant times ϵ. For the last one, we use the identity

$$U_y(t,0)y_1 - y_1 = \int_0^t U_y(t,s)A_y(s)y_1 \ ds, \tag{150}$$

which follows from (iv) in Theorem V.15. The Y-norm of the expression on the right is bounded by

$$t_0 \max_{0 \leq s \leq t \leq t_0} \|U_y(t,s)\|_Y \max_{0 \leq s \leq t_0} \|A_y(s)y_1\|_Y. \tag{151}$$

Since we can bound the last two factors, this is again less than a constant times t_0. It follows that T maps $Q(t_0, R)$ into itself provided that t_0 is chosen small enough.

To show that T is a contraction, we use Lemma V.18. Let y_1 and y_2 be in $Q(t_0, R)$, and let $\hat{y}_1 = Ty_1$, $\hat{y}_2 = Ty_2$. Then

$$\|\hat{y}_1(t) - \hat{y}_2(t)\|_{\infty,X} \leq \|U_{y_1}\|_{\infty,X} [\|f_{y_1} - f_{y_2}\|_{1,X} + \|(A_{y_1} - A_{y_2})y_2\|_{1,X}]. \tag{152}$$

Lemma V.16 provides a bound for $\|U_{y_1}\|_{\infty,X}$. For the other terms we make use of the fact that $\|\cdot\|_{1,X} \leq t_0 \|\cdot\|_{\infty,X}$. Moreover, it is not hard to show that there is a constant C such that

$$\|f_{y_1} - f_{y_2}\|_{\infty,X} \leq C\|y_1 - y_2\|_{\infty,X}, \tag{153}$$

and

$$\|A_{y_1} - A_{y_2}\|_{\infty,Y,X} \le C\|y_1 - y_2\|_{\infty,X}. \tag{154}$$

From this it is evident that T is a contraction if t_0 is small enough. The mapping T has therefore a unique fixed point in $Q(t_0, R)$ and it follows from Lemma V.17, part c, that this fixed point provides a solution to the differential equation.

The approach used here also works in several space dimensions. To illustrate this we treat the initial-value problem for equation $(27)_1$ of Chapter III, which we repeat for the convenience of the reader

$$\ddot{u}^i = A_{ij}^{\alpha\beta}(\nabla u)\frac{\partial^2 u^j}{\partial x^\alpha \partial x^\beta} + \int_0^t M_{ij}^{\alpha\beta}(t-\tau, \nabla u(\cdot,\tau))\frac{\partial^2 u^j}{\partial x^\alpha \partial x^\beta}(\cdot,\tau)\,d\tau + f^i,$$

$$x \in \mathbb{R}^3,\ t \ge 0, \tag{155}_1$$

$$u(x,0) = u_0(x),\ \dot{u}(x,0) = u_1(x). \tag{155}_2$$

As above, we first differentiate the equation with respect to time. Using the notation $v = \dot{u}$, $w = \ddot{u} - \lambda u$ ($\lambda > 0$ is suitably large), we obtain the system

$$\dot{v} = w,$$

$$\dot{w}^i = -\lambda v^i + A_{ij}^{\alpha\beta}(\nabla u)\frac{\partial^2 v^j}{\partial x^\alpha \partial x^\beta} + C_{ijk}^{\alpha\beta\gamma}(\nabla u)\frac{\partial v^k}{\partial x^\gamma}\frac{\partial^2 u^j}{\partial x^\alpha \partial x^\beta}$$

$$+ M_{ij}^{\alpha\beta}(0, \nabla u)\frac{\partial^2 u^j}{\partial x^\alpha \partial x^\beta} + \int_0^t \dot{M}_{ij}^{\alpha\beta}(t-\tau, \nabla u(\cdot,\tau))\frac{\partial^2 u^j}{\partial x^\alpha \partial x^\beta}(\cdot,\tau)\,d\tau + \dot{f}^i, \tag{156}_1$$

with initial conditions

$$v(x,0) = v_0(x) = u_1(x),\ w(x,0) = w_0(x) = u_2(x), \tag{156}_2$$

where u_2 is the initial value of $\ddot{u} - \lambda u$ as computed from (155), and u in $(156)_1$ is computed from the elliptic problem

$$w^i = -\lambda u^i + A_{ij}^{\alpha\beta}(\nabla u)\frac{\partial^2 u^j}{\partial x^\alpha \partial x^\beta} + \int_0^t M_{ij}^{\alpha\beta}(t-\tau, \nabla u(\cdot,\tau))\frac{\partial^2 u^j}{\partial x^\alpha \partial x^\beta}(\cdot,\tau)\,d\tau + f^i. \tag{156}_3$$

Here $\dot{\mathbf{M}}$ is the derivative of $\mathbf{M}$ with respect to its first argument, and

$$C_{ijk}^{\alpha\beta\gamma}(\mathbf{V}) = \frac{\partial A_{ij}^{\alpha\beta}(\mathbf{V})}{\partial V_\gamma^k}. \tag{157}$$

To solve (156), we use the iteration scheme

$$\dot{v}_{(n+1)} = w_{(n+1)}$$

$$\dot{w}^i_{(n+1)} = -\lambda v^i_{(n)} + A^{\alpha\beta}_{ij}(\nabla u_{(n)})\frac{\partial^2 v^j_{(n+1)}}{\partial x^\alpha \partial x^\beta} + C^{\alpha\beta\gamma}_{ijk}(\nabla u_{(n)})\frac{\partial v^k_{(n)}}{\partial x^\gamma}\frac{\partial^2 u^j_{(n)}}{\partial x^\alpha \partial x^\beta}$$

$$+ M^{\alpha\beta}_{ij}(0,\nabla u_{(n)})\frac{\partial^2 u^j_{(n)}}{\partial x^\alpha \partial x^\beta} + \int_0^t \dot{M}^{\alpha\beta}_{ij}(t-\tau, \nabla u_{(n)}(\cdot,\tau))\frac{\partial^2 u^j_{(n)}}{\partial x^\alpha \partial x^\beta}(\cdot,\tau)\,d\tau + \dot{f}^i, \quad (158)$$

where $u_{(n)}$ is determined from $w_{(n)}$ by the elliptic problem $(156)_3$. This nonlinear elliptic problem can be solved via the implicit function theorem as long as $\|w_{(n)} - u_2\|_{H^1}$ remains sufficiently small. Problem (158) is then solved using Theorem V.15.

We make the following assumptions (it is instructive to compare these with the assumptions of Theorem III.5):

(s1) $\mathbf{A}$ is of class C^{m-1} on an open set $\mathcal{O}$ containing $\mathbf{0}$ $(m \geq 4)$.

(s2) The initial data satisfy $u_0 \in H^m(\mathbb{R}^3)$, $u_1 \in H^{m-1}(\mathbb{R}^3)$ and ∇u_0 takes values in $\mathcal{O}$.

(s3) For some $T > 0$, the forcing term f satisfies $f \in C^1([0,T]; H^{m-2}(\mathbb{R}^3))$.

(s4) All derivatives of $\mathbf{M}$ and $\dot{\mathbf{M}}$ with respect to the second argument of orders up to $m-2$ are continuous on $[0,T] \times \mathcal{O}$.

(e) The following symmetry and strong ellipticity assumptions hold on $\mathcal{O}$:

$$A^{\alpha\beta}_{ij} = A^{\beta\alpha}_{ji}, \tag{159$_1$}$$

$$A^{\alpha\beta}_{ij}\varsigma^i\varsigma^j\eta_\alpha\eta_\beta \geq C|\varsigma|^2|\eta|^2, \tag{159$_2$}$$

where $C > 0$.

Theorem V.25:

Under assumptions (s1)-(s4) and (e), there is a $T' > 0$ such that the initial value problem (155) has a unique solution $u \in C^2([0,T']; H^{m-2}(\mathbb{R}^3)) \cap C^1([0,T']; H^{m-1}(\mathbb{R}^3)) \cap C([0,T']; H^m(\mathbb{R}^3))$.

In contrast to Chapter III, we have assumed less temporal smoothness on f and $\mathbf{M}$. Of course, this is reflected in less temporal smoothness of the solution. The semigroup approach used here is closely related to the energy method where the equations are differentiated only with respect to spatial variables. In fact, we can think of the operator S as being an abstraction of spatial differentiation. This is why the semigroup approach needs to be modified for initial-boundary value problems (see [K8], [K9]).

The proof of Theorem V.25 is based on a contraction argument similar to the one given above for the one-dimensional model, and we shall only sketch the basic ideas. We define the spaces $X := H^1(\mathbb{R}^3) \times L^2(\mathbb{R}^3)$, $Y := H^{m-1}(\mathbb{R}^3) \times H^{m-2}(\mathbb{R}^3)$, and $Z := H^{m-2}(\mathbb{R}^3) \times H^{m-3}(\mathbb{R}^3)$. We set $y := (v,w)$ and denote by y_0 the initial value of y. We now define $Q(T',R,L)$ to be the set of all functions $y: [0,T'] \to Y$ such that $\|y(t) - y_0\|_Y \leq R$ for $t \in [0,T']$ and $\|y(t) - y(t')\|_Z \leq L|t - t'|$ for $t,t' \in [0,T']$. For $y \in Q(T',R,L)$, we define $A_y\hat{y}$ as

$$\left(A_y(t)\hat{y}\right)^i := \left(\hat{w}^i, A^{\alpha\beta}_{ij}(\nabla u)\frac{\partial^2 \hat{v}^j}{\partial x^\alpha \partial x^\beta}\right), \tag{160}$$

where u is determined by the elliptic problem $(156)_3$. By $f_y(t)$ we denote the remaining terms on the right hand side of (156). In applying Theorem V.15 to solve the problem

$$\dot{\hat{y}} = A_y(t)\hat{y} + f_y(t), \tag{161}$$

we now have to use a variable norm in order to verify the stability condition (H1), and it is for this reason that the space Z and the Lipschitz condition in the definition of $Q(T', R, L)$ had to be introduced. This norm is defined as

$$\|(\hat{v}, \hat{w})\|^2_{N_y(t)} := \int_{\mathbb{R}^3} A^{\alpha\beta}_{ij}(\nabla u) \frac{\partial \hat{v}^i}{\partial x^\alpha} \frac{\partial \hat{v}^j}{\partial x^\beta} \, dx + \lambda\|\hat{v}\|^2_{L^2} + \|\hat{w}\|^2_{L^2}. \tag{162}$$

For the operator S, we choose the appropriate power of $(I - \Delta)$. With these choices, Theorem V.15 can be applied to the solution of (161) provided T', L, R are chosen appropriately, and it can be shown that the solution operator is a contraction on $Q(T', R, L)$ in the metric defined by $\|\cdot\|_X$. We omit the details and refer to Hughes, Kato and Marsden [H17] where the equations of nonlinear elasticity are discussed. The memory term does not necessitate any essential changes in the argument.

7. Three-dimensional motions of Jeffreys-type materials

In this section, we consider general three-dimensional motions (with Dirichlet boundary conditions) of a class of viscoelastic materials which we shall call of "Jeffreys type". The constitutive laws for these materials give the stress as the sum of a Newtonian contribution and a memory term which is of lower differential order. Polymer solutions with Newtonian solvents are often modelled in such a fashion. This structure of the constitutive law suggests treating the equations of motion as a perturbation of the Stokes equations. In this section, we shall implement this idea within the abstract context of quasilinear parabolic equations as discussed in Section 2. The results discussed in this section are based on [R8]. We remark that they can be generalized to materials in which the Newtonian viscosity is replaced by a history-dependent and tensor-valued term, such as Curtiss-Bird fluids [C13], but we shall not pursue this point here.

The constitutive law for the materials we consider, expressed in terms of the second Piola-Kirchhoff stress and the right Cauchy-Green tensor, is as follows:

$$\Pi(x,t) = -p\mathbf{C}^{-1}(x,t) - \eta\frac{\partial}{\partial t}(\mathbf{C}^{-1}(x,t)) + \mathcal{H}(\mathbf{C}^t(x,\cdot)). \tag{163}$$

In this equation, the first term on the right is the undetermined pressure, the second is a Newtonian viscous contribution. The third term, which involves the history of the deformation, will be assumed to satisfy appropriate smoothness conditions, so that it can be treated as a perturbation of lower differential order. The constitutive law (163) leads to the following equation of motion in Lagrangian coordinates:

$$\rho \ddot{y}^i = \frac{\partial}{\partial x^\beta}\left(\frac{\partial y^i}{\partial x^\alpha}\Pi^{\alpha\beta}\right) + f^i = -\frac{\partial x^\alpha}{\partial y^i}\frac{\partial p}{\partial x^\alpha} + \eta\left(\frac{\partial x^\alpha}{\partial y^j}\frac{\partial x^\beta}{\partial y^j}\frac{\partial^2 \dot{y}^i}{\partial x^\alpha \partial x^\beta}\right.$$

$$\left. -\frac{\partial x^\alpha}{\partial y^j}\frac{\partial x^\beta}{\partial y^l}\frac{\partial^2 y^l}{\partial x^\alpha \partial x^\gamma}\frac{\partial x^\gamma}{\partial y^j}\frac{\partial \dot{y}^i}{\partial x^\beta}\right) + \frac{\partial}{\partial x^\beta}\left(\frac{\partial y^i}{\partial x^\alpha}\mathcal{H}^{\alpha\beta}\right) + f^i(x,t). \tag{164$_1$}$$

In addition to (164)$_1$, we have the incompressibility constraint

$$\det \mathbf{F} = 1. \tag{164$_2$}$$

We consider these equations on a bounded domain $\Omega \subset \mathbb{R}^3$ subject to Dirichlet boundary conditions (in [R8], traction conditions are also considered)

$$y = g \text{ on } \partial\Omega. \tag{164$_3$}$$

An initial history is prescribed for $t < 0$:

$$y(x,t) = y_0(x,t) \text{ for } t < 0, \tag{164$_4$}$$

and the initial data for y and $\dot{y}$ as $t \to 0+$ must be compatible with this given initial history. Finally, we normalize the pressure in order to make it unique:

$$\int_\Omega p(x,t) \, dx = 0. \tag{164$_5$}$$

The following function spaces will be relevant in the analysis. We denote by C^{lim} the space of all continuous functions defined on the interval $[0,\infty)$ which converge to a limit at ∞. By $C^{k,lim}$ we denote the space of all functions whose first k derivatives lie in C^{lim}. In order to reduce the number of symbols needed in our notation, we shall not distinguish between spaces of scalar-, vector- or matrix-valued functions; it will be clear from context which is intended. By $C^{lim}(X)$, where X is a Banach space, we denote the space of all continuous functions $[0,\infty)$ which take values in X and have a limit at ∞. For the spatial dependence, we shall work in L^p-spaces. In what follows, p denotes any real number such that $3 < p \le 6$. This guarantees the Sobolev imbeddings $W^{1,p}(\Omega) \subset C(\bar{\Omega})$ and $W^{1,2}(\Omega) \subset L^p(\Omega)$. We think of all quantities, including the body force f and the boundary data g, as being defined for all values of $t \in \mathbb{R}$. We use a superscript t to denote the history as in Section 3.

We shall make the following assumptions:

(S1) Ω is a bounded domain in $\mathbb{R}^3$ with a boundary of class $C^{3,1}$.

S2) $\mathcal{H}$ is defined on the space C^{lim} and has four continuous Fréchet derivatives. Moreover, $\mathcal{H}$, regarded as acting pointwise with respect to the spatial variable, has three continuous Fréchet derivatives as a mapping from $C^{lim}(W^{1,p}(\Omega))$ into $W^{1,p}(\Omega)$.

S3) The function $(x,s) \to y_0(x,-s)$ lies in $C^{lim}(W^{4,p}(\Omega)) \cap C^{2,lim}(W^{2,p}(\Omega))$.

S4) For every t, f^t lies in $C^{lim}(W^{2,p}(\Omega)) \cap C^{1,lim}(L^p(\Omega))$. Moreover, in the norm of this space, f^t is a uniformly Hölder continuous function of t.

S5) For every t, g^t lies in $C^{1,lim}(W^{4-1/p,p}(\partial\Omega)) \cap C^{3,lim}(W^{2-1/p,p}(\partial\Omega))$, and in the norm of this space g^t is a uniformly Hölder continuous function of t.

C1) The boundary condition $(164)_3$ is compatible with the incompressibility constraint.

C2) The given history y_0 satisfies the equation of motion $(164)_1$, the incompressibility constraint $(164)_2$ and the boundary condition $(164)_3$. Moreover, for each $t \le 0$, the function $x \to y_0(x,t)$ is globally invertible.

(E) $\eta > 0$.

The main result of this section is the following.

Theorem V.26:

Assume that (S1)-(S5), (C1), (C2) and (E) hold. Then there exists some $T > 0$ such that (164) has a unique solution on $\Omega \times [0,T]$ satisfying $y \in C^1([0,T]; W^{4,p}(\Omega)) \cap C^2([0,T]; W^{2,p}(\Omega)) \cap C^3([0,T]; L^p(\Omega))$ and $p \in C([0,T]; W^{3,p}(\Omega)) \cap C^1([0,T]; W^{1,p}(\Omega))$.

The proof will be based on an application of Theorem V.12, and the history dependence will be dealt with according to the recipe prescribed in equations (61) and (62). A number of preparatory steps are required in order to write (164) in a form suitable for applying Theorem V.12. We use the notations

$$F_\alpha^i := \frac{\partial y^i}{\partial x^\alpha}, \quad \chi_i^\alpha := \frac{\partial x^\alpha}{\partial y^i}, \quad G_{\alpha\beta}^i := \frac{\partial^2 y^i}{\partial x^\alpha \partial x^\beta},$$

$$v := \dot{y}, \quad z := \ddot{y}, \quad q := \dot{p}. \tag{165}$$

With these notations, we can rewrite $(164)_1$ in the form

$$\rho z^i = \eta \left(\chi_j^\alpha \chi_j^\beta \frac{\partial^2 v^i}{\partial x^\alpha \partial x^\beta} - \chi_j^\alpha \chi_k^\beta G_{\alpha\gamma}^k \chi_j^\gamma \frac{\partial v^i}{\partial x^\beta} \right)$$

$$- \chi_i^\alpha \frac{\partial p}{\partial x^\alpha} + \frac{\partial}{\partial x^\beta}(F_\alpha^i \mathcal{H}^{\alpha\beta}) + f^i. \tag{166}_1$$

By differentiating $(164)_2$ with respect to time and using $(164)_5$, we obtain

$$\chi_i^\alpha \frac{\partial v^i}{\partial x^\alpha} - \int_\Omega p \, dx = 0. \tag{166}_2$$

231

These two equations, combined with the boundary condition

$$v = \dot{g} \text{ on } \partial\Omega, \tag{166}$$

will be used to solve for v and p with y^t and z considered known. They form an elliptic system, which in fact transforms to the Stokes equation when Eulerian coordinates are introduced.

Henceforth we regard v and p as determined from (166) in terms of z and y and we formulate a system of equations for y and z. The first equation in this system is simply given by

$$\dot{y} = v. \tag{167}$$

An equation describing the evolution of z is obtained from differentiating $(166)_1$ One obtains an equation of the form

$$\rho \dot{z}^i = \eta \left(\chi_j^\alpha \chi_j^\beta \frac{\partial^2 z^i}{\partial x^\alpha \partial x^\beta} - \chi_j^\alpha \chi_k^\beta G_{\alpha\gamma}^k \chi_j^\gamma \frac{\partial z^i}{\partial x^\beta} \right)$$

$$- \chi_i^\alpha \frac{\partial q}{\partial x^\alpha} + f^i + \Phi^i (\nabla y^t, \nabla^2 y^t, \nabla v^t, \nabla^2 v^t, \nabla p), \tag{167}_2$$

where Φ is an expression involving spatial derivatives of y^t and v^t up to the second order and first order spatial derivatives of p. By differentiating $(166)_2$ with respect to time, we obtain

$$\chi_i^\alpha \frac{\partial z^i}{\partial x^\alpha} - \int_\Omega q \, dx = \chi_j^\alpha \frac{\partial v^j}{\partial x^\beta} \chi_i^\beta \frac{\partial v^i}{\partial x^\alpha}. \tag{167}_3$$

The appropriate boundary condition for z is given by

$$z = \ddot{g} \text{ on } \partial\Omega. \tag{167}_4$$

The system (167) is still not in a form suitable for applying Theorem V.12. First, equation $(167)_2$ involves delay terms. Secondly, the boundary conditions have to be reduced to a homogeneous form because they will be incorporated in the definition of the domain of a linear operator. Finally, we have to eliminate the term involving q. Let us recall that in classical Navier-Stokes theory the pressure is eliminated by using the Hodge projection. In the problem studied here, we work in Lagrangian coordinates, and the Hodge projection therefore becomes variable; it depends on the deformation y. We shall now discuss some of the basic properties of this variable Hodge projection.

Assume that the deformation $y \in W^{4,p}(\Omega)$ satisfies the incompressibility condition $(164)_2$ and is globally invertible on $\bar{\Omega}$. Let $\tilde{\Omega}$ denote the image of Ω under the mapping y. Then every vector field $\tilde{w}$ in $L^p(\tilde{\Omega})$ can be decomposed uniquely into a divergence-free part which has zero normal component on the boundary and a gradient. Let us call the two parts $\tilde{w}_1$ and $\tilde{w}_2$. A natural isometry between $L^p(\Omega)$ and $L^p(\tilde{\Omega})$ is given by

$$w(x) = \tilde{w}(y(x)). \tag{168}$$

232

The decomposition of $\tilde{w}$ into $\tilde{w}_1$ and $\tilde{w}_2$ induces a decomposition of w into w_1 and w_2, where w_1 satisfies

$$\frac{\partial}{\partial x^\alpha}\{\chi_i^\alpha w_1^i\} = 0, \quad n_\alpha \chi_i^\alpha w_1^i = 0 \text{ on } \partial\Omega, \tag{169}_1$$

and w_2 is of the form

$$w_2^i = \chi_i^\alpha \frac{\partial r}{\partial x^\alpha}. \tag{169}_2$$

We denote by $P(y)$ the projection of w onto w_1. With X_1 and X_2 defined as the range and nullspace of the ordinary Hodge projection in $L^p(\Omega)$ (henceforth denoted by P), $P(y)$ is a projection along $\mathbf{F}^{-T}X_2$ onto $\mathbf{F}X_1$.

The following lemma relates $P(y)$ to P and the deformation gradient $\mathbf{F}$.

Lemma V.27:

Let Q denote $I - P$. The operator $Q\mathbf{F}^{-1}\mathbf{F}^{-T}Q$ is a bijection from X_2 onto itself. Let L denote its inverse. Then we have

$$P(y) = 1 - \mathbf{F}^{-T}LQ\mathbf{F}^{-1}. \tag{170}$$

Proof:
We need to show that, for any $w \in X_2$, there is a unique $z \in X_2$ such that

$$Q\mathbf{F}^{-1}\mathbf{F}^{-T}z = w, \tag{171}$$

or equivalently, that there is a unique $z \in X_2$ and $z^* \in X_1$ such that

$$\mathbf{F}^{-1}\mathbf{F}^{-T}z = w - z^*. \tag{172}$$

This can be written in the form

$$\mathbf{F}^{-T}z + \mathbf{F}z^* = \mathbf{F}w. \tag{173}$$

The unique solvability of this equation is immediate from the decomposition $L^p(\Omega) = \mathbf{F}^{-T}X_2 \oplus \mathbf{F}X_1$.

Some simple algebra shows that the right hand side of (170) is a projection operator which reduces to the identity on $\mathbf{F}X_1$ and to zero on $\mathbf{F}^{-T}X_2$. This completes the proof of the lemma.

∎

We next introduce a substitution that reduces $(167)_3$ and $(167)_4$ to a homogeneous form. For this, we define b as the solution of the (stationary) Stokes problem

$$\Delta_x b = \nabla_x \pi, \tag{174}_1$$

233

$$\text{div}_x b - \int_\Omega \pi \, dx = \chi_j^\alpha \frac{\partial v^j}{\partial x^\beta} \chi_i^\beta \frac{\partial v^i}{\partial x^\alpha}, \tag{174$_2$}$$

$$b = \mathbf{F}^{-1}\dot{g} \quad \text{on } \partial\Omega. \tag{174$_3$}$$

The problems (166) and (174) are both of Stokes type. Standard elliptic theory can be used to establish the following lemma.

Lemma V.28:

Let $y^t \in C^{lim}(W^{4,p}(\Omega))$ be such that $y^t(0)$ is globally invertible on $\bar{\Omega}$ and satisfies the incompressibility condition. Then, for every $f \in L^p(\Omega)$, $z \in L^p(\Omega)$ and $\dot{g} \in W^{2-1/p,p}(\partial\Omega)$, the system (166) has a unique solution $v(y^t,z,f,\dot{g}) \in W^{2,p}(\Omega)$, $p(y^t,z,f,\dot{g}) \in W^{1,p}(\Omega)$. If $f,z \in W^{2,p}(\Omega)$, $\dot{g} \in W^{4-1/p,p}(\partial\Omega)$, then $v(y^t,z,f,\dot{g}) \in W^{4,p}(\Omega)$, $p(y^t,z,f,\dot{g}) \in W^{3,p}(\Omega)$. Here v and p are continuously differentiable functions of their arguments. Moreover, the system (174) has a unique solution $b \in W^{2,p}(\Omega)$, $\pi \in W^{1,p}(\Omega)$ which depends smoothly (of class C^∞ on $y \in W^{4,p}(\Omega)$, $v \in W^{2,p}(\Omega)$, and $\dot{g} \in W^{2-1/p,p}(\partial\Omega)$.

By differentiating (174) with respect to time and noting that $\dot{v} = z$, we obtain a system of the form

$$\Delta\dot{b} = \nabla\dot{\pi}, \tag{175$_1$}$$

$$\text{div } \dot{b} - \int_\Omega \dot{\pi} \, dx = \Gamma(\nabla y, \nabla v, \nabla z), \tag{175$_2$}$$

$$\dot{b} = \mathbf{F}^{-1}\frac{\partial^3 g}{\partial t^3} - \mathbf{F}^{-1}(\nabla v)\mathbf{F}^{-1}\dot{g} \quad \text{on } \partial\Omega. \tag{175$_3$}$$

Here the dependence of Γ on ∇z is linear. From this system we can find $\dot{b} \in W^{2,p}(\Omega)$ as a unique function of $y \in W^{4,p}(\Omega)$, $v \in W^{3,p}(\Omega)$, $z \in W^{2,p}(\Omega)$ and $\dot{g}, \frac{\partial^3 g}{\partial t^3} \in W^{2-1/p,p}(\Omega)$. We write $\dot{b} = \chi(y,v,z,\dot{g},\frac{\partial^3 g}{\partial t^3})$ to indicate this dependence. We note that to estimate the norm of $\dot{b}$ in $W^{1,2}(\Omega)$ (and hence in $L^p(\Omega)$) we only need a bound on the norm of z in $W^{1,2}(\Omega)$ rather than $W^{2,p}(\Omega)$.

To save on notation, we define

$$\left[L_1(y)w\right]^i := \eta\left(\chi_j^\alpha \chi_j^\beta \frac{\partial^2 w^i}{\partial x^\alpha \partial x^\beta} - \chi_j^\alpha \chi_k^\beta G_{\alpha\gamma}^k \chi_j^\gamma \frac{\partial w^i}{\partial x^\beta}\right), \tag{176$_1$}$$

$$\left[L_2(y)r\right]^i := -\chi_i^\alpha \frac{\partial r}{\partial x^\alpha}, \tag{176$_2$}$$

and we set $c := \mathbf{F}^{-1}z - b$. We now replace (164) with the system

$$\dot{y} = v(y^t,z,f,g), \tag{177$_1$}$$

$$\dot{b} = \chi\left(y,v,\mathbf{F}(c+b),\ddot{g},\frac{\partial^3 g}{\partial t^3}\right), \tag{177$_2$}$$

$$\rho\mathbf{F}\dot{c} = -\rho(\nabla v)c - \rho\mathbf{F}\chi\left(y,v,\mathbf{F}(c+b),\ddot{g},\frac{\partial^3 g}{\partial t^3}\right) - \rho(\nabla v)b$$

$$+L_1(y)\big(\mathbf{F}(c+b)\big) + L_2(y)q + \Phi + \dot{f}, \tag{177}_3$$

$$\operatorname{div} c = 0, \tag{177}_4$$

$$c = 0 \text{ on } \partial\Omega. \tag{177}_5$$

Next we eliminate the term $L_2(y)q$ from equation $(177)_3$ by applying the variable Hodge projection $P(y)$ introduced above. We obtain in this way

$$\rho P(y)\mathbf{F}\dot{c} = -\rho P(y)\Big[(\nabla v)(c+b) + \mathbf{F}\chi\big(y,v,\mathbf{F}(c+b),\ddot{g},\frac{\partial^3 g}{\partial t^3}\big)\Big]$$

$$+ P(y)L_1(y)\big(\mathbf{F}(c+b)\big) + P(y)\Phi + P(y)\dot{f}. \tag{177}_3^*$$

Moreover, we note that $(177)_{4,5}$ imply $\operatorname{div}\dot{c} = 0$, $\dot{c} = 0$ on $\partial\Omega$ and hence $P(y)\mathbf{F}\dot{c} = \mathbf{F}\dot{c}$. Therefore we can write $(177)_3^*$ in the form

$$\rho\dot{c} = \mathbf{F}^{-1}P(y)\Big[-\rho(\nabla v)(c+b) - \rho\mathbf{F}\chi + L_1(y)\big(\mathbf{F}(c+b)\big) + \Phi + \dot{f}\Big]. \tag{177}_3^{**}$$

Equations (177) form a system describing the evolution of the variables y, b and c. A dependence on the history occurs in $(177)_1$, where v depends on y^t and in the term Φ in $(177)_3^{**}$, which involves the histories of y and v (and hence the histories of all the variables). We now define new functions $\hat{v}$, $\hat{\chi}$ etc. according to the prescription given in equation (61), i.e. $\hat{v}$ is the history of v,

$$\hat{v}(y^t, z^t, f^t, g^t)(s) = v(T, y^t, z^t(s), f(t-s), g(t-s)), \tag{178}$$

etc. We can now write (177) in the form

$$\dot{y}^t = \hat{v}(y^t, \mathbf{F}^t(c^t + b^t), f^t, g^t), \tag{179}_1$$

$$\dot{b}^t = \hat{\chi}(y^t, \hat{v}(...), \mathbf{F}^t(c^t + b^t), \ddot{g}^t, \frac{\partial^3 g^t}{\partial t^3}), \tag{179}_2$$

$$\rho\dot{c}^t = (\mathbf{F}^t)^{-1}\hat{P}(\mathbf{F}^t)\Big[-\rho(\nabla\hat{v}(...))(c^t + b^t) - \rho\mathbf{F}^t\hat{\chi}(...)$$

$$+ \hat{L}_1(y^t)\big(\mathbf{F}^t(c^t + b^t)\big) + \hat{\Phi}(...) + \dot{f}^t\Big], \tag{179}_3$$

$$\operatorname{div} c^t = 0, \tag{179}_4$$

$$c^t = 0 \text{ on } \partial\Omega. \tag{179}_5$$

We seek a solution to (179) with the initial condition $y^t(x,s) = y_0(x,-s)$ at $t = 0$ and appropriate initial conditions for b^t and c^t which are obtained by solving (174) at all times prior to 0. It easily follows from the smoothness assumptions we made on the data that at $t = 0$ we have $y^t \in C^{lim}(W^{4,p}(\Omega))$ and $b^t, c^t \in C^{lim}(W^{2,p}(\Omega))$.

We shall now apply Theorem V.12 to the system (179). For this we have to specify a function space in which solutions are considered and we have to identify

235

the terms that are denoted by A and f in the abstract equation (17). For the space, we choose

$$X := C^{lim}\left(W^{4,p}(\Omega)\right) \times C^{lim}\left(W^{2,p}(\Omega)\right) \times C^{lim}\left(\check{L}^p(\Omega)\right), \qquad (180)$$

where $\check{L}^p(\Omega)$ denotes the space of all vectorfields in $L^p(\Omega)$ which are divergence-free and have vanishing normal component on the boundary. Moreover, we define the operator A as follows:

$$A(y^t, b^t, c^t, t)((y')^t, (b')^t, (c')^t) := \left(D_2\,\hat{v}(y^t, ...)\mathbf{F}^t(c')^t,\right.$$

$$D_3\,\hat{\chi}(y^t, ...)\mathbf{F}^t(c')^t + D_2\,\hat{\chi}(y^t, ...)D_2\,\hat{v}(y^t, ...)\mathbf{F}^t(c')^t, \qquad (181)$$

$$\left. \rho^{-1}(\mathbf{F}^t)^{-1}\hat{P}(y^t)\{-\rho\mathbf{F}^t(D_3\,\hat{\chi}(...) + D_2\,\hat{\chi}(...)D_2\,\hat{v}(...))\mathbf{F}^t(c')^t + \hat{L}_1(y^t)\mathbf{F}^t(c')^t\}\right).$$

The remaining terms on the right hand side of $(179)_1$-$(179)_3$ are included in the f-term in the abstract setting. Equation $(179)_4$ is already incorporated in the definition of the space X, and $(179)_5$ is incorporated in the definition of the domain of A:

$$D(A) := \{((y')^t, (b')^t, (c')^t) \in C^{lim}\left(W^{4,p}(\Omega)\right) \times C^{lim}\left(W^{2,p}(\Omega)\right) \times C^{lim}\left(W^{2,p}(\Omega)\right) \mid$$

$$\text{div } (c')^t = 0, \ (c')^t = 0 \text{ on } \partial\Omega\}. \qquad (182)$$

With these identifications, we can verify the assumptions of Theorem V.12 with $\alpha = 0$, $\beta = 1$ and ϵ equal to the Hölder constant in assumptions (S4) and (S5) of the theorem. We omit the straightforward verification of the smoothness conditions (31) and (32) and focus only on the resolvent estimate (30). For this we note first that the operator

$$A_1(y^t) := (\mathbf{F}^t)^{-1}\hat{P}(y^t)\hat{L}_1(y^t)\mathbf{F}^t \qquad (183)$$

is simply the Stokes operator transformed to Lagrangian coordinates. It is well known that the Stokes operator generates an analytic semigroup in $\check{L}^p(\Omega)$, see Solonnikov [S9]-[S11], McCracken [M11] and Giga [G4]. This immediately leads to an estimate of the form (30) for the resolvent of the operator A_1. The remaining terms in the third component of A are defined for $(c')^t \in C^{lim}\left(W^{1,p}(\Omega)\right)$ and hence they are bounded relative to A_1 with relative bound zero. Hence the third component of the operator A generates an analytic semigroup. The first two components can be treated as a perturbation in a fashion analogous to the argument following equation (111) in Section 5. If, for example, we consider the equation

$$D_2\,\hat{v}(...)\mathbf{F}^t(c')^t + \lambda(y')^t = (w')^t, \qquad (184)$$

then we immediately obtain an estimate of the form

$$\|(y')^t(\cdot, s)\|_{W^{4,p}(\Omega)} \le \frac{1}{|\lambda|}\|(w')^t(\cdot, s)\|_{W^{4,p}(\Omega)} + \frac{C}{|\lambda|}\|(c')^t(\cdot, s)\|_{W^{2,p}(\Omega)}, \qquad (185)$$

and we already know how to estimate the last term on the right hand side of this inequality. This concludes the proof of Theorem V.26.

8. Three-dimensional motions of a K-BKZ fluid

In this section we shall discuss the initial-value problem for motions of a K-BKZ fluid on all of $\mathbb{R}^3$. We shall demonstrate how this problem can be treated within the framework of Theorem V.19. To handle the history terms, we shall use the approach described following equation (66). In contrast to the approach used in the previous section, this does not require that the given history satisfy the equation; in fact, we shall not need any compatibility conditions between the history and the initial data. The results discussed in this section were obtained in [R10].

As usual, $\mathbf{F}$ denotes the deformation gradient, and χ_i^α denote the components of $\mathbf{F}^{-1}$. Moreover, we set

$$\Phi(s, x, t) := \mathbf{F}(x, t)\mathbf{F}^{-1}(x, s), \tag{186}_1$$

and

$$\Psi := \Phi^{-1}. \tag{186}_2$$

We remark that Φ is the same quantity that was named Φ in Section 5 of Chapter IV, but we now regard it as a function of the Lagrangian coordinate x rather than the Eulerian coordinate y. As in Section 5 of Chapter IV, the stored energy function W is regarded as a function of Φ. The equation of motion for a K-BKZ fluid, written in Lagrangian coordinates, has the form

$$\rho \ddot{y}^i(x, t) = -\frac{\partial p}{\partial x^\alpha}(x, t)\chi_i^\alpha(x, t) + \int_{-\infty}^t m(t - s)\frac{\partial^2 W}{\partial \Phi_m^i \, \partial \Phi_n^j}(s, x, t)$$

$$\times \left[\frac{\partial^2 y^j}{\partial x^\alpha \partial x^\beta}(x, t)\chi_n^\alpha(x, s)\chi_m^\beta(x, s) + \frac{\partial y^j}{\partial x^\alpha}(x, t)\frac{\partial}{\partial x^\beta}(\chi_n^\alpha(x, s))\chi_m^\beta(x, s)\right] ds + f^i(x, t). \tag{187}_1$$

In addition, we have the incompressibility constraint

$$\det \mathbf{F} = 1. \tag{187}_2$$

We prescribe an initial history for y,

$$y(x, t) = \tilde{y}(x, t), \ t < 0, \tag{187}_3$$

and initial data for $t \to 0+$, which need not be compatible with the history,

$$y(x,0+) = y_0(x), \quad \dot{y}(x,0+) = y_1(x). \tag{187}_4$$

We make the following assumptions.

(S1) The function W is of class C^9.

(S2) $m \in W^{3,1}[0,\infty)$.

(S3) $f \in \bigcap_{k=0}^{3} C^k([0,T]; H^{7-k}(\mathbb{R}^3))$.

(S4) $\tilde{y} - x \in C_b((-\infty,0]; H^8(\mathbb{R}^3))$.

(S5) $y_0 \in H^8(\mathbb{R}^3)$, $y_1 \in H^7(\mathbb{R}^3)$.

(C) The history $\tilde{y}$ and the initial data y_0 and y_1 are compatible with the incompressibility constraint.

(E) The strong ellipticity condition

$$\frac{\partial^2 W}{\partial \Phi_n^i \, \partial \Phi_p^j} \lambda^i \lambda^j \mu_n \mu_p \geq \kappa |\lambda|^2 |\mu|^2 \quad \forall \lambda, \mu \in \mathbb{R}^3 \tag{188}$$

holds with a positive constant κ. Moreover, $m > 0$.

Theorem V.29:

Let assumptions (S1)-(S5), (C) and (E) hold. Then there exists some $T' \in (0,T]$ such that problem (187) has a unique solution $y \in L^2([0,T']; H^8(\mathbb{R}^3)) \cap H^3([0,T']; H^5(\mathbb{R}^3))$.

Remarks V.30:

1. It is clear that this result is not optimal, and that an existence and uniqueness result can be obtained assuming less smoothness of the data, e.g. by using the methods of Chapter III.

2. It is not necessary to assume that (E) holds globally; only strong ellipticity in a neighborhood of the data is required. Also, W need not be globally defined.

3. From a physical point of view, the function W is meaningful only on the manifold $\det \Phi = 1$. Condition (E) should be regarded as valid for an appropriate extension of W. It is possible to substitute a condition for (E) which is independent of the choice of the extension, see [R10].

4. The condition (E) is in general not easy to verify. In [R10], it is shown that (E) holds if the function $\hat{W}(I_1, I_2)$ (as in equation $(189)_1$ of Chapter IV) is monotone increasing with respect to each argument and convex as a function of $\sqrt{I_1}$ and $\sqrt{I_2}$. In the neighborhood of $\Phi = 1$, (E) assumes a simple form, see Section 5 of Chapter IV.

The proof of the theorem will be based on a reformulation of the equations which fits into the abstract framework of Theorem V.19. We introduce the following notations:

$$v := \dot{y}, \quad z = \ddot{y} - \lambda(y-x), \quad w = \dot{z}, \quad q = \dot{p}, \quad r = \ddot{p}, \tag{189}$$

where λ is a positive constant which will later be chosen sufficiently large. Also we set

$$A_{ij}^{mn} := \frac{\partial^2 W}{\partial \Phi_m^i \, \partial \Phi_n^j}. \tag{190}$$

With this notation, we rewrite the first two equations of (187) as follows:

$$\rho z^i - f^i = -\frac{\partial p}{\partial x^\alpha} \frac{\partial x^\alpha}{\partial y^i} + \int_{-\infty}^t m(t-s) A_{ij}^{mn} \left(\nabla y(x,t), \nabla y(x,s) \right)$$

$$\times \frac{\partial}{\partial x^\beta} \left[\frac{\partial y^j}{\partial x^\alpha}(x,t) \frac{\partial x^\alpha}{\partial y^n}(x,s) \frac{\partial x^\beta}{\partial y^m}(x,s) \right] ds - \lambda \rho (y^i - x^i), \tag{191$_1$}$$

$$\det(\nabla y) = 1. \tag{191$_2$}$$

By differentiating (191) once with respect to time, we obtain a system of the form

$$\rho w^i - \dot f^i = -\frac{\partial q}{\partial x^\alpha} \frac{\partial x^\alpha}{\partial y^i} + \int_{-\infty}^t m(t-s) A_{ij}^{mn} \frac{\partial}{\partial x^\beta}$$

$$\times \left[\frac{\partial v^j}{\partial x^\alpha}(x,t) \frac{\partial x^\alpha}{\partial y^n}(x,s) \frac{\partial x^\beta}{\partial y^m}(x,s) \right] ds - \lambda \rho v^i + U^i (\nabla y^t, \nabla^2 y^t, \nabla v, \nabla p), \tag{192$_1$}$$

$$\frac{\partial v^i}{\partial x^\alpha} \frac{\partial x^\alpha}{\partial y^i} = 0, \tag{192$_2$}$$

where U is an expression involving spatial derivatives of y^t up to the second order and first order spatial derivatives of v and p. Further differentiation with respect to time leads to the system

$$\dot z = w, \tag{193$_1$}$$

$$\rho \dot w^i = -\frac{\partial r}{\partial x^\alpha} \frac{\partial x^\alpha}{\partial y^i} + \frac{\partial}{\partial x^\beta} \int_{-\infty}^t m(t-s) A_{ij}^{mn}$$

$$\times \left[\frac{\partial z^j}{\partial x^\alpha}(x,t) \frac{\partial x^\alpha}{\partial y^n}(x,s) \frac{\partial x^\beta}{\partial y^m}(x,s) \right] ds$$

$$- \lambda \rho z^i + V^i (y, \nabla y^t, \nabla^2 y^t, \nabla v, \nabla^2 v, \nabla z, \nabla p, \nabla q) + \ddot f^i, \tag{193$_2$}$$

$$\frac{\partial w^i}{\partial x^\alpha} \frac{\partial x^\alpha}{\partial y^i} = W(\nabla y, \nabla v, \nabla z). \tag{193$_3$}$$

In a fashion similar to Section 4 of Chapter III, we regard (191) and (192) as elliptic problems which are to be solved for y and p and, respectively, v and q. By inserting these solutions in (193), we obtain a problem for z, w and r, from which we eliminate r by using the variable Hodge projection defined in the previous section.

Initial data for z and w are found by the same procedure as in Section 4 of Chapter III. First we note that initial data for y and v as well as the history of y are given. We can then apply the operator $\frac{\partial x^7}{\partial y^i} \frac{\partial}{\partial x^7}$ to equation (191)$_1$. The

239

incompressibility condition can be used to express $\frac{\partial x^\gamma}{\partial y^i}\frac{\partial z^i}{\partial x^\gamma}$ in terms of ∇y and ∇v. This yields an elliptic equation for $p_0(x) := p(x,0)$, which, up to an undetermined constant, has a unique solution such that $\nabla p_0 \in H^6(\mathbb{R}^3)$. From $(191)_1$, we then determine the initial value $z_0(x) := z(x,0) \in H^6(\mathbb{R}^3)$. Similarly we use $(192)_1$ and $(193)_3$ to determine initial data for q and w. These satisfy $\nabla q_0, w_0 \in H^5(\mathbb{R}^3)$.

We now introduce functions $\hat{y}(x,\sigma,t) := y(x,\sigma t)$, etc. for $\sigma \in [0,1]$. With this, we can rewrite (191) as follows

$$\rho\hat{z}^i(x,\sigma,t) - \hat{f}^i(x,\sigma,t) = -\frac{\partial\hat{p}}{\partial x^\alpha}(x,\sigma,t)\frac{\partial x^\alpha}{\partial\hat{y}^i}(x,\sigma,t)$$

$$+ \int_{-\infty}^0 m(\sigma t - s)A_{ij}^{mn}\left(\nabla\hat{y}(x,\sigma,t),\nabla\tilde{y}(x,s)\right)\frac{\partial}{\partial x^\beta}$$

$$\left[\frac{\partial\hat{y}^i}{\partial x^\alpha}(x,\sigma,t)\frac{\partial x^\alpha}{\partial\tilde{y}^n}(x,s)\frac{\partial x^\beta}{\partial\tilde{y}^m}(x,s)\right]ds$$

$$+t\int_0^\sigma m(\sigma t - \sigma' t)A_{ij}^{mn}\left(\nabla\hat{y}(x,\sigma,t),\nabla\hat{y}(x,\sigma',t)\right)\frac{\partial}{\partial x^\beta}$$

$$\left[\frac{\partial\hat{y}^i}{\partial x^\alpha}(x,\sigma,t)\frac{\partial x^\alpha}{\partial\hat{y}^n}(x,\sigma',t)\frac{\partial x^\beta}{\partial\hat{y}^m}(x,\sigma',t)\right]d\sigma'$$

$$-\rho\lambda\left(\hat{y}^i(x,\sigma,t)-x^i\right). \tag{194}_1$$

$$0 = \det\left(\nabla\hat{y}(x,\sigma,t)\right) - 1. \tag{194}_2$$

For $t = 0$, $\hat{z}(x,\sigma,0) = z_0(x)$, we have the solution $\hat{y}(x,\sigma,0) = y_0(x)$, $\hat{p}(x,\sigma,0) = p_0(x)$. We shall apply the implicit function theorem to show the existence of a unique solution $\hat{y},\hat{p}$ if $\hat{z}$ and t are close to z_0 and 0. In choosing appropriate function spaces for applying the implicit function theorem, we must be careful to note that not every function in $H^s(\mathbb{R}^3)$ is the divergence of a vector field in $H^{s+1}(\mathbb{R}^3)$. We therefore introduce the space

$$\tilde{H}^s(\mathbb{R}^3) := \{\phi \in H^s(\mathbb{R}^3) \mid \phi = \mathrm{div}\, u \text{ for some } u \in L^2(\mathbb{R}^3)\}. \tag{195}$$

(It is clear that we can actually take $u \in H^{s+1}(\mathbb{R}^3)$ in (195).) We note that equation (89) of Chapter III can be used to write expressions such as $\frac{\partial v^i}{\partial x^\alpha}\frac{\partial x^\alpha}{\partial y^i}$ in divergence form. Using some simple algebra, it is also possible to write $\det \nabla y - 1$ as a divergence.

We also have to specify a topology for the σ-dependence. The theory of Kato on which we base our existence result requires the spaces to be reflexive, so we cannot use a space of continuous functions. On the one hand, the presence of nonlinear terms requires that the spaces used are Banach algebras, so we cannot use the L^2-norm. On the other hand, we shall need use estimates that are of a pointwise nature with respect to σ, so we cannot use the H^1-norm either. It turns out, however, that the following mixed type of space can be used: For any integer n, let

$$V_n := H^1\left([0,1]; H^n(\mathbb{R}^3)\right) \cap L^2\left([0,1]; H^{n+3}(\mathbb{R}^3)\right). \tag{196}$$

We note that V_n is a Banach algebra for $n \geq 2$.

By using standard elliptic theory and the implicit function theorem we can prove the following lemma (cf. Section 4 of Chapter III for a similar argument).

Lemma V.31:

Let $\lambda > 0$ be chosen sufficiently large (relative to the data of the problem). Let $n = 2$ or $n = 3$. Assume that $t > 0$ is small and that $\hat{z} \in V_n$ is given such that $\|\hat{z} - z_0\|_{V_n}$ is small. Then equation (194) has a solution $(\hat{y}, \hat{p})$, such that $\hat{y} - x \in V_{n+2}$, $\nabla \hat{p} \in V_n$. This solution is unique within a neighborhood of (y_0, p_0), and it depends smoothly on $\hat{z}$. If, moreover $\hat{w} \in V_{n-1}$, then equation (192) can be solved for $\hat{v} \in V_{n+1}$ and $\hat{q}$ such that $\nabla \hat{q} \in V_{n-1}$.

We define the Hodge projection $P(y)$ as in Section 7. By applying $P(y)$ to equation (193)$_2$, we eliminate the term $-\frac{\partial r}{\partial x^\alpha} \frac{\partial x^\alpha}{\partial y^i}$. By differentiating (192)$_2$ once with respect to time, we obtain an equation of the form

$$\frac{\partial z^i}{\partial x^\alpha} \frac{\partial x^\alpha}{\partial y^i} = \Gamma(\nabla y, \nabla v), \tag{197}_1$$

and by differentiating (193)$_3$ with respect to time, we obtain an equation of the form

$$\frac{\partial \dot{w}^i}{\partial x^\alpha} \frac{\partial x^\alpha}{\partial y^i} = \Delta(\nabla y, \nabla v, \nabla z, \nabla w). \tag{197}_2$$

We can use these equations to solve for $(I - P(y))z$ and $(I - P(y))\dot{w}$, respectively. Let us therefore write

$$(I - P(y))z = G(y, v), \quad (I - P(y))\dot{w} = H(y, v, z, w), \tag{198}$$

or equivalently

$$(I - P(\hat{y}))\hat{z} = G(\hat{y}, \hat{v}), \quad (I - P(\hat{y}))\dot{\hat{w}} = H(\hat{y}, \hat{v}, \hat{z}, \hat{w}). \tag{199}$$

For $n = 2$ or $n = 3$, G is a smooth mapping from $V_{n+2} \times V_{n+1}$ into V_{n+1} and H is a smooth mapping from $V_{n+2} \times V_{n+1} \times V_n \times V_{n-1}$ into V_{n-1}.

We now make the definition

$$\left(\tilde{A}(\hat{y}, \sigma, t) \hat{z}(x, \sigma, t) \right)^i := \frac{\partial}{\partial x^\alpha} \int_{-\infty}^{0} m(\sigma t - s) A_{ij}^{mn} \left(\nabla \hat{y}(x, \sigma, t), \nabla \tilde{y}(x, s) \right)$$

$$\times \left[\frac{\partial \hat{z}^j}{\partial x^\beta}(x, \sigma, t) \frac{\partial x^\beta}{\partial \tilde{y}^n}(x, s) \frac{\partial x^\alpha}{\partial \tilde{y}^m}(x, s) \right] ds$$

$$+ t \frac{\partial}{\partial x^\alpha} \int_{0}^{\sigma} m(\sigma t - \sigma' t) A_{ij}^{mn} \left(\nabla \hat{y}(x, \sigma, t), \nabla \hat{y}(x, \sigma', t) \right)$$

$$\times \left[\frac{\partial \hat{z}^j}{\partial x^\beta}(x, \sigma, t) \frac{\partial x^\beta}{\partial \hat{y}^n}(x, \sigma', t) \frac{\partial x^\alpha}{\partial \hat{y}^m}(x, \sigma', t) \right] d\sigma'. \tag{200}$$

241

Using this notation and taking account of equation (199), we can rewrite (193) as follows

$$\dot{\hat{z}}(x,\sigma,t) = \sigma\hat{w}(x,\sigma,t), \tag{201}_1$$

$$\dot{\hat{w}}(x,\sigma,t) = \sigma H(\hat{y}(\cdot,\sigma,t),\hat{v}(\cdot,\sigma,t),\hat{z}(\cdot,\sigma,t),\hat{w}(\cdot,\sigma,t))(x)$$

$$+\frac{\sigma}{\rho}\Big\{P(\hat{y}(\cdot,\sigma,t)\tilde{A}(\hat{y},\sigma,t)P(\hat{y}(\cdot,\sigma,t)\hat{z}(x,\sigma,t)$$

$$+(I - P(\hat{y}(\cdot,\sigma,t)))\Delta_x(I - P(\hat{y}(\cdot,\sigma,t)))\hat{z}(x,\sigma,t)\Big\}$$

$$+\frac{\sigma}{\rho}\Big\{P(\hat{y}(\cdot,\sigma,t)\tilde{A}(\hat{y},\sigma,t)G(\hat{y}(\cdot,\sigma,t),\hat{v}(\cdot,\sigma,t))(x)$$

$$-(I - P(\hat{y}(\cdot,\sigma,t)))\Delta_x G(\hat{y}(\cdot,\sigma,t),\hat{v}(\cdot,\sigma,t))(x)\Big\}$$

$$+\sigma P(\hat{y}(\cdot,\sigma,t))\Big\{-\lambda\hat{z} + \frac{1}{\rho}V(\hat{y},...) + \frac{1}{\rho}\hat{f}\Big\}(x,\sigma,t). \tag{201}_2$$

The system (201) is in a form suitable for applying Theorem V.19. In doing so, we make the following identifications: We set $X = Z = Z' = V_2 \times V_1$ and $Y = V_3 \times V_2$. The operator $A(\hat{z},t)$ is defined by

$$A(\hat{z},t)(\hat{z}',\hat{w}')(x,\sigma) := \sigma(\hat{w}', A^*(\hat{z},t)\hat{z}'), \tag{202}_1$$

where

$$A^*(\hat{z},t)\hat{z}' := \frac{1}{\rho}\Big\{P(\hat{y}(\cdot,\sigma)\tilde{A}(\hat{y},\sigma,t)P(\hat{y}(\cdot,\sigma)\hat{z}'(x,\sigma)$$

$$+(I - P(\hat{y}(\cdot,\sigma)))\Delta_x(I - P(\hat{y}(\cdot,\sigma)))\hat{z}'(x,\sigma) - \lambda\hat{z}'(x,\sigma)\Big\}. \tag{202}_2$$

Here λ is chosen large enough so that A^* is invertible and $\hat{y}$ is determined in terms of $\hat{z}$ according to Lemma V.31. The remaining terms on the right hand side of (202) are included in the "f-term" in Theorem V.19. Using Lemma V.31, it is not hard to verify assumption (A3) and (f). We have to define a norm N and an operator S such that the remaining assumptions hold. For S we choose

$$S(\hat{z},t)(\hat{z}',\hat{w}') = (\hat{w}', A^*(\hat{z},t)\hat{z}'). \tag{203}$$

The nontrivial part of assumption (S) is the existence of S^{-1} which follows from standard elliptic theory. Assumption (A2) is obvious because A and S commute. To define N, we proceed as follows: On the space $L^2([0,1];H^1(\mathbb{R}^3) \times L^2(\mathbb{R}^3))$, define

$$\|(\hat{z}',\hat{w}')\|_{N^*(\hat{z},t)}^2 := \Big(\hat{z}', -A^*(\hat{z},t)\hat{z}'\Big)_{L^2} + \|\hat{w}'\|_{L^2}^2. \tag{204}_1$$

On $L^2([0,1];H^{k+1}(\mathbb{R}^3) \times H^k(\mathbb{R}^3))$ we define

$$\|(\hat{z}',\hat{w}')\|_{N_k^*(\hat{z},t)} := \|S^k(\hat{z},t)(\hat{z}',\hat{w}')\|_{N^*(\hat{z},t)}, \tag{204}_2$$

and finally we define the following norm on X:

$$\|(\hat{z}',\hat{w}')\|^2_{N(\hat{z},t)} := \|\frac{\partial}{\partial\sigma}(\hat{z}',\hat{w}')\|^2_{N^*_1(\hat{z},t)} + \|(\hat{z}',\hat{w}')\|^2_{N^*_1(\hat{z},t)}. \tag{205}$$

Assumption (N) is easily verified. Finally, to verify (A1), we use the Lumer-Phillips theorem. We have

$$\left((\hat{z}',\hat{w}'),A(\hat{z},t)(\hat{z}',\hat{w}'))\right)_{N(\hat{z},t)} = \left((\hat{z}',\hat{w}'),(\sigma\hat{w}',\sigma A^*(\hat{z},t)\hat{z}'))\right)_{N(\hat{z},t)}$$

$$= \left((\hat{z}',\hat{w}'),(\sigma\hat{w}',\sigma A^*(\hat{z},t)\hat{z}'))\right)_{N^*_1(\hat{z},t)}$$

$$+\left((\frac{\partial\hat{z}'}{\partial\sigma},\frac{\partial\hat{w}'}{\partial\sigma}),\frac{\partial}{\partial\sigma}(\sigma\hat{w}',\sigma A^*(\hat{z},t)\hat{z}'))\right)_{N^*_1(\hat{z},t)}. \tag{206}$$

In this expression the first term on the right hand side vanishes, and the second reduces to

$$\left((\frac{\partial\hat{z}'}{\partial\sigma},\frac{\partial\hat{w}'}{\partial\sigma}),(\hat{w}',A^*(\hat{z},t)\hat{z}'+\sigma\frac{\partial}{\partial\sigma}[A^*(\hat{z},t)]\hat{z}'))\right)_{N^*_1(\hat{z},t)}. \tag{207}$$

This term can easily be estimated in terms of $\|(\hat{z}',\hat{w}')\|^2_N$. This completes the verification of the assumptions for Theorem V.19.

Finally, we remark that it is not entirely trivial that the solution to the transformed problem (201) actually yields a solution to the original equation (187). A proof of this relies on uniqueness results. We shall not present the details of the argument but refer instead to Section 4 of Chapter III for a similar situation.

VI Steady flows of viscoelastic fluids

1. Introduction

While there are well known existence theorems for steady flows of Newtonian fluids, even at high Reynolds numbers [L1], there are very few results for viscoelastic fluids. Even the case of slow flows perturbing a state of rest, with which the discussion in this chapter is exclusively concerned, is far from fully understood. For such slow flows it is natural to introduce a small parameter ε, which measures the size of the velocity, and to expand the velocity, pressure and extra stress in powers of ε. If the flow is steady and the velocity is of order ε, then it follows from the recurrence relation (36) of Chapter I that the nth Rivlin-Ericksen tensor $\mathbf{A}_n$ is of order ε^n. Let us now assume that the history of the relative deformation gradient is sufficiently smooth so that it can be expanded in a Taylor series about the present time $\tau = t$:

$$\mathbf{C}_r(\tau, y, t) = \mathbf{1} + \sum_{i=1}^{n} \frac{(\tau - t)^i}{i!} \mathbf{A}_i(y, t) + O((\tau - t)^{n+1}). \tag{1}$$

Since $\mathbf{A}_i$ is of order ε^i, this Taylor series also leads to a series in powers of ε. If the constitutive functional satisfies appropriate smoothness conditions (see Coleman and Noll [C5]), then the expansion of $\mathbf{C}_r$ in powers of ε leads to a corresponding expansion of the stress tensor, which can be used in the equations of motion. If the expansions for $\mathbf{C}_r$ and for the constitutive functional are truncated at first order, then the stress becomes an isotropic linear functional of $\mathbf{A}_1$, which leads to the constitutive law of the Newtonian fluid. The approximate constitutive laws obtained by proceeding to higher orders are known as "fluids of grade n" or "nth order fluids".

This motivates the following procedure. On a bounded domain Ω, consider the problem

$$\rho(v \cdot \nabla)v = \operatorname{div} \mathbf{T} - \nabla p + \varepsilon f,$$

$$\operatorname{div} v = 0, \tag{2}$$

$$v|_{\partial \Omega} = \varepsilon v_0.$$

A formal expansion for the solution is obtained by the ansatz

$$v = \varepsilon v_1 + \varepsilon^2 v_2 + \dots,$$

$$p = \varepsilon p_1 + \varepsilon^2 p_2 + \dots \tag{3}$$

The expansion for v leads to a corresponding expansion for $\mathbf{T}$ by the procedure described above. A formal "solution" to the equations is then obtained by equating corresponding powers of ε. At each order, one must solve an inhomogeneous Stokes problem.

There is, in general, no proof that such formal expansions actually converge to or approximate asymptotically a solution. The mathematical problem lies in the singular perturbation introduced by the recurrence formula for the Rivlin-Ericksen tensors

$$\mathbf{A}_{n+1} = (v \cdot \nabla)\mathbf{A}_n + \mathbf{L}^T \mathbf{A}_n + \mathbf{A}_n \mathbf{L}. \tag{4}$$

The term $(v \cdot \nabla)\mathbf{A}_n$ introduces higher order derivatives at each step of the expansion. Hence the rigorous justification of such expansions is not merely a matter of applying the implicit function theorem; in fact one must generally require the existence of infinitely many spatial derivatives even to make formal sense of the expansion to arbitrary order. Convergence results can therefore at best be expected if the data of the problem are analytic. The singularly perturbed nature of the problem also raises questions about the appropriate number of boundary conditions. Equations with higher order spatial derivatives generally require more conditions at the boundary, and in fact, it is heuristically to be expected that materials with memory will require extra conditions at inflow boundaries, since the history of the flow before the fluid enters the domain Ω should influence its motion inside Ω. However, the expansion procedure outlined above leads to a well-posed problem at every order if only the velocity is prescribed on the boundary.

Niggemann [N3] has studied a class of one-dimensional problems which contain derivatives of all orders. He considers equations of the form

$$v''(x) = \varepsilon f(x) + \sum_{N=1}^{\infty} \sum_{j_1,\dots j_N = 0}^{\infty} a_{j_1 \dots j_N} v(x) v^{(j_1)}(x) \dots v^{(j_N)}(x) \tag{5}$$

with appropriate boundary conditions. His conditions on the coefficients $a_{j_1 \dots j_N}$ are such that this becomes a regular perturbation problem when posed in an appropriate space of analytic functions. The basic idea is that, although the derivative operator is unbounded, certain infinite sums of products of derivatives may in fact turn out to be bounded multilinear operators. Unfortunately, it does not appear that analogues of Niggemann's assumptions apply to problems in viscoelasticity. As explained above, the higher derivatives result from terms involving $(v \cdot \nabla)$. This means that each time a higher derivative enters, it comes with a factor v, and the terms involving a given power of v contain only derivatives up to a given order. If the above problem is to model viscoelastic fluids, the inner sum should therefore involve finitely many terms for each given value of N, but this case is excluded by Niggemann's assumptions.

Another possible approach to studying the existence slow steady flows of viscoelastic fluids might be through the use of Nash-Moser type implicit function theorems (for a review on such theorems, see [H2]). Such an approach requires the solvability not only of the linearized problem at the rest state, but of the linearization about arbitrary velocity fields which are small in an appropriate sense; such a condition is clearly not easy to verify, and even if it can be verified it may well be possible to proceed more directly using a contraction argument based on some iterative scheme. For fluids with differential constitutive equations such an approach was developed in [R11]. These results will be discussed in the next section.

2. Slow steady flows of a Maxwell fluid

In this section, we study slow steady flows of an upper convected Maxwell fluid in a bounded domain Ω with smooth boundary. The equations are

$$\rho(v \cdot \nabla)v - \operatorname{div} \mathbf{T} + \nabla p - f = 0,$$

$$\operatorname{div} v = 0, \tag{6}$$

$$v|_{\partial\Omega} = v_0.$$

We assume that the body force f and the velocity v_0 on the boundary are suitably small; moreover, we assume that v_0 is tangent to the boundary of Ω (see the next section for the discussion of inflow boundaries). The extra stress $\mathbf{T}$ is given by the constitutive law of an upper convected Maxwell fluid (see (59) in Chapter I). In the case of steady flow this means

$$(v \cdot \nabla)\mathbf{T} - (\nabla v)\mathbf{T} - \mathbf{T}(\nabla v)^T + \lambda\mathbf{T} = \eta\lambda(\nabla v + (\nabla v)^T). \tag{7}$$

The specific form of the terms $(\nabla v)\mathbf{T}$ and $\mathbf{T}(\nabla v)^T$ is not important for our discussion, and they could be replaced by other nonlinear combinations of ∇v and $\mathbf{T}$. The methods employed below can therefore be carried over to other differential models.

We shall show that for sufficiently small data f and v_0 a solution can be constructed by using an iteration scheme which alternates between solving a "Stokes-like" problem and a hyperbolic equation whose characteristics are streamlines. Similar iterations are used in numerical calculations [C12]. To introduce this iteration scheme, we apply the divergence operator to the constitutive equation (7) and obtain

$$(v \cdot \nabla)\operatorname{div} \mathbf{T} - (\nabla v)\operatorname{div} \mathbf{T} + \lambda\operatorname{div} \mathbf{T} = \mathbf{T} : \partial^2 v + \eta\lambda\Delta v. \tag{8}$$

Here we have set $\mathbf{T} : \partial^2 = \sum_{j,k} T^{jk} \frac{\partial^2}{\partial y^j \partial y^k}$. If we substitute div $\mathbf{T}$ from the equation of motion, we find

$$\nabla[(v \cdot \nabla)p + \lambda p] - [\nabla v + (\nabla v)^T]\nabla p - (v \cdot \nabla)f + (\nabla v)f - \lambda f$$

$$= \mathbf{T} : \partial^2 v + \eta\lambda\Delta v - \rho(v \cdot \nabla)(v \cdot \nabla)v + \rho(\nabla v)(v \cdot \nabla)v - \lambda\rho(v \cdot \nabla)v, \qquad (9)$$

and we introduce a new quantity $q := (v \cdot \nabla)p + \lambda p$. Since we are interested in small solutions, we can use 0 as a starting value for the iteration. The scheme is

$$v^0 = 0, \; p^0 = q^0 = 0, \; \mathbf{T}^0 = 0, \qquad (10)_1$$

$$\mathbf{T}^n : \partial^2 v^{n+1} + \eta\lambda\Delta v^{n+1} - \rho(v^n \cdot \nabla)(v^n \cdot \nabla)v^{n+1} - \nabla q^{n+1}$$

$$= -[(\nabla v^n) + (\nabla v^n)^T]\nabla p^n - (v^n \cdot \nabla)f + (\nabla v^n)f$$

$$-\lambda f - \rho(\nabla v^n)(v^n \cdot \nabla)v^n + \lambda\rho(v^n \cdot \nabla)v^n,$$

$$\text{div } v^{n+1} = 0, \; v^{n+1} = v_0 \text{ on } \partial\Omega, \int\int\int_\Omega q^{n+1} = 0, \qquad (10)_2$$

$$(v^{n+1} \cdot \nabla)p^{n+1} + \lambda p^{n+1} = q^{n+1}, \qquad (10)_3$$

$$(v^{n+1} \cdot \nabla)\mathbf{T}^{n+1} - (\nabla v^{n+1})\mathbf{T}^{n+1} - \mathbf{T}^{n+1}(\nabla v^{n+1})^T + \lambda\mathbf{T}^{n+1}$$

$$= \eta\lambda[(\nabla v^{n+1}) + (\nabla v^{n+1})^T]. \qquad (10)_4$$

The essential point in setting up this iteration is that the singularly perturbed operator $(v \cdot \nabla) + \lambda$ is inverted rather than evaluated during each step of the iteration, one thus avoids the repeated application of unbounded operators and associated loss of differentiability. We can state the following theorem.

Theorem VI.1:

Assume that $\|f\|_{H^2(\Omega)}$ and $\|v_0\|_{H^{5/2}(\partial\Omega)}$ are sufficiently small and that v_0 is tangent to $\partial\Omega$. Then the iteration (10) converges to a solution of (6), (7). This solution satisfies $v \in H^3(\Omega)$, $\mathbf{T}$, $p \in H^2(\Omega)$, and it is the only solution for which these norms are small. If higher regularity of the data is assumed, the solution also has higher regularity, i.e. if s is any integer greater than 2 and $f \in H^s(\Omega)$, $v_0 \in H^{s+1/2}(\partial\Omega)$, then $v \in H^{s+1}(\Omega)$, $\mathbf{T}$, $p \in H^s(\Omega)$.

Proof:

We shall proceed in two steps. First we show that all iterates remain small in the norms indicated in the theorem, and then we shall prove convergence in a norm with the Sobolev index decreased by one (this procedure is similar to that used to obtain existence results for initial value problems for hyperbolic systems, see Chapter III). The step $(10)_2$ of the iteration is simply a perturbation of the Stokes problem, and we find that, as long as $\|\mathbf{T}^n\|_2$ and $\|v^n\|_3$ are sufficiently small, an estimate of the form

$$\|v^{n+1}\|_3 + \|q^{n+1}\|_2 \leq C_1\{\|p^n\|_2\|v^n\|_3 + \|v^n\|_2\|f\|_2$$

247

$$+\|f\|_1 + \|v_0\|_{5/2} + \|v^n\|_3^2\|v^n\|_1 + \|v^n\|_2^2\} \tag{11}$$

holds with some constant C_1. The equation $(10)_3$ is solved by integration along the characteristic streamlines: Let $x(y,\tau)$ be defined as the solution of the equation

$$\frac{d}{d\tau}x(y,\tau) = -v^{n+1}(x(y,\tau)), \quad x(y,0) = y. \tag{12}$$

Then the solution to $(10)_3$ is given by

$$p^{n+1}(y) = \int_0^\infty e^{-\lambda\tau}q^{n+1}(x(y,\tau))\,d\tau. \tag{13}$$

If v is Lipschitz, then (12) is uniquely solvable, and it is clear that continuity of q^{n+1} implies continuity of p^{n+1}. Higher regularity of q^{n+1} implies the same degree of regularity for p^{n+1} provided derivatives of $x(y,\tau)$ with respect to y do not increase too rapidly with τ. This is guaranteed if ∇v^{n+1} is small in the supremum norm. The operator $(v^{n+1}\cdot\nabla)$ is skew-adjoint in $L^2(\Omega)$ and we immediately obtain the estimate

$$\|p^{n+1}\|_0 \leq \frac{1}{\lambda}\|q^{n+1}\|_0. \tag{14}$$

By differentiating equation $(10)_3$, we can obtain estimates for higher derivatives, and we find that, as long as $\|v^{n+1}\|_3$ is small, there is a constant C_2 such that

$$\|p^{n+1}\|_2 \leq C_2\|q^{n+1}\|_2. \tag{15}$$

Since we can regard $(10)_4$ as a perturbation of $(10)_3$, we also obtain

$$\|\mathbf{T}^{n+1}\|_2 \leq C_3\|v^{n+1}\|_3. \tag{16}$$

By combining (11), (15) and (16), it is easy to see that $\|v^n\|_3$, $\|q^n\|_2$, $\|p^n\|_2$ and $\|\mathbf{T}^n\|_2$ remain uniformly small if $\|f\|_2$ and $\|v_0\|_{5/2}$ are small enough.

To show convergence of the iteration, we simply have to subtract the equations at stage n and $n-1$, and then use the estimates already obtained. For example, if we subtract $(10)_3$ at stage $n-1$ from $(10)_3$ at stage n, we find

$$(v^n\cdot\nabla)p^n - (v^{n-1}\cdot\nabla)p^{n-1} + \lambda(p^n - p^{n-1}) = q^n - q^{n-1}, \tag{17}$$

which we rewrite in the form

$$((v^{n-1}\cdot\nabla)+\lambda)(p^n - p^{n-1}) = q^n - q^{n-1} - (v^n - v^{n-1})\cdot\nabla p^n. \tag{18}$$

From (18) and the bounds already established for $\|p^n\|_2$, $\|v^n\|_3$, we easily find an estimate of the form

$$\|p^n - p^{n-1}\|_1 \leq C_4(\|q^n - q^{n-1}\|_1 + \|v^n - v^{n-1}\|_2). \tag{19}$$

248

A similar estimate for $\|\mathbf{T}^n - \mathbf{T}^{n-1}\|_1$ follows from $(10)_4$, and by using it in $(10)_2$, we find the required contraction estimate. Thus we find that v^n converges in H^2 and p^n, q^n, $\mathbf{T}^n$ converge in H^1. Even though the convergence is in the weaker norms, the solution satisfies the higher regularity claimed in the theorem because we have uniform bounds for the higher norms of the iterates. The uniqueness claimed in the theorem also follows from the contraction estimate.

Higher regularity of the solution follows from similar estimates for the derivatives of the iterates.

If f and v_0 are proportional to a small parameter ε and have sufficient regularity, then it is possible, by using similar procedures as above, to obtain estimates for difference quotients of the solution with respect to ε. In this way, formal expansions as described in Section 1 can be justified in an asymptotic sense.

3. Problems with inflow boundaries

At inflow boundaries, additional boundary conditions will be required because of the history dependence of the constitutive equation. This leads to two new problems. The required additional boundary conditions have to be characterized, and problems of compatibility arise at those points where inflow and outflow boundaries join. As a model problem which has an inflow boundary but avoids the latter complication, we consider flow through a domain bounded by two infinite parallel planes, with periodic boundary conditions imposed in the unbounded directions. Instead of looking for flows perturbing a state of rest, we now look for flows perturbing a uniform flow with velocity V transverse to the boundaries. The results discussed in this section are based on [R13].

We consider transverse flow through the strip $0 \le x \le 1$ with periodic boundary conditions imposed in the y- and z-directions. The periods are denoted by L_y and L_z. For $f = 0$, a solution of (6) and (7) is given by $v = (V, 0, 0)$, $p = 0$, $\mathbf{T} = 0$, where V is a positive constant. We are looking for flows which are small perturbations of such a flow. We prescribe a small body force f and boundary conditions for the velocity which are small perturbations of the constant velocity: $v(0, y, z) = (V, 0, 0) + v_1(y, z)$, $v(1, y, z) = (V, 0, 0) + v_2(y, z)$. These boundary conditions must be consistent with incompressibility,

$$\int_0^{L_z} \int_0^{L_y} e_x \cdot v_1(y, z) \, dy \, dz = \int_0^{L_z} \int_0^{L_y} e_x \cdot v_2(y, z) \, dy \, dz, \qquad (20)$$

where e_x is the unit vector pointing in the x-direction. In addition, we shall need conditions on the stresses at the inflow boundary $x = 0$. The nature of these conditions will now be discussed.

249

As in Section 2, we base our analysis on (7) and (9). The iteration scheme (10) requires a minor modification because we are perturbing uniform flow rather than a state of rest. The modified scheme is as follows:

$$v^0 = (V,0,0), \ p^0 = q^0 = 0, \ \mathbf{T}^0 = \mathbf{0}, \tag{21}_1$$

$$\mathbf{T}^n : \partial^2 v^{n+1} + \eta\lambda\Delta v^{n+1} - \rho(v^n \cdot \nabla)(v^n \cdot \nabla)v^{n+1} - \lambda\rho(v^n \cdot \nabla)v^{n+1} - \nabla q^{n+1}$$

$$= -[(\nabla v^n) + (\nabla v^n)^T]\nabla p^n - (v^n \cdot \nabla)f + (\nabla v^n)f - \lambda f - \rho(\nabla v^n)(v^n \cdot \nabla)v^n,$$

$$\text{div } v^{n+1} = 0, \ v^{n+1} = (V,0,0) + v_1 \text{ on } x = 0, \ v^{n+1} = (V,0,0) + v_2 \text{ on } x = 1,$$

$$\int_0^1 \int_0^{L_y} \int_0^{L_z} q^{n+1} \ dz \ dy \ dx = 0, \tag{21}_2$$

$$(v^{n+1} \cdot \nabla)p^{n+1} + \lambda p^{n+1} = q^{n+1}, \tag{21}_3$$

$$(v^{n+1} \cdot \nabla)\mathbf{T}^{n+1} - (\nabla v^{n+1})\mathbf{T}^{n+1} - \mathbf{T}^{n+1}(\nabla v^{n+1})^T + \lambda\mathbf{T}^{n+1}$$

$$= \eta\lambda[(\nabla v^{n+1}) + (\nabla v^{n+1})^T]. \tag{21}_4$$

As long as $V^2 < \lambda\eta/\rho$, and v^n and $\mathbf{T}^n$ are small perturbations of $(V,0,0)$ and $\mathbf{0}$, equation $(21)_2$ is an elliptic system for v^{n+1} and q^{n+1}. If $V^2 > \lambda\eta/\rho$, a change of type occurs (cf. Section II.2), and this case will not be considered here. Equations $(21)_3$ and $(21)_4$ are hyperbolic equations with the streamlines as characteristics. If there are no inflow or outflow boundaries, as in Section 2 above, characteristics do not cross the boundary, and $(21)_3$ and $(21)_4$ can be solved without any boundary conditions imposed. In the present situation, however, we have to prescribe p and $\mathbf{T}$ at the inflow boundary $x = 0$ in order to solve $(21)_3$ and $(21)_4$.

It can be shown along lines similar to Section 2 that with such inflow boundary conditions the iteration will converge, provided that the data are sufficiently small. However, it will in general not converge to a solution of the original problem. The limit obtained from the iteration will satisfy (7) and (9), but the original problem is (6) and (7). Let us recall that (9) was obtained by substituting $\rho(v \cdot \nabla)v + \nabla p - f = \text{div } \mathbf{T}$ into (8), and (8) was obtained by differentiating (7). Hence the fact that (9) is satisfied means that if we set

$$h := \rho(v \cdot \nabla)v + \nabla p - f - \text{div } \mathbf{T}, \tag{22}$$

then

$$(v \cdot \nabla)h - (\nabla v)h + \lambda h = 0. \tag{23}$$

In order to have a solution of the original problem, we need to know that $h = 0$. If streamlines do not cross the boundary (as in Section 2), then (23) does in fact guarantee this. In the present case (23) implies that $h = 0$ provided that $h = 0$ at the inflow boundary. This condition has to be viewed as a constraint on the possible inflow data for p and $\mathbf{T}$, and the iteration (21) has to be modified in order to accommodate this constraint. In three dimensions $\mathbf{T}$ has six components, p has one and h has three, thus suggesting that four inflow boundary conditions can be prescribed. In two dimensions, $\mathbf{T}$ has three components, p has one and h

as two, suggesting that one needs two inflow boundary conditions. This is also suggested by the analysis of characteristics (see Section II.2). It contradicts the belief of many rheologists that it is possible to prescribe all components of the extra stress $\mathbf{T}$ at an inflow boundary.

We want to restrict the choice of inflow boundary conditions in $(21)_3$, $(21)_4$ by the constraint

$$\operatorname{div} \mathbf{T}^{n+1} - \nabla p^{n+1} = \rho(v^{n+1} \cdot \nabla)v^{n+1} - f \quad \text{at } x = 0. \tag{24}$$

On the left hand side of (24), the x-derivatives of p^{n+1} and $\mathbf{T}^{n+1}$ can be expressed using $(21)_3$ and $(21)_4$, e.g.

$$\frac{\partial p^{n+1}}{\partial x} = -\frac{\lambda}{V} p^{n+1} + \frac{q^{n+1}}{V} + \text{nonlinear terms}, \tag{25}$$

and (24) assumes the form

$$\frac{\lambda}{V}(p^{n+1} - T_{11}^{n+1}) + T_{12,y}^{n+1} + T_{13,z}^{n+1} = \dots$$

$$-\frac{\lambda}{V}T_{12}^{n+1} + (T_{22}^{n+1} - p^{n+1})_y + T_{23,z}^{n+1} = \dots \tag{26}$$

$$-\frac{\lambda}{V}T_{13}^{n+1} + T_{23,y}^{n+1} + (T_{33}^{n+1} - p^{n+1})_z = \dots$$

The right hand sides indicated by dots contain terms involving f, v^{n+1}, q^{n+1} and nonlinear terms which also involve $\mathbf{T}^{n+1}$ and p^{n+1}.

We have to use equation (26) in order to express some stress components at $x = 0$ in terms of others. To obtain a convergent iteration, we want to do this in such a way that no loss of regularity occurs when solving for the undetermined stress components. Unfortunately, if we are to avoid a loss of regularity, it is not possible to prescribe some of the stress components at inflow and leave the others to be determined. Rather, we have to prescribe stress components "partially"; more precisely, let each component of the stress be expanded in a Fourier series in y and z, e.g.

$$T_{11}(0, y, z) = \sum_{k,l} t_{11}^{kl} e^{2\pi i(ky/L_y + lz/L_z)}. \tag{27}$$

We can then prescribe the Fourier components of T_{11}, T_{22}, T_{13} and T_{33} for $|k| \geq |l|$, $k \neq 0$ and solve for those of p, T_{12} and T_{23}. For $|l| > |k|$, we prescribe the Fourier components of T_{11}, T_{12}, T_{22} and T_{33} and solve for those of p, T_{13} and T_{23}. For $k = l = 0$, we prescribe the Fourier components of T_{11}, T_{22}, T_{23} and T_{33} and solve for those of p, T_{12} and T_{13}. In the two-dimensional case, this amounts to prescribing T_{11} and T_{22} and solving for p and T_{12}.

The solution procedure is now as follows. We start the iteration with initial data $(21)_1$. Then at each step we compute a new v and q using $(21)_2$. Then we compute a new p and $\mathbf{T}$ from $(21)_3$ and $(21)_4$ with inflow boundary conditions

which are in part prescribed and in part computed from (26) in the manne
outlined above. We show now that such an iteration converges under appropriat
smallness conditions for the body force f, the velocity boundary data v_1 and v
and the prescribed part of the stress boundary data.

As usual, we denote by H^s the space of all functions on the strip $0 \le x \le 1$
which are periodic with periods L_y and L_z in the y- and z-directions and have
derivatives which are square integrable over one period. Sobolev spaces of periodi
functions living on one of the boundaries $x = 0$ or $x = 1$ are denoted by $H^{(s)}$
The corresponding norms are denoted by $\| \cdot \|_s$ and $\| \cdot \|_{(s)}$.

In the following, let s be any integer ≥ 1. We assume that the body forc
and the velocity boundary data satisfy the bound

$$\|f\|_{s+1} \le \gamma, \quad \|v_1\|_{(s+3/2)} \le \gamma, \quad \|v_2\|_{(s+3/2)} \le \gamma, \tag{28}$$

where γ is a positive number which will later be chosen small. The stress at th
inflow boundary consists of a prescribed part $\mathbf{T}_p$ and an unknown part $\mathbf{T}_u$, whic
must be determined from (26) at each stage of the iteration. We assume that

$$\|\mathbf{T}_p\|_{(s+1)} \le \gamma. \tag{29}$$

As in Section 2 above, the convergence proof consists of two parts: First w
show that all iterates remain bounded in a certain norm, and then we use thi
fact to prove convergence in a weaker norm. In order to carry out the first par
let us assume that

$$\|v^n - (V,0,0)\|_{s+2} \le \varepsilon, \quad \|p^n\|_{s+1} \le \varepsilon, \quad \|\mathbf{T}^n\|_{s+1} \le \varepsilon, \tag{30}$$

where ε is small. As long as $\rho V^2 < \eta \lambda$ and ε is sufficiently small, $(21)_2$ is a.
elliptic system for v^{n+1} and q^{n+1}, and a standard argument shows that there i
a constant C_1 such that

$$\|v^{n+1} - (V,0,0)\|_{s+2} + \|q^{n+1}\|_{s+1} \le C_1 (\gamma + \varepsilon^2) =: \delta. \tag{31}$$

In the next step, $\mathbf{T}_u^{n+1}$ and p^{n+1} on the inflow boundary are determined fror
(26). Using the trace theorem, we see that q^{n+1} and ∇v^{n+1} are in $H^{(s+1/2)}$. B
using this in (26), it is easy to show that for small δ there is a constant C_2 suc
that

$$\|\mathbf{T}_u^{n+1}\|_{(s+1)} + \|p^{n+1}\|_{(s+1)} \le C_2 (\gamma + \delta) =: \sigma. \tag{32}$$

We now turn to the hyperbolic equations $(21)_3$ and $(21)_4$. As in Section 2, th
solution to these equations can be obtained by integrating along streamlines. T
obtain estimates for the solution, let us multiply equation $(21)_3$ by p and integrat
over the domain:

$$\int_0^1 \int_0^{L_y} \int_0^{L_z} p^{n+1}(v^{n+1} \cdot \nabla)p^{n+1} + \lambda(p^{n+1})^2 \; dz \; dy \; dx$$

252

$$= \int_0^1 \int_0^{L_y} \int_0^{L_z} p^{n+1} q^{n+1} \ dz \ dy \ dx. \tag{33}$$

The left hand side is equal to

$$\frac{1}{2} \int_0^{L_y} \int_0^{L_z} e_x \cdot v^{n+1}(1, y, z)(p^{n+1}(1, y, z))^2 \ dz \ dy$$

$$-\frac{1}{2} \int_0^{L_y} \int_0^{L_z} e_x \cdot v^{n+1}(0, y, z)(p^{n+1}(0, y, z))^2 \ dz \ dy$$

$$+\lambda \int_0^1 \int_0^{L_y} \int_0^{L_z} (p^{n+1})^2 \ dz \ dy \ dx. \tag{34}$$

The first term in (34) is positive, and hence we obtain

$$\lambda \int_0^1 \int_0^{L_y} \int_0^{L_z} (p^{n+1})^2 \ dz \ dy \ dx$$

$$\leq \int_0^1 \int_0^{L_y} \int_0^{L_z} p^{n+1} q^{n+1} \ dz \ dy \ dx$$

$$+\frac{1}{2} \int_0^{L_y} \int_0^{L_z} e_x \cdot v^{n+1}(0, y, z)(p^{n+1}(0, y, z))^2 \ dz \ dy. \tag{35}$$

That is, we have an estimate of the L^2-norm for p in terms of the L^2-norm of q and the L^2-norm of the inflow boundary data. By differentiating $(21)_3$ with respect to y or z, we can obtain estimates for derivatives of p in these directions; estimates for x-derivatives can be obtained from the equation itself. The same argument can be applied to $(21)_4$. If δ is sufficiently small, we obtain estimates of the form

$$\|p^{n+1}\|_{s+1} + \|\mathbf{T}^{n+1}\|_{s+1} \leq C_3(\sigma + \gamma + \delta). \tag{36}$$

If we now choose ε small enough and γ sufficiently small relative to ε, we will have $\delta \leq \varepsilon$ and $C_3(\sigma + \gamma + \delta) \leq \varepsilon$. This implies that (30) holds with n replaced by $n + 1$. By induction, we see that (30) holds for all values of n.

The fact that all the iterates are bounded and in fact small can now be used to show convergence of the iteration in a weaker norm. More specifically, we can show that u^n converges in H^{s+1} and that p^n, q^n, $\mathbf{T}^n$ converge in H^s. The argument is exactly as for the case without inflow boundaries discussed above, and we omit the details.

4. The case of several relaxation modes

The method discussed in Section 2 can be modified for the case where **T** is the sum of a finite number of terms, each of which satisfies an equation of the form (7). We may allow one of the relaxation times to be zero, leading to a Newtonian contribution to the stress. We consider the constitutive law

$$\mathbf{T} = 2\eta_0 \mathbf{D} + \sum_{k=1}^{N} \mathbf{T}_k, \tag{37}$$

where

$$(v \cdot \nabla)\mathbf{T}_k - (\nabla v)\mathbf{T}_k - \mathbf{T}_k(\nabla v)^T + \lambda_k \mathbf{T}_k = 2\eta_k \lambda_k \mathbf{D}. \tag{38}$$

Here $\eta_0 \geq 0$, $\eta_k, \lambda_k > 0$. We consider again the situation of Section 2, i.e. small perturbations of the rest state with a tangential velocity prescribed on the boundary. The basic ideas are as before, and we omit details; what is different now is how the "Stokes-like" part of the iteration is constructed. As before, we take the divergence of (38), leading to

$$(v \cdot \nabla)\text{div } \mathbf{T}_k - (\nabla v)\text{div } \mathbf{T}_k + \lambda_k \text{div } \mathbf{T}_k = \eta_k \lambda_k \Delta v + \mathbf{T}_k : \partial^2 v, \tag{39}$$

which we rewrite as

$$\text{div } \mathbf{T}_k = ((v \cdot \nabla) + \lambda_k)^{-1} [\eta_k \lambda_k \Delta v + \mathbf{T}_k : \partial^2 v + \nabla v \text{ div } \mathbf{T}_k]. \tag{40}$$

Inserting (40) into the equation of motion, we obtain

$$\rho(v \cdot \nabla)v = \eta_0 \Delta v + \sum_k ((v \cdot \nabla) + \lambda_k)^{-1} \cdot$$

$$[\eta_k \lambda_k \Delta v + \mathbf{T}_k : \partial^2 v + \nabla v \text{ div } \mathbf{T}_k] - \nabla p + f. \tag{41}$$

In (41), Δv is multiplied by the linear operator

$$\mathcal{L}[v] := \eta_0 + \sum_k \eta_k \lambda_k ((v \cdot \nabla) + \lambda_k)^{-1}. \tag{42}$$

Clearly $\mathcal{L}[v]$ is a bounded operator in $L^2(\Omega)$. If $\eta_0 \neq 0$, then $\mathcal{L}[v]$ also has a bounded inverse. This follows easily from the Lax-Milgram lemma and $(\mathcal{L}[v]u, u) \geq \eta_0(u, u)$. If v has sufficient regularity and the L^∞-norm of ∇v is small, then $\mathcal{L}[v]$ also maps higher Sobolev spaces bijectively onto themselves.

After multiplying (41) by $\mathcal{L}[v]^{-1}$, we get the term Δv on the right hand side. We also obtain a term $\mathcal{L}[v]^{-1}\nabla p$, which is further manipulated as follows

$$\mathcal{L}[v]^{-1}\nabla p = \nabla(\mathcal{L}[v]^{-1}p)$$

$$-\mathcal{L}[v]^{-1} \sum_k \eta_k \lambda_k ((v \cdot \nabla) + \lambda_k)^{-1} \{(\nabla v)^T \nabla[((v \cdot \nabla) + \lambda_k)^{-1} \mathcal{L}[v]^{-1} p]\}. \quad (43)$$

We now set $q = \mathcal{L}[v]^{-1} p$ and construct the Stokes part of the iteration as follows

$$\Delta v^{n+1} - \nabla q^{n+1} = \rho \mathcal{L}[v^n]^{-1} (v^n \cdot \nabla) v^n$$

$$-\mathcal{L}[v^n]^{-1} \sum_k ((v^n \cdot \nabla) + \lambda_k)^{-1} [\mathbf{T}_k^n : \partial^2 v^n + \nabla v^n \ \mathrm{div} \ \mathbf{T}_k^n] - \mathcal{L}[v^n]^{-1} f \quad (44)$$

$$-\mathcal{L}[v^n]^{-1} \sum_k \eta_k \lambda_k ((v^n \cdot \nabla) + \lambda_k)^{-1} \{(\nabla v^n)^T \nabla[((v^n \cdot \nabla) + \lambda_k)^{-1} q^n]\}.$$

If η_0 vanishes, this approach fails because $\mathcal{L}[v]$ is not invertible. In this case, we proceed as follows. We choose any positive number $\tilde{\lambda}$. Then we apply the operator $(v \cdot \nabla) + \tilde{\lambda}$ to equation (41), and obtain

$$\rho((v \cdot \nabla) + \tilde{\lambda})(v \cdot \nabla) v = ((v \cdot \nabla) + \tilde{\lambda})\{\sum_k ((v \cdot \nabla) + \lambda_k)^{-1} \cdot$$

$$[\eta_k \lambda_k \Delta v + \mathbf{T}_k : \partial^2 v + \nabla v \ \mathrm{div} \ \mathbf{T}_k] - \nabla p + f\}. \quad (45)$$

We can now set

$$\mathcal{L}[v] = ((v \cdot \nabla) + \tilde{\lambda}) \sum_k \eta_k \lambda_k ((v \cdot \nabla) + \lambda_k)^{-1}, \quad q = \mathcal{L}[v]^{-1}((v \cdot \nabla) + \tilde{\lambda}) p. \quad (46)$$

The operator $\mathcal{L}[v]$ is bounded, and we can again use the Lax-Milgram lemma to show it is invertible. With these new definitions of $\mathcal{L}[v]$ and q we can proceed as above.

References

Numbers in curly brackets following a reference indicate the pages where it is referred to.

[A1] M. Abramowitz and I.A. Stegun, *Handbook of Mathematical Functions*, Dover 1965 {48}

[A2] J.D. Achenbach and D.P. Reddy, Note on wave propagation in linearly viscoelastic media, *Z. Angew. Math. Phys.* **18** (1967), 141-144 {38}

[A3] S. Agmon, A. Douglis and L. Nirenberg, Estimates near the boundary for solutions of elliptic partial differential equations satisfying general boundary conditions, *Comm. Pure Appl. Math.* **12** (1959), 623-727 and **17** (1964), 35-92 {109, 110}

[A4] G. Andrews, On the existence of solutions to the equation $u_{tt} = u_{xxt} + \sigma(u_x)_x$, *J. Diff. Eq.* **35** (1980), 200-231 {134}

[A5] G. Andrews and J.M. Ball, Asymptotic behaviour and changes of phase in one-dimensional nonlinear viscoelasticity, *J. Diff. Eq.* **44** (1982), 306-341 {134}

[A6] S.S. Antman, The theory of rods, in: S. Flügge (ed.), *Handbuch der Physik VIa/2*, Springer 1972, 641-703 {26}

[A7] S.S. Antman and R. Malek-Madani, Dissipative mechanisms, in: S.S. Antman, J.L. Ericksen, D. Kinderlehrer and I. Müller (eds.), *Metastability and Incompletely Posed Problems*, Springer Lecture Notes in Mathematics 1987 {134}

[A8] S.S. Antman and R. Malek-Madani, Existence and uniqueness of solutions to an initial-boundary value problem of nonlinear viscoelasticity, preprint {134}

[A9] S.S. Antman, Material constraints in continuum mechanics, *Atti Acc. Naz. Lincei, Rendiconti, Cl. Sci. fis. mat. nat., Ser. VIII,* **70** (1981), 256-264 {18}

[B1] I. Babuška and I. Hlaváček, On the existence and uniqueness of solution in the theory of viscoelasticity, *Arch. Mech. Stos.* **18** (1966), 47-83 {82}

[B2] J.M. Ball, Global invertibility of Sobolev functions and the interpenetration of matter, *Proc. Roy. Soc. Edinburgh* **88A** (1981), 315-328 {12, 104}

[B3] Bateman Project, *Tables of Integral Transforms*, Vol. 1, McGraw-Hill 1954 {48}

[B4] A.N. Beris, R.C. Armstrong and R.A. Brown, Finite element calculation of viscoelastic flow in a journal bearing: II. Moderate eccentricity, *J. Non-Newtonian Fluid Mech.* **19** (1986), 323-347 {38}

[B5] B. Bernstein, Hypo-elasticity and elasticity, *Arch. Rat. Mech. Anal.* **6** (1960), 89-104 {75}

[B6] B. Bernstein, E.A. Kearsley and L.J. Zapas, A study of stress relaxation with finite strain, *Trans. Soc. Rheology* **7** (1963), 391-410 {22}

[B7] D.S. Berry, A note on stress pulses in viscoelastic rods, *Phil. Mag., Ser. 8,* **3** (1958), 100-102 {38}

[B8] R.B. Bird, R.C. Armstrong and O. Hassager, *Dynamics of Polymeric Liquids, Vol. 1: Fluid Mechanics*, John Wiley 1977 (new edition in press) {11,21}

[B9] R.B. Bird, O. Hassager, R.C. Armstrong and C.F. Curtiss, *Dynamics of Polymeric Liquids, Vol. 2: Kinetic Theory*, John Wiley 1977 (new edition in press) {21}

[B10] R.B. Bird, Kinetic theory and constitutive equations for polymeric liquids, *J. Rheology* **26** (1982), 277-299 {21}

[B11] L. Boltzmann, Zur Theorie der elastischen Nachwirkung, *Ann. Phys. Chem.* **7** (1876), Ergänzungsband, 624-654 [=*Sitzungsber. Akad. Wiss. Wien* **70** (1874), 275-306] {3,4,5}

[B12] H. Brézis, *Opérateurs Maximaux Monotones*, North Holland 1973 {192}

[C1] C. Chen and W. von Wahl, Das Rand-Anfangswertproblem für quasilineare Wellengleichungen in Sobolevräumen niedriger Ordnung, *J. Reine Angew. Math.* **337** (1982), 77-112 {84,85}

[C2] P.J. Chen, Growth and decay of waves in solids, in: S. Flügge (ed.), *Handbuch der Physik VIa/3*, Springer 1973, 303-402 {61}

[C3] R.M. Christensen, *Theory of Viscoelasticity*, Academic Press 1971 {11,38}

[C4] B.T. Chu, Stress waves in isotropic linear viscoelastic materials, *J. Mécanique* **1** (1962), 439-446 {38,51}

[C5] B.D. Coleman and W. Noll, An approximation theorem for functionals with applications in continuum mechanics, *Arch. Rat. Mech. Anal.* **6** (1960), 355-370 {20,22,244}

[C6] B.D. Coleman, H. Markovitz and W. Noll, *Viscometric Flows of Non-Newtonian Fluids*, Springer 1966 {11}

[C7] B.D. Coleman, M.E. Gurtin and I.R. Herrera, Waves in materials with memory I, *Arch. Rat. Mech. Anal.* **19** (1965), 1-19 {60}

[C8] B.D. Coleman and M.E. Gurtin, Waves in materials with memory II, *Arch. Rat. Mech. Anal.* **19** (1965), 239-265 {7,60}

[C9] B.D. Coleman, R.J. Duffin and V.J. Mizel, Instability, uniqueness and nonexistence theorems for the equation $u_t = u_{xx} - u_{xtx}$ on a strip, *Arch. Rat. Mech. Anal.* **19** (1965), 100-116 {22}

[C10] B.D. Coleman and V.J. Mizel, Norms and semi-groups in the theory of fading memory, *Arch. Rat. Mech. Anal.* **23** (1967), 87-123 {20}

[C11] B.D. Coleman and V.J. Mizel, On the general theory of fading memory, *Arch. Rat. Mech. Anal.* **29** (1968), 18-31 {20}

[C12] M.J. Crochet, A.R. Davies and K. Walters, *Numerical Simulation of Non-Newtonian Flow*, Elsevier 1984 {246}

[C13] C.F. Curtiss and R.B. Bird, A kinetic theory for polymer melts, *J. Chem. Phys.* **74** (1981), 2016-2033 {23,217,229}

[D1] C.M. Dafermos, The mixed initial-boundary value problem for the equations of one-dimensional nonlinear viscoelasticity, *J. Diff. Eq.* **6** (1969), 71-86 {134}

[D2] C.M. Dafermos, An abstract Volterra equation with applications to linear viscoelasticity, *J. Diff. Eq.* **7** (1970), 554-569 {89}

[D3] C.M. Dafermos and J.A. Nohel, Energy methods for nonlinear hyperbolic Volterra integrodifferential equations, *Comm. PDE* **4** (1979), 219-278 {151}

[D4] C.M. Dafermos and J.A. Nohel, A nonlinear hyperbolic Volterra equation in viscoelasticity, *Amer. J. Math.* Supplement (1981), 87-116 {130,152,154, 156}

[D5] C.M. Dafermos and W.J. Hrusa, Energy methods for quasilinear hyperbolic initial-boundary value problems. Applications to elastodynamics, *Arch. Rat. Mech. Anal.* **87** (1985), 267-292 {84,85,87,115}

[D6] C.M. Dafermos, Dissipation in materials with memory, in: A.S. Lodge, M. Renardy and J.A. Nohel (eds.), *Viscoelasticity and Rheology*, Academic Press 1985, 221-234 {62}

[D7] C.M. Dafermos, Development of singularities in the motion of materials with fading memory, *Arch. Rat. Mech. Anal.* **91** (1986), 193-205 {62,69}

[D8] W. Desch, R. Grimmer and M. Zeman, Wave propagation for abstract integrodifferential equations, in: F. Kappel and W. Schappacher (eds.), *Infinite-Dimensional Systems*, Springer Lecture Notes in Mathematics 1076 (1984), 62-70 {99}

[D9] W. Desch and R. Grimmer, Smoothing properties of linear Volterra integrodifferential equations, preprint {49}

[D10] G. Doetsch, *Introduction to the Theory and Application of the Laplace Transformation*, Springer 1974 {44}

[D11] M. Doi and S.F. Edwards, Dynamics of concentrated polymer systems, *J. Chem. Soc. Faraday* **74** (1978), 1789-1832 and **75** (1979), 38-54 {6}

[D12] N. Dunford and J.T. Schwartz, *Linear Operators I*, John Wiley 1958 {198}

[D13] F. Dupret, J.M. Marchal and M.J. Crochet, On the consequence of discretization errors in the numerical calculation of viscoelastic flow, *J. Non-Newtonian Fluid Mech.* **18** (1985), 173-186 {38}

[E1] D.G. Ebin and R.A. Saxton, The initial value problem for elastodynamics of incompressible bodies, *Arch. Rat. Mech. Anal.* **94** (1986), 15-38 {104}

[E2] H. Engler, On the dynamic shear flow problem for viscoelastic liquids, *SIAM J. Math. Anal.*, to appear {134,184,185,190}

[E3] H. Engler, Weak solutions of a class of quasilinear hyperbolic integro-differential equations describing viscoelastic materials, submitted to *Arch. Rat. Mech. Anal.* {135}

[F1] X.-J. Fan and R.B. Bird, Configuration-dependent friction coefficients and elastic dumbbell rheology, *J. Non-Newtonian Fluid Mech.* **18** (1985), 255-272 {17}

[F2] G.M.C. Fisher and M.E. Gurtin, Wave propagation in the linear theory of viscoelasticity, *Q. Appl. Math.* **23** (1965), 257-263 {38}

258

[F3] A. Friedman, *Partial Differential Equations*, Holt, Rinehart and Winston 1969 {200, 201, 202}

[G1] P.G. de Gennes, Coil-stretch transition of dilute flexible polymers under ultrahigh velocity gradients, *J. Chem. Phys.* **60** (1974), 5030-5042 {17}

[G2] M. Giaquinta and G. Modica, Non-linear systems of the type of the stationary Navier-Stokes system, *J. reine angew. Math.* **330** (1982), 173-214 {109}

[G3] H. Giesekus, A unified approach to a variety of constitutive models for polymer fluids based on the concept of configuration dependent molecular mobility, *Rheol. Acta* **21** (1982), 366-375 {24}

[G4] Y. Giga, Analyticity of the semigroup generated by the Stokes operator in L_r-spaces, *Math. Z.* **178** (1981), 297-329 {236}

[G5] I.S. Gradshteyn and I.M. Ryzhik, *Table of Integrals, Series and Products*, Academic Press 1980 {45}

[G6] A.E. Green and R.S. Rivlin, The mechanics of non-linear materials with memory I, *Arch. Rat. Mech. Anal.* **1** (1957/58), 1-21 {22}

[G7] A.E. Green, R.S. Rivlin and A.J.M. Spencer, The mechanics of non-linear materials with memory II, *Arch. Rat. Mech. Anal.* **3** (1959), 82-90 {22}

[G8] A.E. Green and R.S. Rivlin, The mechanics of non-linear materials with memory III, *Arch. Rat. Mech. Anal.* **4** (1960), 387-404 {22}

[G9] J.M. Greenberg, The existence of steady shock waves in nonlinear materials with memory, *Arch. Rat. Mech. Anal.* **24** (1967), 1-21 {29, 76}

[G10] J.M. Greenberg, R.C. MacCamy and V.J. Mizel, On the existence, uniqueness, and stability of solutions of the equation $\sigma'(u_x)u_{xx} + \lambda u_{xxt} = \rho_0 u_{tt}$, *J. Math. Mech.* **17** (1968), 707-728 {134}

[G11] J.M. Greenberg, On the existence, uniqueness, and stability of the equation $\rho_0 X_{tt} = E(X_x)X_{xx} + \lambda X_{xxt}$, *J. Math. Anal. Appl.* **25** (1969), 575-591 {134}

[G12] J.M. Greenberg and R.C. MacCamy, On the exponential stability of solutions of $E(u_x)u_{xx} + \lambda u_{xtx} = \rho u_{tt}$, *J. Math. Anal. Appl.* **31** (1970), 406-417 {134}

[G13] J.M. Greenberg, A priori estimates for flows in dissipative materials, *J. Math. Anal. Appl.* **60** (1977), 617-630 {149}

[G14] J.M. Greenberg and L. Hsiao, The Riemann problem for the system $u_t + \sigma_x = 0$, $(\sigma - \hat{\sigma}(u))_t + \frac{1}{\epsilon}(\sigma - \mu\hat{\sigma}(u)) = 0$, *Arch. Rat. Mech. Anal.* **82** (1983), 87-108 {73}

[G15] E.M. Griest, W. Webb and R.W. Schiessler, Effect of pressure on viscosity of higher hydrocarbons and their mixtures, *J. Chem. Phys.* **29** (1958), 711-720 {18}

[G16] R.C. Grimmer and A.J. Pritchard, Analytic resolvent operators for integral equations in Banach space, *J. Diff. Eq.* **50** (1983), 234-259 {55, 211}

[G17] G. Gripenberg, Nonexistence of smooth solutions for shearing flows in a nonlinear viscoelastic fluid, *SIAM J. Math. Anal.* **13** (1982), 954-961 {62}

[G18] M.E. Gurtin, *An Introduction to Continuum Mechanics*, Academic Press 1981 {11, 13}

[H1] J.K. Hale and J. Kato, Phase space for retarded equations with infinite delay, *Funk. Ekv.* **21**, 11-41 {208}

[H2] R.S. Hamilton, The inverse function theorem of Nash and Moser, *Bull. Amer. Math. Soc.* **7** (1982), 65-222 {245}

[H3] K.B. Hannsgen and R.L. Wheeler, Behavior of the solutions of a Volterra equation as a parameter tends to infinity, *J. Integral Eq.* **7** (1984), 229-237 {49}

[H4] H. Hattori, Breakdown of smooth solutions in dissipative nonlinear hyperbolic equations, *Quart. Appl. Math.* **40** (1982/83), 113-127 {62}

[H5] M.L. Heard, An abstract parabolic Volterra integrodifferential equation, *SIAM J. Math. Anal.* **13** (1982), 81-105 {208}

[H6] M.L. Heard, A class of hyperbolic Volterra integrodifferential equations, *Nonlinear Analysis* **8** (1984), 79-93 {170, 208}

[H7] E.J. Hinch, Mechanical models of dilute polymer solutions in strong flows, *Phys. Fluids* **20** (1977), S22-S30 {17}

[H8] W.J. Hrusa, A nonlinear functional differential equation in Banach space with applications to materials with fading memory, *Arch. Rat. Mech. Anal.* **84** (1983), 99-137 {96, 169}

[H9] W.J. Hrusa, Global existence of classical solutions to the equations of motion for materials with fading memory, in: J.H. Lightbourne III and S.M. Rankin III (eds.), *Physical Mathematics and Nonlinear Partial Differential Equations*, Marcel Dekker 1985, 97-109 {169}

[H10] W.J. Hrusa, Global existence and asymptotic stability for a semilinear hyperbolic Volterra equation with large initial data, *SIAM J. Math. Anal.* **16** (1985), 110-134 {170}

[H11] W.J. Hrusa and J.A. Nohel, Global existence and asymptotics in one-dimensional nonlinear viscoelasticity, in: P.G. Ciarlet and M. Roseau (eds.), *Trends and Applications of Pure Mathematics to Mechanics*, Springer Lecture Notes in Physics 195 (1984), 165-187 {149}

[H12] W.J. Hrusa and J.A. Nohel, The Cauchy problem in one-dimensional nonlinear viscoelasticity, *J. Diff. Eq.* **59** (1985), 388-412 {154, 155, 156, 164, 166}

[H13] W.J. Hrusa and M. Renardy, On a class of quasilinear partial integrodifferential equations with singular kernels, *J. Diff. Eq.* **64** (1986), 195-220 {118, 119, 121, 169}

[H14] W.J. Hrusa and M. Renardy, On wave propagation in linear viscoelasticity, *Quart. Appl. Math.* **43** (1985), 237-254 {38, 49}

[H15] W.J. Hrusa and M. Renardy, An existence theorem for the Dirichlet problem in the elastodynamics of incompressible materials, *Arch. Rat. Mech. Anal.*, to appear {101, 117}

[H16] W.J. Hrusa and M. Renardy, A model equation for viscoelasticity with a nonintegrable memory function, preprint {118, 133}

[H17] T.J.R. Hughes, T. Kato and J.E. Marsden, Well-posed quasi-linear second-order hyperbolic systems with applications to nonlinear elastodynamics and general relativity, *Arch. Rat. Mech. Anal.* **63** (1976), 273-284 {84, 200, 206, 207, 229}

[H18] R.R. Huilgol, *Continuum Mechanics of Viscoelastic Liquids*, John Wiley 1975 {11}

[H19] J.K. Hunter and M. Slemrod, Unstable viscoelastic flow exhibiting hysteretic phase changes, *Phys. Fluids* **26** (1983), 2345-2351 {37}

[I1] E.F. Infante and J.A. Walker, A stability investigation for an incompressible simple fluid with fading memory, *Arch. Rat. Mech. Anal.* **72** (1980), 203-218 {173}

[J1] F. John, Formation of singularities in one-dimensional nonlinear wave propagation, *Comm. Pure Appl Math.* **27** (1974), 377-405 {63}

[J2] M.W. Johnson and D. Segalman, A model for viscoelastic fluid behavior which allows non-affine deformation, *J. Non-Newtonian Fluid Mech.* **2** (1977), 255-270 {24, 35}

[J3] G.S. Jordan, O.J. Staffans and R.L. Wheeler, Local analyticity in weighted L^1-spaces and application to stability problems for Volterra equations, *Trans. Amer. Math. Soc.* **174** (1982), 749-782 {54, 55, 217}

[J4] D.D. Joseph, Instability of the rest state of fluids of arbitrary grade larger than one, *Arch. Rat. Mech. Anal.* **75** (1981), 251-256 {22}

[J5] D.D. Joseph, M. Renardy and J.C. Saut, Hyperbolicity and change of type in the flow of viscoelastic fluids, *Arch. Rat. Mech. Anal.* **87** (1985), 213-251 {29, 30, 34, 35}

[J6] D.D. Joseph, Hyperbolic phenomena in the flow of viscoelastic fluids, in: A.S. Lodge, M. Renardy and J.A. Nohel (eds.), *Viscoelasticity and Rheology*, Academic Press 1985, 235-321 {62}

[J7] D.D. Joseph, O. Riccius and M. Arney, Shear wave speeds and elastic moduli for different liquids, II. Experiments, *J. Fluid Mech.* **171** (1986), 309-338 {6, 38}

[J8] D.D. Joseph, J.E. Matta and K. Chen, Delayed die swell, preprint {38}

[K1] Ya. I. Kanel', A model system of equations for the one-dimensional motion of a gas, *Diff. Equations* **4** (1968), 374-380 {134}

[K2] F. Kappel and W. Schappacher, Some considerations to the fundamental theory of infinite delay equations, *J. Diff. Eq.* **37** (1980), 141-183 {208}

[K3] T. Kato, *Perturbation Theory for Linear Operators*, Springer 1966 {198}

[K4] T. Kato, Linear evolution equations of "hyperbolic" type, *J. Fac. Sci. Univ. Tokyo, Sec. I,* **17** (1970), 241-258 {200, 204}

[K5] T. Kato, Linear evolution equations of "hyperbolic" type II, *J. Math. Soc. Japan* **25** (1973), 648-666 {200, 204, 205}

[K6] T. Kato, The Cauchy problem for quasi-linear symmetric hyperbolic systems, *Arch. Rat. Mech. Anal.* **58** (1975), 181-205 {90, 200}

[K7] T. Kato, Quasi-linear equations of evolution with application to partial differential equations, in: W.N. Everitt (ed.), *Spectral Theory of Differential Equations*, Springer Lecture Notes in Mathematics 448 (1975), 25-70 {200, 207, 225}

[K8] T. Kato, Linear and quasi-linear equations of hyperbolic type, in: G. DaPrato and G. Geymonat (eds.), *Hyperbolicity*, Centro Internazionale Matematico Estivo, II ciclo, Cortona 1976, 125-191 {81, 84, 200, 207, 223, 228}

[K9]   T. Kato, Abstract differential equations and nonlinear mixed problems, pre-print {81, 85, 200, 223, 228}

[K10] A. Kaye, *Non-Newtonian flow in incompressible fluids*, College of Aeronautics, Cranfield, England, Tech. Note 134 (1962) {22}

[K11] J.Y. Kazakia and R.S. Rivlin, Run-up and spin-up in a viscoelastic fluid, *Rheol. Acta* **20** (1981), 111-127 {38}

[K12] J.U. Kim, Global smooth solutions for the equations of motion of a nonlinear fluid with fading memory, *Arch. Rat. Mech. Anal.* **79** (1982), 97-130 {135, 171, 174, 175}

[K13] S. Klainerman and A. Majda, Singular limits of quasilinear hyperbolic systems with large parameters and the incompressible limit of compressible fluids, *Comm. Pure Appl. Math.* **34** (1981), 481-524 {104}

[K14] S. Klainerman and A. Majda, Compressible and incompressible fluids, *Comm. Pure Appl. Math.* **35** (1982), 637-656 {104}

[K15] W. Kosiński, Gradient catastrophe in the solution of nonconservative hyperbolic systems, *J. Math. Anal. Appl.* **61** (1977), 672-688 {60}

[L1]   O.A. Ladyženskaya, *The Mathematical Theory of Viscous Incompressible Flow*, Gordon and Breach 1969 {213, 215, 244}

[L2]   H.M. Laun, Description of the non-linear shear behaviour of a low density polyethylene melt by means of an experimentally determined strain dependent memory function, *Rheol. Acta* **17** (1978), 1-15 {5}

[L3]   P.D. Lax, Development of singularities of solutions of nonlinear hyperbolic partial differential equations, *J. Math. Phys.* **5** (1964), 611-613 {58}

[L4]   M.J. Leitman and G.M.C. Fisher, The linear theory of viscoelasticity, in: S. Flügge (ed.), *Handbuch der Physik VIa/3*, Springer 1973, 1-123 {21}

[L5]   A.I. Leonov, Nonequilibrium thermodynamics and rheology of viscoelastic polymer media, *Rheol. Acta* **15** (1976), 85-98 {24}

[L6]   J.L. Lions and E. Magenes, *Non-Homogeneous Boundary Value Problems and Applications* , Springer 1972 {iv, 90, 115, 137}

[L7]   A.S. Lodge and J. Meissner, On the use of instantaneous strains, superposed on shear and elongational flows of polymeric liquids, to test the Gaussian network hypothesis and to estimate the segment concentration and its variation during flow, *Rheol. Acta* **11** (1972), 351-352 {75}

[L8]   A.S. Lodge, *Body Tensor Fields in Continuum Mechanics*, Academic Press 1974 {11}

[L9]   A.S. Lodge, A classification of constitutive equations based on stress relaxation predictions for the single-jump shear strain experiment, *J. Non-Newtonian Fluid Mech.* **14** (1984), 67-83 {76}

[L10] S.-O. Londen, An existence result on a Volterra equation in a Banach space, *Trans. Amer. Math. Soc.* **235** (1978), 285-304 {135}

[L11] M. Luskin, On the classification of some model equations for viscoelasticity, *J. Non-Newtonian Fluid Mech.* **16** (1984), 3-11 {30}

[M1]   R.C. MacCamy and V.J. Mizel, Existence and nonexistence in the large of solutions of quasilinear wave equations, *Arch. Rat. Mech. Anal.* **25** (1967), 299-320 {58}

[M2] R.C. MacCamy, Existence, uniqueness and stability of solutions of the equation $u_{tt} = \frac{\partial}{\partial x}\big(\sigma(u_x) + \lambda(u_x)u_{xt}\big)$, *Indiana Univ. Math. J.* **20** (1970), 231-238 {134}

[M3] R.C. MacCamy, A model for one-dimensional nonlinear viscoelasticity, *Quart. Appl. Math.* **35** (1977), 21-33 {151, 154, 164}

[M4] R.C. MacCamy, A model Riemann problem for Volterra equations, *Arch. Rat. Mech. Anal.* **82** (1983), 71-86 {73}

[M5] A. Majda, *Compressible Fluid Flow and Systems of Conservation Laws in Several Space Variables*, Springer 1984 {57, 63}

[M6] R. Malek-Madani and J.A. Nohel, Formation of singularities for a conservation law with memory, *SIAM J. Math. Anal.* **16** (1985), 530-540 {62}

[M7] P. Markowich and M. Renardy, Lax-Wendroff methods for hyperbolic history value problems, *SIAM J. Num. Anal.* **21** (1984), 24-51 (Corrigendum: **22** (1985), 204) {71}

[M8] J.E. Marsden and T.J.R. Hughes, *Mathematical Foundations of Elasticity*, Prentice Hall 1983 {16}

[M9] A. Matsumura, Global existence and asymptotics of the solutions of the second order quasilinear hyperbolic equations with first order dissipation, *Publ. Res. Inst. Math. Sci. Kyoto Univ.* **A 13** (1977), 349-379 {135, 139, 140}

[M10] J.C. Maxwell, On the dynamical theory of gases, *Phil. Trans. Roy. Soc. London* **A 157** (1867), 49-88 {24}

[M11] M. McCracken, The resolvent problem for the Stokes equation on halfspace in L_p, *SIAM J. Math. Anal.* **12** (1981), 201-228 {236}

[M12] G.H. Meisters and C. Olech, Locally one-to-one mappings and a classical theorem on schlicht functions, *Duke Math. J.* **30** (1963), 63-80 {12, 104}

[M13] O.E. Meyer, Theorie der elastischen Nachwirkung, *Ann. Phys. Chem.* **151** (1874), 108-119 {3}

[N1] A. Narain and D.D. Joseph, Linearized dynamics for step jumps of velocity and displacement of shearing flows of a simple fluid, *Rheol. Acta* **21** (1982), 228-250 {38}

[N2] G.A. Nariboli, Asymptotic theory of wave motion in rods, *Z. Angew. Math. Mech.* **49** (1968), 379-392 {26}

[N3] M. Niggemann, A model equation for non-Newtonian fluids, *Math. Meth. Appl. Sci.* **3** (1981), 200-217 {245}

[N4] T. Nishida, Nonlinear hyperbolic equations and related topics in fluid dynamics, Publications Mathématiques d'Orsay 78.02 (1978) {135, 140}

[N5] J.A. Nohel and D.F. Shea, Frequency domain methods for Volterra equations, *Adv. Math.* **22** (1976), 278-304 {151}

[N6] J.A. Nohel and M. Renardy, Development of singularities in nonlinear viscoelasticity, to appear in: *Amorphous Polymers and Non-Newtonian Fluids*, IMA Volumes in Mathematics, Vol. 6, Springer {61}

[N7] W. Noll, A mathematical theory of the mechanical behavior of continuous media, *Arch. Rat. Mech. Anal.* **2** (1958/59), 197-226 {17, 20}

[N8] J.W. Nunziato, E.K. Walsh, K.W. Schuler and L.M. Barker, Wave propagation in nonlinear viscoelastic solids, in: S. Flügge (ed.), *Handbuch der Physik VIa/4*, Springer 1974, 1-108 {80}

[O1] J.G. Oldroyd, On the formulation of rheological equations of state, *Proc. Roy. Soc. London* **A 200** (1950), 523-541 {18,23}

[O2] J.G. Oldroyd, Non-Newtonian effects in steady motion of some idealized elastico-viscous liquids, *Proc. Roy. Soc. London* **A 245** (1958), 278-297 {23,24,68}

[O3] J.G. Oldroyd, Some steady flows of the general elastico-viscous liquid, *Proc. Roy. Soc. London* **A 283** (1965), 115-133 {17}

[O4] A.P. Oskolkov, Certain model nonstationary systems in the theory of non-Newtonian flows IV, *J. Soviet Math.* **25** (1985), 902-917 {134}

[O5] A.P. Oskolkov, Theory of nonstationary flows of Kelvin-Voigt fluids, *J. Soviet Math.* **28** (1985), 751-758 {134}

[P1] A. Pazy, *Semigroups of Linear Operators and Applications to Partial Differential Equations*, Springer 1983 {9,194,198,199,200,201,219}

[P2] R.L. Pego, Phase transitions in one-dimensional nonlinear viscoelasticity: Admissibility and stability, *Arch. Rat. Mech. Anal.* **97** (1987), 353-394 {134}

[P3] C.J.S. Petrie, *Elongational Flows*, Pitman 1977 {11,21,26}

[P4] C.J.S. Petrie and M.M. Denn, Instabilities in polymer processing, *AIChE J.* **22** (1976), 200-236 {37}

[P5] N. Phan-Thien and R.I. Tanner, A new constitutive equation derived from network theory, *J. Non-Newtonian Fluid Mech.* **2** (1977), 353-365 {24}

[P6] A.C. Pipkin, Shock structure in a viscoelastic fluid, *Quart. Appl. Math.* **23** (1965/66), 297-303 {76}

[P7] A.C. Pipkin, *Lectures on Viscoelasticity Theory*, Springer, 2nd edition 1986 {11,49}

[P8] G. da Prato, M. Ianelli and E. Sinestrari, Regularity of solutions of a class of linear integrodifferential equations in Banach spaces, *J. Integral Eq.* **8** (1985), 27-40 {55,211}

[P9] J. Prüß, Positivity and regularity of hyperbolic Volterra equations in Banach spaces, submitted to *Math. Ann.* {49}

[R1] M. Rammaha, Development of singularities of smooth solutions of nonlinear hyperbolic Volterra equations, *Comm. Part. Diff. Eq.*, to appear {62}

[R2] J.W.S. Rayleigh, On the motion of solid bodies through viscous liquid, *Phil. Mag.* **21** (1911), 697-711 {39}

[R3] M. Reiner, The Deborah number, *Physics Today* **17** (1) (1964), 62 {38}

[R4] M. Renardy, A quasilinear parabolic equation describing the elongation of thin filaments of polymeric liquids, *SIAM J. Math. Anal.* **13** (1982), 226-238 {26,136}

[R5] M. Renardy, Some remarks on the propagation and non-propagation of discontinuities in linearly viscoelastic liquids, *Rheol. Acta* **21** (1982), 251-254 {38}

[R6] M. Renardy, A class of quasilinear parabolic equations with infinite delay and application to a problem of viscoelasticity, *J. Diff. Eq.* **48** (1983), 280-292 {210}

[R7] M. Renardy, Singularly perturbed hyperbolic evolution problems with infinite delay and an application to polymer rheology, *SIAM J. Math. Anal.* **15** (1984), 333-349 {210}

[R8] M. Renardy, Local existence theorems for the first and second initial-boundary value problem for a weakly non-Newtonian fluid, *Arch. Rat. Mech. Anal.* **83** (1983), 229-244 {210, 229, 230}

[R9] M. Renardy, On the domain space for constitutive laws in linear viscoelasticity, *Arch. Rat. Mech. Anal.* **85** (1984), 21-26 {4, 22}

[R10] M. Renardy, A local existence and uniqueness theorem for a K-BKZ fluid, *Arch. Rat. Mech. Anal.* **88** (1985), 83-94 {178, 210, 237, 238}

[R11] M. Renardy, Existence of slow steady flows of viscoelastic fluids with differential constitutive equations, *Z. Angew. Math. Mech.* **65** (1985), 449-451 {246}

[R12] M. Renardy, Some remarks on the Navier-Stokes equations with a pressure dependent viscosity, *Comm. Part. Diff. Eq.* **11** (1986), 779-793 {18}

[R13] M. Renardy, Inflow boundary conditions for steady flows of viscoelastic fluids with differential constitutive laws, *Rocky Mt. J. Math.*, to appear {249}

[R14] R.S. Rivlin, Some applications of elasticity theory to rubber engineering, *Proc. 2nd Rubber Technology Conf.*, Heffers, Cambridge 1948, 204-211 {75}

[R15] R.S. Rivlin and J.L. Ericksen, Stress-deformation relations for isotropic materials, *J. Rat. Mech. Anal.* **4** (1955), 323-425 {18, 22}

[R16] R.S. Rivlin, Stress-relaxation in incompressible elastic materials at constant deformation, *Quart. Appl. Math.* **14** (1956), 83-89 {75}

[R17] P.E. Rouse, A theory of the linear viscoelastic properties of dilute solutions of coiling polymers, *J. Chem. Phys.* **21** (1953), 1271-1280 {6}

[R18] I.M. Rutkevich, The propagation of small perturbations in a viscoelastic fluid, *J. Appl. Math. Mech.* **34** (1970), 35-50 {30, 34}

[R19] I.M. Rutkevich, On the thermodynamic interpretation of the evolutionary conditions of the equations of the mechanics of finitely deformable viscoelastic media of Maxwell type, *J. Appl. Math. Mech.* **36** (1972), 283-295 {30}

[S1] J.C. Saut and D.D. Joseph, Fading memory, *Arch. Rat. Mech. Anal.* **81** (1982), 53-95 {20}

[S2] S. Schochet, The incompressible limit in nonlinear elasticity, *Comm. Math. Phys.* **102** (1985), 207-215 {104}

[S3] W.R. Schowalter, *Mechanics of Non-Newtonian Fluids*, Pergamon Press 1978 {11, 21}

[S4] K. Schumacher, Remarks on semilinear partial functional-differential equations with infinite delay, *J. Math. Anal. Appl.* **80** (1981), 261-290 {208}

[S5] M. Slemrod, A hereditary partial differential equation with applications in the theory of simple fluids, *Arch. Rat. Mech. Anal.* **62** (1976), 303-321 {210, 212, 213, 216}

[S6] M. Slemrod, An energy stability method for simple fluids, *Arch. Rat. Mech. Anal.* **68** (1978), 1-18 {217}

[S7] M. Slemrod, Instability of steady shearing flows in a nonlinear viscoelastic fluid, *Arch. Rat. Mech. Anal.* **68** (1978), 211-225 {62}

[S8] P.E. Sobolevskii, Equations of parabolic type in a Banach space, *Amer. Math. Soc. Transl.* **49** (1966), 1-62 {199, 200, 201, 202, 203, 219, 220, 222}

[S9] V.A. Solonnikov, Estimates for the solution of a nonstationary linearized systems of Navier-Stokes equations, *Proc. Steklov Inst.* **70** (1964), 213-31 (Russian) {236}

[S10] V.A. Solonnikov, On the differential properties of the solutions of the first boundary-value problem for a nonstationary system of Navier-Stokes equations, *Proc. Steklov Inst.* **73** (1964), 221-291 (Russian) {236}

[S11] V.A. Solonnikov, Estimates for solutions of nonstationary Navier-Stokes equations, *J. Soviet Math.* **8** (1977), 467-529 {236}

[S12] J.H. Song and J.Y. Yoo, Numerical simulation of viscoelastic flow through sudden contraction using type dependent difference method, *J. Non-Newtonian Fluid Mech.*, to appear {38}

[S13] O.J. Staffans, An inequality for positive definite Volterra kernels, *Proc. Amer. Math. Soc.* **58** (1976), 205-210 {133, 151}

[S14] O.J. Staffans, On a nonlinear hyperbolic Volterra equation, *SIAM J. Math. Anal.* **11** (1980), 793-812 {151, 155}

[S15] G.G. Stokes, On the theories of the internal friction of fluids in motion and of the equilibrium and motion of elastic solids, *Trans. Cambridge Phil. Soc.* **8** (1845), 287-319 {18}

[S16] G.G. Stokes, On the effect of the internal friction of fluids on the motion of pendulums, *Trans. Cambridge Phil. Soc.* **9** (1850), 8 {39}

[T1] R.I. Tanner, Note on the Rayleigh problem for a viscoelastic fluid, *Z. Angew. Math. Phys.* **13** (1962), 573-580 {38}

[T2] R. Temam, *Navier-Stokes Equations*, North Holland, 3rd ed., 1984 {114}

[T3] J.P. Tordella, Unstable flow of molten polymers, in: F. Eirich (ed.), *Rheology Theory and Applications, Vol. 5*, Academic Press 1969, 57-92 {37}

[T4] C.A. Truesdell, Hypo-elasticity, *J. Rat. Mech. Anal.* **4** (1955), 83-133 and 1019-1020 {74}

[T5] C.A. Truesdell and W. Noll, The non-linear field theories of mechanics, in: S. Flügge (ed.), *Handbuch der Physik III/3*, Springer 1965 {11, 74}

[U1] J.S. Ultman and M.M. Denn, Anomalous heat transfer and a wave phenomenon in dilute polymer solutions, *Trans. Soc. Rheology* **14** (1970), 307-317 {30, 38}

[W1] E.K. Walsh, Wave propagation in viscoelastic solids, in: A.S. Lodge, M. Renardy and J.A. Nohel (eds.), *Viscoelasticity and Rheology*, Academic Press 1985, 13-46 {80}

[W2] C.C. Wang, Stress relaxation and the principle of fading memory, *Arch. Rat. Mech. Anal.* **18** (1965), 117-126 {20}

[W3] C.C. Wang, The principle of fading memory, *Arch. Rat. Mech. Anal.* **18** (1965), 343-366 {20}

[W4] J.L. White and A.B. Metzner, Development of constitutive equations for polymeric melts and solutions, *J. Appl. Polym. Sci.* **7** (1963), 1867-1889 {24}

[W5] J.L. White, Dynamics of viscoelastic fluids, melt fracture and the rheology of fiber spinning, *J. Appl. Polym. Sci.* **8** (1964), 2339-2357 {38}

[Y1] J.Y. Yoo, M. Ahrens and D.D. Joseph, Hyperbolicity and change of type in sink flow, *J. Fluid Mech.* **153** (1985), 203-214 {38}

[Z1] B.H. Zimm, Dynamics of polymer molecules in dilute solutions: viscoelasticity, flow birefringence and dielectric loss, *J. Chem. Phys.* **24** (1956), 269-278 {6}

Index